BIG DATA IN ASTRONOMY

BIG DATA IN ASTRONOMY

Scientific Data Processing for Advanced Radio Telescopes

Edited by

LINGHE KONG

Research Professor, Computer Science and Engineering, Shanghai Jiao Tong University, Shanghai, China

TIAN HUANG

Research Associate, Astrophysics Group, Cavendish Lab, Cambridge University, Cambridge, United Kingdom

YONGXIN ZHU

Professor, Shanghai Advanced Research Institute, Chinese Academy of Sciences, Shanghai, China

SHENGHUA YU

Associate Professor, Joint Laboratory for Radio Astronomy Technology, National Astronomical Observatories, Chinese Academy of Sciences, Beijing, China

ELSEVIER

Elsevier
Radarweg 29, PO Box 211, 1000 AE Amsterdam, Netherlands
The Boulevard, Langford Lane, Kidlington, Oxford OX5 1GB, United Kingdom
50 Hampshire Street, 5th Floor, Cambridge, MA 02139, United States

Library of Congress Cataloging-in-Publication Data
A catalog record for this book is available from the Library of Congress

British Library Cataloguing-in-Publication Data
A catalogue record for this book is available from the British Library

ISBN: 978-0-12-819084-5

For information on all Elsevier publications
visit our website at https://www.elsevier.com/books-and-journals

Publisher: Candice Janco
Acquisitions Editor: Amy Shapiro
Editorial Project Manager: Lena Sparks
Production Project Manager: Kumar Anbazhagan
Cover Designer: Christian J. Bilbow

Typeset by SPi Global, India

Contents

Part B Big data processing

Part C Computing technologies

Contributors

Xuelei Chen National Astronomical Observatories, Chinese Academy of Sciences, Beijing, China

Yatong Chen Dalian University of Technology, Dalian, China

Hui Deng Center for Astrophysics, Guangzhou University, Guangzhou Higher Education Mega Center, Guangzhou, China

Sen Du School of Microelectronics, Shanghai Jiao Tong University, Shanghai, China

Siyu Fan Department of Computer Science and Engineering, Shanghai Jiao Tong University, Shanghai, China

Kaiyu Fu Department of Computer Science and Engineering, Shanghai Jiao Tong University, Shanghai, China

Stephen F. Gull Astrophysics Group, Cavendish Lab, Cambridge University, Cambridge, United Kingdom

Peter Hague Astrophysics Group, Cavendish Lab, Cambridge University, Cambridge, United Kingdom

Junjie Hou School of Microelectronics, Shanghai Jiao Tong University, Shanghai, China

Tian Huang Astrophysics Group, Cavendish Lab, Cambridge University, Cambridge, United Kingdom; Institute of High Performance Computing, Agency for Science, Technology and Research (A*STAR), Singapore, Singapore

Linghe Kong Shanghai Jiao Tong University, Shanghai, China

Rui Kong Shanghai Jiao Tong University, Shanghai, China

Jiale Lei Shanghai Jiao Tong University, Shanghai, China

Qiuhong Li School of Computer Science, Fudan University, Shanghai, China

Ting Li Department of Computer Science and Engineering, Shanghai Jiao Tong University, Shanghai, China

Bin Liu National Astronomical Observatories, Chinese Academy of Sciences, Beijing, China

Dongliang Liu National Astronomical Observatories, Chinese Academy of Sciences, Beijing, China

Yuan Luo Department of Computer Science and Engineering, Shanghai Jiao Tong University, Shanghai, China

Ying Mei Center for Astrophysics, Guangzhou University, Guangzhou Higher Education Mega Center, Guangzhou, China

Bojan Nikolic Astrophysics Group, Cavendish Lab, Cambridge University, Cambridge, United Kingdom

Danny C. Price Centre for Astrophysics and Supercomputing, Swinburne University, Hawthorn, VIC, Australia; Department of Astronomy, University of California at Berkeley, Berkeley, CA, United States

Shijin Song School of Microelectronics, Shanghai Jiao Tong University, Shanghai, China

Yuefeng Song School of Microelectronics, Shanghai Jiao Tong University, Shanghai, China

Jinlin Tan Shanghai Jiao Tong University, Shanghai, China

Sze Meng Tan Picarro Inc., Santa Clara, CA, United States

Rodrigo Tobar International Center for Radio Astronomy Research (ICRAR), The University of Western Australia, Crawley, Perth, WA, Australia; Kunming University of Science and Technology, Chenggong District, Kunming, China

Feng Wang Center for Astrophysics, Guangzhou University, Guangzhou Higher Education Mega Center, Guangzhou; Kunming University of Science and Technology, Chenggong District, Kunming, China; International Center for Radio Astronomy Research (ICRAR), The University of Western Australia, Crawley, Perth, WA, Australia

Shoulin Wei Kunming University of Science and Technology, Chenggong District, Kunming, China; International Center for Radio Astronomy Research (ICRAR), The University of Western Australia, Crawley, Perth, WA, Australia

Chen Wu International Center for Radio Astronomy Research (ICRAR), The University of Western Australia, Crawley, Perth, WA, Australia; Kunming University of Science and Technology, Chenggong District, Kunming, China

Huaiguang Wu Zhengzhou University of Light Industry, Zhengzhou, China

Haoyang Ye Astrophysics Group, Cavendish Lab, Cambridge University, Cambridge, United Kingdom

Haihang You Institute of Computing Technologies, Chinese Academy of Sciences, Beijing, China

Shenghua Yu National Astronomical Observatories, Chinese Academy of Sciences, Beijing, China

Yu Zheng School of Microelectronics, Shanghai Jiao Tong University, Shanghai, China

Yongxin Zhu Shanghai Advanced Research Institute, Chinese Academy of Sciences; School of Microelectronics, Shanghai Jiao Tong University, Shanghai; University of Chinese Academy of Sciences, Beijing, China

Preface

In recent years, radio astronomy is experiencing the accelerating explosion of data. Modern telescopes can image enormous portions of the sky. For example, the Square Kilometer Array (SKA), which is the world's largest radio telescope, generates over an Exabyte of data every day. To cope with the challenges and opportunities offered by the exponential growth of astronomical data, the new disciplines and technologies are emerging. For example, in China, the fastest supercomputer, Sunway TaihuLight, is used to undertake the processing task of big data in radio astronomy.

Since the big data era poses many new challenges in radio astronomy, we should think about a series of problems: How to process, calibrate, and clean the astronomical big data; How to optimize and accelerate the algorithms of data processing; How to extract knowledge from big data, and so on.

This book provides a comprehensive review on the latest research developments and results in the interdisciplinary of radio astronomy and big data. It presents recent advances and insights in radio astronomy from the special point of view of data processing. Challenges and techniques in various stages of the life cycle of data science are covered in this book.

In this book, we first have a quick review of the fundamentals of radio astronomy and the big data problems in this field. Then, we introduce the advanced big data processing technologies, including preprocessing, real-time streaming, digitization, channelization, packeting, correlation, calibration, and scale-out. Moreover, we present the state-of-the-art computing technologies such as execution framework, heterogeneous computing platform, high-performance computing, image library, and artificial intelligence in astronomical big data. In the end, we look into the future development, especially mapping the universe with 21-cm observations.

This book will be a valuable resource for students, researchers, engineers, policy makers working in various areas related to big data in radio astronomy.

Linghe Kong

Tian Huang

Yongxin Zhu

Shenghua Yu

Acknowledgments

This work was supported in part by the China Ministry of Science and Technology, China Natural Science Foundation, Chinese Academy of Sciences and China SKA office. Special thanks should be also dedicated to Mr. Linhao Chen on behalf of China Ministry of Science and Technology, Ms. Shuang Liu on behalf of China SKA office, Prof. Bo Peng and Prof. Di Li on behalf of FAST telescope, Chinese Academy of Sciences, for their advice and guidance. This work will not be possible without the discussions and support from many of our collaborators, colleagues, and students. We would especially like to thank Mr. Chris Broekema, Professor Guihai Chen, Professor Xueming Si, Mr. Zhe Wang, and Mr. Shuaitian Wang provided insightful feedbacks and discussions. We are also grateful to our Editorial Project Manager Ms. Lena Sparks, Editor Ms. Sheela Bernardine B. Josy, and the anonymous reviewers of this book for their constructive criticism of the earlier manuscripts.

PART A

Fundamentals

Introduction to radio astronomy

Jinlin Tan, Linghe Kong
Shanghai Jiao Tong University, Shanghai, China

1 The history of astronomy

Astronomy is the science studying celestial objects (including stars, planets, comets, and galaxies) and phenomena (such as auroras and cosmic background radiation). It involves physics, chemistry, and the evolution of the universe.

Astronomy is one of the oldest disciplines, appearing almost simultaneously with ancient science. Recent findings show that prehistoric cave paintings dating back to 40,000 years ago may be considered to be astronomical calendars. Throughout the history of astronomy, every milestone showed the wisdom and courage of human beings. The Copernican revolution made people dare to imagine that the sun is at the center of the planets. Then, Kepler revealed the laws of planetary movement. Newton combined Galileo's experiments with Kepler's laws and established the law of universal gravitation, which has become an important symbol of modern scientific determination and also the basis of physics [1].

1.1 Ancient astronomy

During the ancient astronomy period, amateur astronomers could only observe celestial bodies with the naked eye or through primitive astronomical instruments. The main contribution in that period was the visible position of the celestial body. Ancient Babylon made the calendar by observing the activities of the moon, and determined the leap month. Chaldeans could predict the date of the eclipse of the sun and moon. Ancient Egyptians divided a whole day into day and night, each containing 12 h. Later, the Pythagorean theorem proved that the Earth is round according to the movements of stars. However, it was really difficult to imagine that the universe could be comprehensively observed in the future.

Big Data in Astronomy. https://doi.org/10.1016/B978-0-12-819084-5.00014-6

1.2 Astronomy from the mid-16th century to the mid-19th century

Copernicus's heliocentric theory was an epoch-making revolution that pioneered modern natural science and modern astronomy. Then, the birth of the telescope in the early 17th century provided a new means of observation for astronomy and brought countless new discoveries in astronomy. The birth of the telescope also greatly improved the positioning accuracy of the celestial body, as shown in Fig. 1.1, which brought about the rapid development of astronomy. The discovery of the gravitational top floor in the second half of the 17th century helped astronomy develop from just a simple description of the visual position and visual motion of a celestial body. The interaction between celestial bodies and the stages of their mutual movement, celestial mechanics, has flourished since then. In the second half of the 18th century, the birth of the Kant-Laplace Nebula, the origin of the solar system, strongly impacted the metaphysical view of nature at the time and opened up a new field of research in astronomy called celestial chemistry. In 1785, William Herschel initially established the concept of the Milky Way, extending the horizons of people from the solar system to the Milky Way, and the vision was greatly broadened [3].

Fig. 1.1 The first large-scale sky surveys were carried out by Ryle at Cambridge in the early 1950s. Taken from A. Hewish, Early techniques in radio astronomy, Adv. Imaging Electron Phys. 91 (1995) 285–290. Fig. 1.

1.3 Astronomy since the mid-19th century

Before the middle of the 19th century, people were limited to using telescopes to observe celestial bodies with human eyes. Although this method of observation brought many important astronomical discoveries, it could not reveal the physical nature of celestial bodies. In the middle of the 19th century, spectroscopic techniques, observation techniques, and photographic techniques were applied to astronomy almost simultaneously, leading to the birth of astrophysics. As a result, human understanding of celestial bodies made another leap from the development of mechanical movements of celestial bodies to the study of various physical and chemical movements of celestial bodies. Entering the 20th century, the birth of quantum mechanics provided a powerful theoretical weapon for the further development of astrophysics. Then, the creation of general relativity in 1915 led to the birth of modern cosmology. The discovery of extragalactic galaxies in the 1920s once again expanded people's horizons and opened a new page for human exploration of the universe. From the 1930s through the 1950s, the rise of radio detection technology and space detection technology enabled the detection of celestial spheres from pure optical bands to the entire electromagnetic wave band. This ushered in the era of full-wave astronomy, leading to numerous new discoveries. Now, astronomy is moving forward at an unprecedented rate [2,4].

2 What is radio astronomy

Astronomers around the world use radio telescopes to observe the naturally occurring radio waves that come from stars, planets, galaxies, clouds of dust, and molecules of gas. Most of us are familiar with visible-light astronomy and what it reveals about these objects. Visible light—also known as optical light—is what we see with our eyes. However, visible light doesn't tell the whole story about an object. To get a complete understanding of a distant quasar or a planet, for example, astronomers study it in as many wavelengths as possible, including the radio range. There's a hidden universe out there, radiating at wavelengths and frequencies we can't see with our eyes. Each object in the cosmos gives off unique patterns of radio emissions that allow astronomers to get the whole picture of a distant object. Radio astronomers study emissions from gas giant planets, blasts from the hearts of galaxies, or even precisely ticking signals from a dying star.

Radio waves from space were discovered in 1932, but actual radio astronomy research was conducted during World War II. Initially, alert radars detected strong radio noise emitted from the sun and made people aware that the conditions for studying celestial bodies using radio were mature. This is because the Earth's atmosphere can pass radio waves, and the receiving devices at that time could receive specific signals from space, as shown in Fig. 1.2. The atmospheric window of Earth's radio astronomy contains wavelengths from a few millimeters to three or 40 m.

Radio astronomy has not only become an important auxiliary of optical astronomy, but it has also uniquely opened up a series of new scientific fields. Methodologically, radio astronomy can study celestial bodies by radio waves emitted from the ground. In this sense, it raises astronomy from purely observing science to a

Fig. 1.2 Centaurus A radio image and the moon to scale superimposed on the Australia Telescope Compact Array which made this 1.4 GHz image. From Ekers, R.D. (2014). Non-thermal radio astronomy. Astroparticle Physics, 53 (2), 152–159; a composite image by Ilana Feain, Tim Cornwell & Ron Ekers (CSIRO/ATNF); ATCA northern middle lobe pointing courtesy R. Morganti (ASTRON); Parkes data courtesy N. Junkes (MPIfR); ATCA & Moon photo: Shaun Amy, CSIRO.

certain experiment. At the same time, from the characteristics of work, radio waves have an important growth than light waves. First, some material processes such as the movement of charged particles can generate radio waves but do not emit light; then, radio waves can pass through the light. Dust and clouds enable radio astronomical instruments to work day and night, but in the study of the universe, the vast space is behind dense interstellar material, which was previously inaccessible by optical methods. Now, the radio method has been extensively explored. These characteristics have given radio astronomy a sudden rise in modern science [5].

2.1 How does radio astronomy occur

Karl Jansky, who worked at the Telephone Research Laboratory, discovered the radiation of space radio waves in 1933 when trying to determine the cause of interference in transatlantic telephone communications. However, unexpected noise appeared. The peak signal arrived 4 min earlier every day, and Jansky needed this to be of extraterrestrial origin because that corresponded to a lateral time. The reaction from Bell Laboratories was light rog. As Glowtraber later observed, "Because it was so thin, it was not even an interesting source of radio interference!" When the intervention was judged to be "extraterrestrial" by Bell Telephone Laboratories, there was little support to discover more locations. Sullivan was interested in time by some astronomers, but most of the decibel engineers and super receivers were far away from their world. Jansky's discovery was ignored by other scientists until December 1938.

In 1937, the successful amateur hammer Glow Trevor had a hard time understanding Jansky. In a backyard in Wheaton, Illinois, Ha made $2000 and $32 parabolic dishes and began looking for the radio signal Jansky obtained. Initially, the only type of natural radio radiation, known as thermal heat, and radiation became stronger at shorter wavelengths, resulting in shorter wavelengths than those used by Jansky. But because nothing was visible in the shortwave, Trevor went to the longwave until he found a sign that matched what Jansky saw. The radio program had to be strong in longer wavelengths and nonthermal orthodontics, but it was more obscure because there was no astronomical concept of nonthermal radiation at any wavelength. In 1950, when radiation was synchronized with high-energy cosmetic particles (spaceships), this result was able to be integrated with the world's larger scientific image and strange wireless noise world in 13 years. Some of the astronomical features were made part of astronomy.

2.2 The radio stars, quasars, and black holes

2.2.1 The strongest radio source, Cygnus A, in the sky

Stanley Hey, a cofounder of British radio emissions from the Sun during World War II, found in 1946 that one of the strongest radio emission sources varied in units of 10–30 s. He decided that the diameter of the source should be small. Later, he realized that the fluctuations were ionospheric flashes and that they were not necessary, but claims on a small diameter were still accurate. This is the size of the star, but how was there no optical counterpart? Does every radio star emit such radio [6]?

2.2.2 The discovery of cliff allergens and radio galaxies

In 1946 at Dover Heights near Sydney, a telescope was constructed on the cliff to measure the interference between the direct waves and those reflected by the sea (a Lloyd's mirror). This cliff interferometer was built to locate the origin of the solar radio emission and to identify the radio stars. The idea of a cliff interferometer came from the multiple path interference already seen in shipborne radar in WWII, and was used to improve positional information. John Bolton and his colleagues [7] at CSIRO in Australia were able to measure positions accurately enough to identify three of the strongest of the mysterious discrete sources of radio emissions that, up until this time, were thought to be radio stars. One was the Crab nebula, the remnant of a star that the Chinese saw explode in 1054 CE. The other two were an even greater surprise. Centaurus A and Virgo A (strongest sources of radio emissions in the constellations of Centaurus and Virgo) had conspicuous bright optical identifications that were galaxies, not stars! These were galaxies far outside our own Milky Way that were undergoing such a violent explosion that they were among the brightest objects in the radio sky and became the most luminous sources known in the universe. This discovery, with some help from the now very enthusiastic optical astronomers at Mt. Palomar in the United States, led to the eventual identification of the strongest of all the radio sources, Cygnus A. It was found to be a very faint galaxy so distant that it was obvious that the radio telescopes were already probing the most distant reaches of the universe!

2.2.3 Nonthermal radiation

This is a very confusing story and the misunderstanding of the early radio data exacerbated the confusion. Some wireless power supplies are supposed to have a smaller diameter. They are correct, but it is wrong to think that all broadcasts in the Milky

Way are the sum of all radio stars. It is also recognized that the radio is similar to the sun, but this is not true. They are a mixture of Galaxy Nebula (SNR) and Galaxy Star (AGN) [8].

2.2.4 Synchronous radiation

In 1949, Fermi explained the acceleration of career particles in interstellar media, although Langmuir had observed and explained the synchrotron radiation seen in the General Electric synchrotron in 1947. But none of them associated high energy particles with cosmic radio emissions.

2.2.5 Synchrotron radiation pattern

In 1949, the abnormal nonthermal radio radiation sold by sunlight was interpreted as plasma vibration. Alfven suggested that this abnormal radiation from the sun's radiation was synchrotron radiation. Kiepenhauer conducted further research in 1950, suggesting that galactic radio radiation could be generated in the synchrotron process in the interstellar environment (ISM). He recognized the existence of interstellar magnetic fields and assumed that cosmic rays contained relativistic electrons. In the Western world, this explanation was almost ignored. But in Russia, Ginzburg and Shklovsky enthusiastically accepted this due to the clear evidence of magnetic fields and cosmic radiation particles. At present, most Western astronomers do not understand the importance of cosmic rays [7].

2.2.6 Connect nonthermal radiation and cosmic rays

Ginzburg pointed out in 1951 that the synchrotron radiation of relativistic electrons in the magnetic field of the galaxy "is very natural and attractive as an explanation for the general radio emissions of the galaxy." In 1953, Shkolovsky published his seminal paper explaining that the radiation from the Crab nebula was the radiation of radio and optical synchrotrons. In 1957, Burbidge noted that radio and optical wavelength synchrotrons could explain planes in the M87 radio galaxy. By this time, the radio synchrotron emission was well accepted for galactic supernova remnants and for extragalactic sources, so the pieces of the nonthermal radio synchrotron puzzle were falling into place.

2.2.7 Astrophysics of cosmic rays

Ginzburg said that space astronomy began in the early 1950s, and that during the synchrotron process, nonthermal radiation could be used to notice cosmological rays that were far from the world. The Crab Nebula and the first radio galaxies have been recognized. As the radio waves travel directly, the cosmic rays have access to information about the electronic composition of the cosmic rays far from the Earth, in our galaxy, other galaxies, and cavities. The source of exposure can be monitored at all wavelengths under ray or UHE (ultraenergetic) conditions without offset.

2.2.8 Discovery of quasars

Before 1963, extragalactic radio sources were almost all identified with giant elliptical galaxies. When the 3C273 radio source occulted by the moon, this changed in unexpected ways. Cyril Hazard observed the occultation using CSIRO's Perks radio telescope. The nature of the unresolved plane spectrum of a steep-spectrum 2000 jet was shown. The morphology and position clearly identified this strong but previously unidentified radio source with a bright 13 magnitude star and a wisp (jet) of optical emission in the same location as the radio jet. Martin Schmidt took the star's spectrum and interpreted it as a redshift corresponding to a 0.15 light speed pickup. This meant that the unprecedented size came from something as small as a star, equivalent to the entire galaxy. This was the first quasar. This discovery sparked the first Texas symposium on gravitational collapse and relativistic astrophysics. Only very small black holes provide the energy needed for a small volume. This was a paradigm shift in astronomy and the process of explaining the role of supergiant black holes in the evolution of the universe continues to this day [9].

2.3 The radio astronomy instrument: Radio telescope

Radio astronomy was born in the 1930s, and it is a discipline that studies astronomical phenomena by observing radio waves from celestial bodies.

Due to the Earth's atmospheric disturbance, radio waves from celestial bodies can reach the Earth only at wavelengths of about 1 mm to 30 m. So far, most radio astronomy studies have been conducted in this group. In radio astronomy, radio wave reception technology is used as a means of observation, and the object of observation extends throughout the celestial body. The range extends from the celestial bodies near the solar system to the various celestial bodies in the Milky Way to far beyond the Milky Way. The radio technology in the radio astronomy band didn't really develop until the 1940s.

So what is a radio telescope? A typical astronomical telescope is called an astronomical optical telescope because it can only observe visible light emitted by other objects. Apart from the visible band of light, the human eye cannot directly detect it as a visible image, but there are many other bands of radio waves that can be picked up and measured. Radio telescopes have been used to "observe" all directions from the sky [10–12].

An astronomical tool for sending radio waves. Equipped with highly directional antennas and compatible electronic devices. Therefore, radio telescopes are said to be closer to antennas receiving radar than telescopes. Of course, later technical processing can also process radio waves taken from radio telescopes and convert them into data or images. The visible effects of light can only be seen with ordinary telescopes, but radio telescopes can observe the radio phenomena of astronomical objects. Radio waves can pass through media between dust in space but light waves can't pass, so radio telescopes can pass through interstellar dust.

Egypt observed the unknown universe in the distance. At the same time, radio telescopes operate almost nonstop throughout the day because radio waves are less sensitive to light and weather. Astronomy has evolved rapidly because of the invention of the radio telescope. It reveals many wonderful phenomena in the universe. For example, the Cygnus A radio galaxy was discovered through a radio telescope. It emits more than 100 billion times more radio energy than the sun emits per second. The largest radio and optical radio galaxy ever discovered. The telescope knows nothing about it. Moreover, the four major discoveries of 1960s astronomy—pulsars, quasars, cosmic microwave background radiation, and interstellar organic molecules—are all connected to radio telescopes. In the history of the Nobel Prize, five of the seven awards called the Astronomy Awards are based on observations from radio telescopes, and radio astronomy is the birthplace of the Nobel Prize.

The principle of a radio telescope is to focus on reflecting the pan using a form of antenna, collect signals from a few square meters to thousands of square meters at one point, take radio waves and determine the position and trajectory it is.

In 1931, Bell Labs in the United States received antennas from the Milky Way center using an antenna array. Later, American Glow Treve built a 9.5-ft-long antenna in his backyard. He received radio waves from the center of the Milky Way in 1939, and the first radio map was designed based on observations. The antenna used by Rebec was the world's first telescope dedicated to astronomical observation, and radio astronomy was

Fig. 1.3 The 5-km telescope designed by Ryle and completed in 1972. Taken from A. Hewish, Early techniques in radio astronomy, Adv. Imaging Electron Phys. 91 (1995) 285–290. Fig. 3.

born. In 1972, Ryle designed a 5-km telescope as a new radio telescope to promote the development of radio astronomy, as shown in Fig. 1.3.

2.4 Some achievements of radio astronomy

Basically solving the problem of the material distribution of the Milky Way defined the shape of the vortex arm and the position of the core of the Milky Way, especially the gas that was found near the center of the Milky Way (within 100 million light years) had significant radial motion. This provides important clues to the salinization of the Milky Way.

More than a thousand cosmic radio sources have been discovered, but most of them have not yet been identified, and the identified parts are all special targets except for normal extragalactic galaxies and ionized hydrogen clouds in the Milky Way. These goals all show unusually intense movements and extreme instability. This fact makes us think that the true face of the entire universe actually contains more dramatic movements than previously thought [13].

It was discovered that the radio radiation of the Crab Nebula was generated by the so-called "synchronous accelerator" mechanism, which led to the hypothesis that cosmic radio radiation and original cosmic radiation have the same cause. This hypothesis of great attraction is enriched from the further observation and exploration of radio astronomy, and the results obtained are likely to play a major role in the progress of the entire physics.

The results of radio astronomy added important materials for the study of solar physics. In particular, a series of different types of solar "radiation bursts" have been discovered. These bursts generally have higher power (the highest on record has reached 10 million times the usual solar radiation), and we believe that at least some of them are generated by a plasma mechanism. These phenomena discovered by radio astronomy in the sun and in space can be seen as new and important revelations in nature: there is a kind of institution that can produce such huge energy, and the law that governs it will inevitably be in human life and have important applications.

Some of the embarrassing solar radio bursts are closely related to the bursting outbursts. As soon as they appear, there is a magnetic storm on the Earth, and short-wave radio communication is disturbed at the same time. Observing the solar radio phenomenon makes up an important aspect of the "Japan-Traditional Relationship." Using these observations, we may make predictions about communication interference and prepare the telecommunications sector.

Radio astronomy is not only great for basic research, but its development makes it closely related to practical applications. In the voyage of the universe, communication, monitoring, and remote control are all top priority issues, and these works require the application of radio astronomy. Humans now have powerful radio transmission technology. The signals emitted on a single frequency far exceed the solar radio waves of the same frequency, so radio astronomy can be used for communication and monitoring. It can interfere with wall cosmic radiation, which is not possible with optical or other methods. A huge radio telescope can receive very weak signals, and this instrument is actually a very sensitive ear. The radio astronomy method can also be used for rocket navigation. In half of navigation and aviation, the "radio sextant" positioning can compensate for the defects of the optical method of hand and rain. The radio astronomy method is revolutionizing ground communication technology, and the use of the moon as a radio relay station can solve the problem of long-distance communication on the Earth. It also proposes the use of the ionization residuals left by meteors in the ionosphere as a reflector for long-distance communication on the ground. According to research, this communication can be carried out without interruption, and the power required for communication is small. The working frequency is stable and is not affected by the Earth's atmosphere.

2.5 Astronomical research nowadays

Astronomy has been amazing since the 1960s. As a result, astronomy has written an excellent chapter in the development of human natural science. The most exciting and fascinating discoveries of astronomy depend increasingly on a larger scale. The collaboration of scientific research equipment relies on an increasing amount of mining data and analysis. At the same time, human transparency, diversity, and interdisciplinary integration make human life more scientific and technical. Astronomical learning has really entered the era of multiband, multifaith. People use multiple observation devices to detect the same celestial object at the same time and receive almost all electromagnetic waves. Yet, they also receive full information about the spectrum. You can also use nonelectromagnetic radiation sources such as neutrinos and gravity waves to study celestial bodies. One of the most representative examples is neutron star bonding, which two astronomers discovered in August 2017. The ground laser gravitational wave station and the VIRGO gravitational wave detector first detected the time and space of the neutron star fusion process, followed by the most powerful spatial and terrestrial telescopes. In addition to improving the recognition of gravitational waves, it was confirmed by observations of short gamma ray bursts and giant supernovae. Strange celestial bodies provide a powerful force for collaborative research in a new understanding of astronomy. Observation-based astronomy has long suffered from data shortages, and astronomy has already undergone a revolutionary shift in the information age of the 21st century. Currently, astronomical observation is gradually entering the era of big data. Research methods and communication methods have also undergone significant changes. To give an example, the Boy Supernova is a wonderful firework in the universe, and the earliest astronomical record of a supernova [14,15]. Supernovae are being studied at leading positions in astrophysics, and the 2011 Nobel Prize in Physics was awarded to three astronomical houses whose contributions were due to supernova observations that the universe is expanding velocity. A supernova is a very rare event, and one was captured 10 years ago. Whenever a supernova is observed, it is very difficult. You will have to rely on a lot of research to inevitably lead to the telescope tracking competition in the world. Numerical simulation and theoretical calculation. Today, optical surveys can be sent annually. Currently, more than 1000 supernovae are unusual, deep, and ineffective for mines. The data collected from these large surveys may generate more new findings. Ideal for astronomical observation provided by next-generation super telescopes, such as SKA Ascension, and is still a rare celestial object, and will become a regular customer in 5 to 10 years. Statistics, information science, and astronomy are closely combined. Create, organize, analyze, and investigate macrocosm truths and astronomical rules that provide data analysis tools for astronomers based on large space data acquisition [16].

3 Advanced radio telescope

3.1 The square kilometer array (SKA)

In recent years, in order to promote astronomy, the international community has sought to capture the history of the world. To this end, communities and people around the world are gathering resources and experiences to conduct powerful observations exploring the full electromagnetic spectrum as well as gravitational waves, cosmic rays, and gamma rays. This includes the design and construction of the site. The Square Kilometer Array (SKA) is one of these telescopes with up to a million square feet. The SKA was originally created as an international astronomical initiative. In 1993, the International Union of Radio Sciences (URSI) created a working group to study next-generation radio. Since then, 19 countries and 55 laboratories took part in this. Seven different concepts of early SKA technology were selected mainly. Through rigorous agreement, two sites were identified in the Central African region and in Western Australia in the Kalus region, which are suitable for many SKA target areas. Currently, 15 funding agencies regularly discuss financing and development opportunities for SKA. During the project, the telescope will operate for 10 GHz. As a rule, at higher levels, all telescopes are designed to solve the most important problems in astronomy, in particular, problems with clouds in the wavelength range. The US 10-Year Review Commission, 2000–2010, outlined these goals in its New Astronomy and Astrophysics reports:

- Identify the large-scale characteristics of the universe (quantity, distribution, and nature).
- Problems and energy, times, long history.
- Explore the beginning of the modern universe in which the first stars and galaxies were created.
- Understanding black hole formation of any size.
- Study the formation of stars and planets as well as the birth and development of giant planets and Earthlike planets.
- Understand how the astronomical environment affects the planet.

Similar targets have recently been identified in similar reports in other countries and regions, such as the European Astro Net process.

Radio observation solves this goal differently than other wavelength bonds. It has won several physical awards for its observations on radio astronomy and technology development. This success was almost simultaneous in new technology applications. In addition, powerful centimeter forces and measuring waves give rise to phenomena, hidden objects, and properties that generate radio waves almost everywhere. In addition, this generation of telescopes can be designed with a large area of imaging, polymers, and special features with wide spatial and spectral resolution as well as high sensitivity. All these characteristics can be achieved at the same time using the latest technology. The exact wavelength range is not yet fixed, but SKA provides images and other data at wavelengths of 1 cm (30 GHz) at 4.3 m (70 MHz) [17].

The scientific significance of radio astronomy is the ability to use thin optical spectrum lines with a large number of components that do not respond to the universe, especially the 21 cm hydrogen lines, the most common feature. The ability to study physics under extremes, for example, due to high synchronization accuracy when pulses come from a radio tube; Use of celestial physical trends found in molecular spectrum lines; Solving magnetic fields in space; Indeed, ubiquitous interactions of high-energy electrons and magnetic fields are tracked to generate synchrotron radiation from stars, galaxies, and galaxy clusters.

As modern astronomy develops, the SKA should be designed to monitor these phenomena around the world. Cosmic expansion reduces radiation and receives an observation wavelength. The change in red is defined as the observed radiation, if the view is the wavelength measured by the observer, and the radical world is the wavelength at which long objects are released in space. In the neighboring world, Hubble changes the red color greatly according to the rule $v = c$, where v is the reduced rate and c is the speed of light as the world expands. The world usually has similar but more complex relationships.

In radio astronomy, it has the following scientific significance: many components that do not respond to the Earth, especially the 21 cm hydrogen lines, the ability to use thin spectrum lines with the most common characteristics and so on.

Physics are in extreme conditions. For example, keeping accurate time when the pulses leave wireless coverage; the exploitation of celestial physical trends found in the spectral lines of molecules; solving magnetic fields in space; and almost everywhere interactions in high-energy electrons and magnetic fields are traced to produce synchrotron radiation from stars, galaxies, and galaxy clusters.

The main element of a radio telescope is priced. In astronomical physical situations, especially in the early world, radio emissions are weak, requiring a very sensitive telescope to observe the phenomena described above. To achieve the key science goals of early space research, the SKA must be much more sensitive than wireless telescopes in current and meter wave gauges. This goal is specified in the case of SKA science. This is part of the way in which new science is expressed in potential and technical needs. Department of design solutions available; Articles on key technical issues and costs to be met; Share how your project makes technical and design decisions. At this stage of the project, there is no direct way to make a simple solution. It must be considered that many other aspects are parallel and repeated [18].

The SKA uses radio telescope design methods developed over the last 40–50 years. The concepts of aperture synthesis are more sophisticated than Van Cittert-Zernike's theorem and space, incoming radio emission spectrum, expected field structure, and time.

The rejection of external RFI signals (radio frequency interference) should be refused. To illustrate these concepts with terrestrial radio telescopes, an antenna and set of receivers are configured to cover a large area of land and to provide the required spatial sampling or telescope. The amplitude and diagrams are correctly shown, interconnected in pairs, and integrated to reduce noise. You can use this data to reproduce the original brightness, reproduce the spectrum of each point in the sky, and reproduce the time fluctuations in the spectrum and in space. For decades, telescopes and radio telescopes have been limited to opinions of about 104 m2 (the Arecibo radio telescopes are a special objective consisting of a single stroke with very limited resolution, other than radio telescopes in applications) to achieve mass-mass production of large-convex antenna, optical fiber for massive data transmission, and high-speed digital signal processing to improve computer signal analysis.

3.2 Fast

As shown in Fig. 1.4, the 500-m spherical radio telescope (FAST) is a major scientific project in China to create the world's largest single plate telescope. The innovative design concept and engineering solves the most effective way in the best way. FAST is the Chinese contribution to international efforts to create a

Fig. 1.4 FAST 3-D model. Taken from D. Li, The early science opportunities for the five-hundred-meter aperture spherical radio telescope (fast), Proc. Int. Astron. Union 8 (S291) (2012), 325–330. Fig. 1.

Square Root Collection (SRA). As a single-plate radio telescope, FAST can explore the neutral hydrogen of the Milky Way and other galaxies, detect size pulses, search for the first gas star, and hear possible signs. You can start many scientific goals from other civilizations.

The idea of placing a large spherical disc in a karst depression began with the Arecibo telescope. FAST is an Arecibo aerial with three distinct elements: a depression similar to karst, a large area to a 500 m telescope, and an antiaircraft angle of 40 degrees. The main reflective active spherical cushioning can be adjusted to the ground to achieve full polarization and a wide band, without the need for a complex feed system. The light feeding room is controlled by cables and servers, and in conjunction with a parallel robot as an auxiliary control system, can be moved accurately. With the support of China and the global astronomical community, the FAST feasibility study has been running for 14 years. The National Development and Reform Commission approved rapid funding in July 2007 with a budget of 700 million Yuan. The project commenced 5.5 years from March 2011, and launched in 2016 [18,19].

Fig. 1.4 shows the FAST optical geometry and its three characteristic elements: a large chamber of karst 2 located in the southern part of Guizhou, a reflector with an active core of 500 m, which may have a spherical correction 3, gondola 4, cable-driven and Correct correction servos, and robot parallel as auxiliary controller system, are used to make the most accurate parts of the receiver. Multiband receivers will be installed in the cab, covering a 70 MHz–3 GHz frequency range. The telescope shall be equipped with different tools and terminals for different scientific purposes.

With the cabin suspension system with a deep cut and feed there, a large open angle can be at the reflector fast, a large angle up to 40 degrees and a total light surface of 300 m. Using some special feeding technologies (such as feeding-shaped string (PAF)), you can extend a large zenith angle to 60 degrees south from FAST outside the center of the galaxy. Due to the large collection area and the latest intake system, the initial sensitivity of the band L (the most scientifically important FAST band) reaches $2000\,m^2\,K^{-1}$. In this band, it is proposed to launch 19 horn-based receivers to increase the speed of measurement. The maximum turning time is fixed to 10 min, which is limited by the power of the high-voltage engine. The coverage needs to meet the main part of the fastest-growing scientific problem. When 3 GHz was installed as the upper limit of the first phase of the telescope design, compared to the ceiling of 8 GHz established in the previous model in 2000, the control accuracy requirements were reduced by 2 h. Both construction phases reduced construction risks, project time, and capital budget. The main scientific motivation for creating the largest radio telescope is the unused FAST sensitivity and the high speed of measurement to measure the accuracy of the radio. FAST should improve our understanding of cosmology, the evolution of galaxies, the life-sized fat (ISM), a star formation, and an exoplanet. Scientific goals include:

- HI ISM galactic measure with a resolution comparable to the large-scale CO survey.
- Find out that there are 44,000 new galactic recesses and find out the first galactic wicker.
- Find thousands of HI galaxies and open one big galaxy with a maximum $z \sim 1$.
- Spectral study of the radio frequency spectrum of a rich galaxy source with continuous coverage from 70 MHz to 3 GHz.
- Search radio signals from exoplanets.

Radio frequency interference (RFI) often affects radio surveillance. To achieve high FAST sensitivity, the radio environment must be very quiet. Dawodang has a very quiet radio environment as well as a karst size and shape cavity suitable for FAST's construction. The surrounding mountains offer excellent protection against RFI. Guizhou is located on a geologically stable plate with a very small earthquake. The climate is mild. This makes all the Dawodang sites suitable for rapid radiology.

FAST science sets strict limits on the surface deformation of the principal active reflector. You must check the location of each node accurately. This not only depends on the cable network structure, rear frames, and reflective elements, but also on the condition of thousands of cables and assets. Faster tracking is done by adjusting the reflector in real time. Therefore, the rate of pressure relative to the speed of the accelerator should be high enough to track an empty target (in extreme cases, about 15 h). The shape and size of the panel were carefully chosen to minimize the deformation required by the sphere to make a paraboloid and reduce internal polarization.

Because of the large size of the telescope, there is no strong connection between the reflector and the feed cabin. When the Arecibo model is adopted, 10,000 t of metal will depend on the reflector, but this is not practical. In contrast, FAST power cables support and control cables and servomotors. A secondary adjustment system is used in the cab to achieve the required accuracy.

The design has three main components, including a cable grate that supports and controls the feed cabin, an adjustable auxiliary cabin device that holds the most accurate part of the receiver, and feedback control. A number of models have been developed to test the complex problems of the cabin suspension subsystem. Because of the large size and complex dynamics of cabin suspension, the law of similarity cannot guarantee the accuracy of the reduced model. Only a few inches can move the feed cab after an initial adjustment check. The accuracy of the feed, accessible with secondary stabilizers, was found to be a few millimeters that meet the requirements of the telescope pilot.

4 The challenge of radio astronomy

4.1 System noise

The initial sensitivity and the measurement speed (Ae/Tsys) are proportionate to the results. Since radio astronomy was born, the function of radio telescopes has increased significantly, increasing by 10 times every 10 years. This is almost entirely due to the improved noise system of the existing telescope system, in particular the noise component generated in the first amplifier. A low noise amplifier (LNA) is attached to each antenna in the telescope. To achieve this performance, LNAs must be loaded at 10–20 K. Because systems using this technology are large, complex, and expensive, it is sensible to use them as little as possible. A block assembly area is according to other design constraints. At a wavelength of about 30 cm, LNAs have enough prospects operating at

300 K (room temperature) so that they can balance the future at the longest edge of the wavelength range and use unused SKA frequency amplifiers.

It is clear that with the successful replacement of SPF solutions with AA design solutions, an increased 2 coefficient for SSFoM (Ae/Tsys) is sufficient to compensate for the highest Tsys values that can be generated because AA cannot be achieved by using the translator. In both cases, the initial costs are assumed to be the same. PAF has a similar argument. Therefore, before any of the scaling technologies are used, the performance ratio/cost threshold (SSFoM/cost unit) needs to be crossed [17,20].

As for the actual costs per unit area, AA and PAF may be higher than SPF; the upgrading of the room temperature receiver system appears to be the most likely route for the successful use of AA and PAF. Because the RNA in the AA scheme is distributed evenly over the wells, there is only one design option for LNAs: an LNA room temperature development with a very low noise level. Now, at room temperature in the 700–1400 MHz band, a 20 K noise temperature can be achieved at the LNA input connector on the stand. The PAF model has greater design flexibility because there is a smaller number of spatially concentrated LNAs (i.e., in the central region of the reflector), and real cooling can be done. Therefore, there are two LNA design options: a) room temperature LNA and b) LNA cooling sufficient to exceed the above noise threshold. For option b, also note that as the ambient temperature decreases, the LNA noise improves unilaterally. Significant improvements can be achieved at a physical temperature of 102 K. One of the limiting factors for the best combinations of LNA products at present is that the best receivers have a 2:1 coverage ratio between short and long waves. This means that approximately five LNA feeder combinations are required for each reflector. Very broad channels are tuned for SKA, and routes for Allen telescopic arrays have been developed covering a 15:1 wavelength ratio. It will be a significant improvement in the cost of solutions based on SKA reflectors.

The design of the LNA and antenna as integrated optimized units has a common design material to achieve low noise and high throughput. This applies to an SPF reflector as well as AA and PAF. This is the practical task of design and measurement. Typically, people with the ability to design antennas and transmitters create standardization of input barriers. In the case of AA and PAF, another design parameter is the associated noise between the elements of an adjacent antenna. In addition, special measurement methods should be used to measure and compare the noise characteristics of embedded devices that are difficult to measure individually.

At wavelengths greater than 1 m, the noise from the "sky" begins to increase significantly from the noise of the system. At these wavelengths, LNAs operating at room temperature (300 K) are suitable for SKA. Although they may have to be used in large quantities, they are very cheap to manufacture.

4.2 Antennas and collecting area

The SKA design telescope noise reduction has a major impact on SKA design, preventing further (violent) landings. SKA's science goal requires higher sensitivity than the prescribed amounts available in existing telescopes. Therefore, the only way to increase sensitivity is to increase the totals (ηA), where η is the open efficiency. As the existing telescopes, totals are more effective than 0.5, there is little space left for a significant increase in efficiency, but efficiency must be maintained in new designs. In the case of AA, efficiency means the coefficient of average predicted loss, which is $\cos(Z)$, where Z is the angle of rotation framework. To compensate for this factor, it is necessary to establish the largest AA collection area. Large parabolic electric reflectors are the only design available for wavelengths less than 20 cm, although they can work on longer waves. The version with the primary focus of these antennae can retain high efficiency with diameters greater than 10. The aerial view is reflective according to (λ/d) 2 pounds (increases by wavelength).

SKA's reflective antenna design is suitable for complex and cost-effective balance. The limits are as follows [16,19]:

(a) The collection area should be sufficient to meet the sensitivity characteristics 104 $m^2 = K$ (use the lower edge of sensitivity shown in Table 1.1). Using a 40 K Tsys score and 0.7 aperture efficiency, the total collection area was $5.7 \times 105\,m^2$.

(b) The diameter may not be less than 10 λ_{max}, where λ_{max} is the maximum wavelength, which we believe will be useful for reflective technology. This wavelength shall be determined by the relative success of AA technology, but is unlikely to exceed 1 m.

(c) The shortest operating wavelength is <3 cm (>10 GHz).

This decision was made after several years of engineering and astronomical exchanges due to very high price pressures and many basic SKA sciences, which can be achieved by 10 GHz. An important factor is to decide that SKA's overall target is to achieve a wavelength under 1 cm, but the minimum RFI wavelength should be highlighted when selecting a telescope position, particularly in wavelength systems. Taking all constraints into account, a preliminary study of the optimization problem gave estimates for a 15 m antenna with 2000–3000 diameters.

One of the biggest challenges for SKA is to make these apartments affordable. In the current small market for large reflector antennae, there is no incentive to highlight technologies to produce thousands of antennae. Traditional models for large wireless reflectors use a steel or aluminum space frame structure that supports many adjustable reflection panels. This provides the performance required by SKA, but production, installation, and maintenance costs are significant constraints. The development of new materials and manufacturing technologies based on compost can significantly reduce the cost of gross reflector production. The molded-based method enables the production of a reproducible mirror. Depreciation is reduced in a large number of parts and the exact costs of mold. In addition, the reflector may contain one or more components, which can be installed on the site without any adjustments. The mold-based reflector can be of metal or a composite material. Apart from the advantages in weight and stiffness, the thermal expansion coefficient of its components is almost zero. Rarely feel the desired shape. The reflectors are offset or shaped like symmetrical shapes. The advantage of molded metal antennas is that they are not given much attention to protect them from sunlight, but the compound material must be protected with a UV-resistant coating. While the comparative costs of the two new technologies are not fully estimated, it is clear that each method of manufacture contributes a nearly flat wavelength curve, with a reflection wavelength of 3 cm. It should be noted that the reflector is only part of the reflective antenna model. Manufacturing technology should apply to components, towers, and foundations. AA production has similar challenges. AA still has plenty of space for mass production of thousands of antennae and receiver chains. For example, the technology has been developed to print a series of antennas, and robotic assembly technology can be used to use active AA distribution components. Because AA reveals many active ingredients, the natural challenge is how to prevent damage to the electromagnetic pulses from nearby lightning strikes. This requires design and testing of a lightning protection system for the entire area of the group. As noted in the previous section, the challenge of AA and PAF is to increase research costs at a competitive cost.

4.3 Data transmission

No SKA type could be developed without facilitating the transmission of data by the use of optical fiber transmission technology. However, data transfer limits the performance of SCA, especially those used in long-range antenna clusters (long buses). Fortunately, the creation of a large range of detached views (for example,

10 square phases) with high resolution is not a scientific priority. The rates are R coefficient fees, which refer to the environmental sector and transmission distances. In the simplest case, when an antenna is connected to the total SKA, $B \propto Nbeam$, where B is the total instantaneous bandwidth of the antennae connected to the same units, N is the number of antennae and the number of antennae. The beams in the field of vision. Note that the radius of a single-angle antenna or cluster is consistent with 2/Ae and $Nbeam \propto (Atot/N\lambda 2)\ \Omega$, then $R \propto BAtot\Omega/\lambda 2$, where the EU is an effective surface of the antenna cluster. Atot is the sum of the effective catch area of the telescope and volume field of vision. This proportionality is independent of aperture technology (e.g., SPF, AA, or PAF). In the case of AA and PAF, this is a separate design parameter, and to reduce its cost, it can, if necessary, be limited by the available FoV determined by an antenna model.

The length of the transmission depends on the configuration of the antenna on the ground. In practice, different remote systems require different data transfer technologies and involve different costs. The distance should be at least 30–3106 m. Optical transmission is cheaper than digital transmission over several kilometers, but output system performance has not yet been fully established. Data transmitted over an area must be more than a few kilometers in a digital format. It is clear that the cost depends directly on the number of bits used to encode each sample. In the absence of disturbed signs, radio astronomy can be performed at 2 bits per sample without much impact on performance. Even in the countries where they chose SKA radio stations, RF interference is likely to require at least 4 bits from air and space sources. The data rates for maximum bandwidth, FoV, and total area are as follows: 8 GHz SPF bands creating 160 Gbit/s (assuming 25% higher) for each disk, or 480 T/total for one band 3000; PAF with polarization bandwidth of 700 MHz to be generated at 840 t/s with 2000 nut per container; and a total of 250 AA nuts covering a range of 250 degrees with bandwidth generates 700 MHz of full data at 4.1 Pbps. The cost of these very high data rates is likely to limit bandwidth, FoV, or bandwidth.

The problem with the "storage wall" in SKA scientific computing applications is that the problem is with bandwidth I/O, one of the main bottlenecks in the system. Even if Tian He # 2 is not used properly, there is not enough space for large data such as SKA, and it is difficult to observe and analyze the accident, new architecture is essential based on this data. Audit. intensive processing of scientific data. As mentioned above, the rapid implementation of SKA should be reduced by the following real-time processing workloads. In the real case, when processing data in real time, there are certain requirements for designing software and hardware systems for the overall system architecture for integrated installation as well as the centralized application of computer systems. Software algorithms have many problems such as data center monitoring, refrigeration cabinet, full control. To fund the construction of the ceiling, the default computer requirements and the real-time and low-energy requirements as well as corresponding operating costs are required. In addition, storage, archiving, searching, and calculating a large amount of data offers very high requirements on the entire network of ecological computers.

The task of maintaining the radio astronomy sky is growing rapidly, as it did over the past 50 years or more. The increase in electronic devices, the increase in power, the capture of digital and wireless electronics, and increasing the transmitter power and frequency flexibility create challenges for successful research telescopes. In addition, there is an increasing desire to monitor a range of protected radio astronomy, for example, broadband pulse surveys or HI translated to red.

The National Radio Astronomy Observatory, located in Green Bank, West Virginia, operates two quiet zones wireless community, 13,000 square feet of radio silence, and 10 miles west of astronomy radio and television. If society participates more actively in electronic devices and communication technologies, astronomers find it difficult to maintain calm zone airwaves. Examples include the transition to LED lighting, authorized by the federal government, and the wider use of electronic devices, such as radar and WiFi in cars. Increasing the number of high-power satellites and increase the frequency capacity of many transmitters.

These problems have been resolved to some extent, as a new reception technology (phased array) and advanced RF liquefaction support astronomy observation in the presence of these electronic transmissions. Here, astronomers around the world allocate future and possible future problems, even in a quiet area of protective airwaves.

5 The development tendency of radio astronomy

5.1 Mid-frequency aperture arrays

Astronaut technology is at the forefront of new developments and discoveries in astronomy. LOFAR currently exhibits dense and rugged AA capabilities at low frequency. For AAs between 450 and 1450 MHz, they must prove their scientific value in existing technology. Their appearance and flexibility put them in a great position. The Extension Cluster Control Program is committed to showing a general interest in science, in particular. For the central region, this is the state of AIA. And it has developed an EMBRACE that has already demonstrated high variability in pulsar visualization systems at the same time. It also serves as a test to prove the AIA's technical reliability and stability. The next step is AA-level technology that can be used for advanced science.

For decades now, technology has reached the stage of preparation for the development of future scientific instruments, especially square kilometers arrays (SKA). In this section, we explore dense mass gaps optimized to control the frequency range of individual receptors at frequencies between 450 and 1450 MHz. This technology is part of the program to improve SKA devices.

There are many benefits to a narrow diet. Individual members are basically broadband recipients, so they are strong candidates for individual antenna members. Due to these high-frequency problems, the cost of the required performance is reduced a lot. If the EU has no body to move, they can be repeated for a few seconds. This makes the technology most suitable for fast-tracking events. The unique nature of AA can be seen in many directions at once. More than one pixel in each field is a convenient technology for creating the cheapest and most effective research tool from AA. The field width is limited to the basic level of processing power, which can create unprecedented speed capability. The main problem with using this is to reduce the costs of woodworking and processing.

AA works in a multibeam format. First, the elements of individual wavelengths are mixed in a small radius, which is equal to the wavelength of the material. This specifies the type of device in the field.

5.2 Entering a near future

We are having a very interesting time in radio astronomy as the first results begin in the next-generation radio telescopes. Instead of using large plates, these radio telescopes use suction sensors to combine the signals from several small antennas. This method allows for greater space and larger assembly space. It can also produce more dynamic equipment. Antenna subarrays can be sacrificed to simultaneously view multiple targets or to maximize the viewing area. Alternatively, we can use the sub editor to see different frequencies and add bandwidth to one source view. This type of telescope module is as easy as it can be improved, allowing us to add faster computers or more recipients.

The goal of transition science is to explore the space available in these new tools and as many new tracks as possible. This requires a monitoring mechanism. The sensitivity of the new radio telescope is increased so we can find very harsh but very common radio changes. When complete, the Meer KAT will be a very attractive radio telescope with a 20-km vessel of 6413.5 m. Another equally important Islamic strategy is the most unusual but wonderful material. With the significant increase in various tools, such as ASKAP, this type of observation can do this. ASKAP's unique 1.4 GHz scene is square feet. The current field of visual arts is the 13-wave receiver of Parks' telescope, which includes frequencies in squared steps. Not a mysterious telescope, but the majestic view of the park is the only telescope to recognize the barrel. With ASKAP you can get more out of these events [19,20,21].

In fact, the easiest way to explore a new measurement space is to look at it in a new frequency band. It was impossible to make high-quality observations at low frequencies before design and when presenting composite compositions. The quality and sensitivity of conventional telescopes depends on the size of the dishwasher. We are looking for ships that are technically impossible to construct a reasonable solution at low frequencies. However, current tires replace long tires, which can be viewed using LOFAR (Neutral Miniature), LWA (Long Wave Array), and MWA (Broadband Array). If the incidence is low, wireless observation can't be done on the reflector. The incredible feel of this instrument is very useful for searching for new types of transitions in uncaptured frequency bands.

Each of these telescopes is considered a vehicle, a small project that uses new alternative and usability technologies to develop the telescope. The flow of information is stored on one disk but is expected to be reduced in multiple folders before it is sent to the network operator. Real-time searches for unique GPUs and functional devices are currently the only two solutions available to address these challenges.

It's too early to guess whether a new generation of radio devices will cause a new generation of floods. We can be sure that some assumptions must be partially met, as the efficiency and power of the survey increases. It's definitely a secret to a strange new cross. If the past shows the future, you may need something special, such as a tree, to start a new golden age.

References

[1] R.D. Ekers, Non-thermal radio astronomy, Astropart. Phys. 53 (2) (2014) 152–159.

[2] A. Hewish, Early techniques in radio astronomy, Adv. Imaging Electron Phys. 91 (1995) 285–290.

[3] J. Bennett, Instruments in the history of astronomy, Endeavour 23 (3) (1999) 98–99.
[4] A. Hadravová, P. Hadrava, Astronomy in Prague: from the past to the present, Proc. Int. Astron. Union 2 (14) (2006) 3–13.
[5] Chengalur & Jayaram, N., Radio astronomy and the giant meter-wave radio telescope, Resonance 23 (2) (2018) 165–182.
[6] P.E. Dewdney, P.J. Hall, R.T. Schilizzi, T.J.L.W. Lazio, The square kilometer array, Proc. IEEE 97 (8) (2009) 1482–1496.
[7] K. O'Neil, J. Ford, M. Mccarty, T. Minter, R. Prestage, Future challenges for radio astronomy, in: Radio Science Meeting, IEEE, 2013.
[8] D. Li, The early science opportunities for the five-hundred-meter aperture spherical radio telescope (fast), Proc. Int. Astron. Union 8 (S291) (2012) 325–330.
[9] D.L. Jones, K. Wagstaff, D.R. Thompson, L. D'Addario, U. Rebbapragada, Big data challenges for large radio arrays, in: Aerospace Conference Proceedings, IEEE, 2012.
[10] S.J. Wijnholds, R. Nijboer, K.J.B. Grainge, J.D. Bregman, Overview of SKA calibration challenges and impact of design decisions, in: General Assembly & Scientific Symposium, IEEE, 2011.
[11] A. van Ardenne, New generations of radio telescopes: antenna concepts and technologies, in: IEEE Twelfth International Conference on Antennas and Propagation, 2003, pp. 526–529.
[12] B. Juswardy, F. Schlagenhaufer, P.J. Hall, Radio interference evaluations of photovoltaic modules for radio astronomy active antenna, in: 2013 Asia-Pacific Symposium on Electromagnetic Compatibility (APEMC), IEEE, 2015.
[13] R.V. Kozhyn, V.V. Vynogradov, D.M. Vavriv, Low-noise, high dynamic range digital receiver/spectrometer for radio astronomy applications, in: The Sixth International Kharkov Symposium on Physics and Engineering of Microwaves, Millimeter and Submillimeter Waves and Workshop on Terahertz Technologies, IEEE, 2007.
[14] R.D. Norrod, J.R. Fisher, B.D. Jeffs, K.F. Warnick, Development of cryogenic phased array feeds for radio astronomy antennas, in: IEEE International Symposium on Phased Array Systems & Technology, IEEE, 2010.
[15] T.S. Bird, Role of radio astronomy as a testbed for future wireless applications, in: International Conference on Electromagnetics in Advanced Applications, IEEE, 2016.
[16] A. Soliman, S. Weinreb, Optimization of small reflector antennas for radio astronomy, in: IEEE 2016 United States National Committee of URSI National RADIO Science Meeting (USNC-URSINRSM), 2016, pp. 1–2.
[17] I.M. Van Bemmel, A. Van Ardenne, J.G.B. De Vaate, A.J. Faulkner, R. Morganti, Mid-frequency aperture arrays: the future of radio astronomy, in: Proceedings of the Meeting "Resolving the Sky – Radio Interferometry: Past, Present and Future", 2012.
[18] R.P. Breton, T. Hassall, The future for radio astronomy, Astron. Geophys. 54 (6) (2013) 6.36–6.39.
[19] P.A. Abiola, F.O. Emmanuel, Intelligent cognitive radio models for enhancing future radio astronomy observations, Adv. Astron. 2016 (2016) 1–15.
[20] K. Michael, Radio astronomy in the future: impact on relativity, Proc. Int. Astron. Union 5 (S261) (2009) 366–376.
[21] P. Sarti, M. Negusini, S. Montaguti, F. Mantovani, F. Buffa, G.L. Deiana, An overview of the sardinia radio telescope geodetic potential at national and international levels, Mem. S. A. It. Suppl. 10 (2006) 107.

Fundamentals of big data in radio astronomy

Jiale Lei, Linghe Kong
Shanghai Jiao Tong University, Shanghai, China

1 Big data and astronomy

1.1 Background of big data

Today, there is no doubt that we are living in an era of big data. It is apparent that data have increased rapidly in many mainstream fields over these years. Fig. 2.1 shows the global data volume from 2010 to 2025 published by International Data Corporation (IDC), and the data from 2020 to 2025 is predicted in some statistical way. As shown in Fig. 2.1, in 2018 the total data generated and copied all over the world was about 33 zetabytes (ZB) while the number was only 2 ZB in 2010. IDC also predicted that the figure will grow to an incredible 175 ZB by the year 2025. It is true that data is growing explosively.

With the exponential increase of data, the term "big data" has been more of a concern in many fields. At the beginning, big data was mainly used to refer to enormous datasets. Unlike traditional datasets, big data generally includes great amounts of unstructured data. The unstructured data typically needs more real-time processing and analysis. Big data usually contains hidden values that need to be revealed through appropriate big data technology, which brings about both opportunities and challenges.

The foremost industry that has confronted big data challenges is probably the Internet companies. It is reported that Google processes hundreds of PB (about 10^{15} Bytes) of data per month, and Facebook generates log data at the PB level every month. Baidu, a Chinese search engine company, processes tens of PBs of data per day, and Alibaba's subsidiary Taobao generates up to tens of TB data for online trading. More incredible is Taobao's "double eleven" shopping day on Nov. 11, which is regarded

Big Data in Astronomy. https://doi.org/10.1016/B978-0-12-819084-5.00010-9

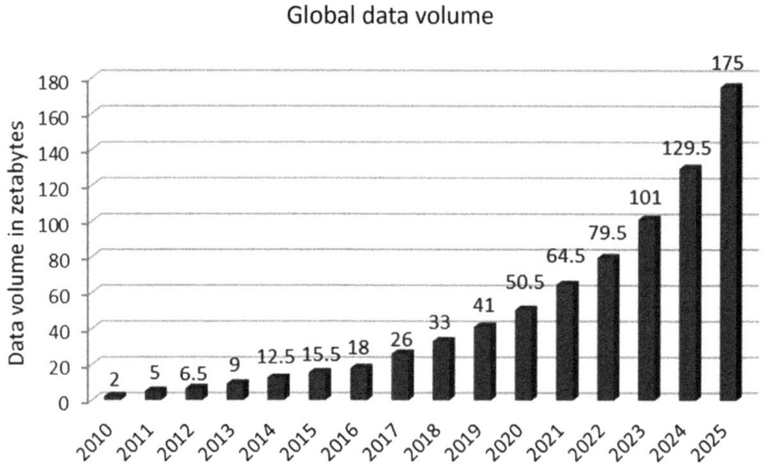

Fig. 2.1 The continuously increasing global data volume published by IDC.

as a shopping festival just like Black Friday. On that day in 2019, the total volume of trade reached about $28.2 billion, which smashed records. The peak of orders that day was 544,000 orders per second, and over the whole day, the total data volume reached 970 PB.

Not long ago, almost all these decision makers began to get interested in big data because of its high potential. For example, many government agencies have announced major projects to accelerate big data research and applications. In addition, big data has also became famous in the academy. Two premier scientific journals, Nature and Science, have opened special columns for big data.

While the volume of large datasets is growing, this begs the following questions: How can we collect and integrate massive data from various data sources? How do we store and manage such vast heterogeneous datasets? How can we process and analyze the datasets at different levels effectively?

1.2 Definitions and features of big data

Nowadays, big data contains many meanings. People still hold different viewpoints on the definition of big data, though its importance has been well recognized. In general, big data means the datasets that can hardly be handled by traditional architectures within an acceptable time. From different perspectives, scientific and technological enterprises, research scholars, data analysts, and technical practitioners give different definitions of big data. Some mainstream definitions of big data are listed in the following to help us get a better understanding of the profound social, economic, and technological connotations of big data.

The very first definition of big data may be traced back to 2011, when Doug Laney, an analyst at META (presently Gartner), defined challenges and opportunities brought about by increased data with a 3Vs model, that is, **the increase of volume, velocity, and variety**, in a research report. Although such a model was not originally used to define big data, Gartner and many other enterprises, including IBM and some research departments of Microsoft, still used the "3Vs" model to describe big data within the following several years. One of the most popular and formal definitions of big data was given by Apache Hadoop. It defined big data as "datasets (that) could not be captured, managed, and processed by general computers within an acceptable scope." And based on this definition, in May 2011, McKinsey and Company, a global consulting agency, defined big data as "the next frontier for innovation, competition, and productivity." This definition includes two connotations: First, dataset volumes that conform to the standard of big data may grow over time or with technological advances; and second, dataset volumes that conform to the standard of big data in different applications differ from each other. At present, the volume of big data generally ranges from several TB to several PB, and it is possible that the volume can still rise. From the definition by McKinsey and Company, it can be seen that the volume of a dataset is not the only criterion for big data. The increasingly growing data scale and its management that could not be handled by traditional database technologies are the next two key features.

Others also have different opinions, including IDC. In 2011, an IDC report announced big data as "big data technologies describe a new generation of technologies and architectures, designed to economically extract value from very large volumes of a wide variety of data, by enabling the high-velocity capture, discovery, and/or analysis." With this definition, the characteristics of big data may be summarized as four Vs, that is, **volume (great volume), variety (various modalities), velocity (high-velocity generation), and value (huge value but very low density)**. Because of its emphasis on the meaning and necessity of big data, such a 4V definition was widely recognized. This definition indicates the most critical problem in big data, which is how to explore values from datasets with an enormous scale, various types, and rapid generation. As Jay Parikh, Deputy Chief Engineer of Facebook, said, "You could only own a bunch of data other than big data if you do not utilize the collected data." In addition, NIST defines big data as "Big data shall mean the data of which the data volume, acquisition speed, or data representation limits the capacity of using traditional relational methods to conduct effective analysis or the data (that) may be effectively processed with important horizontal zoom technologies," which emphasizes the technological aspect of big data. It indicates that efficient methods or technologies need to be developed and applied to analyze and process big data.

Moreover, Krik Borne, Principal Data Scientist and Executive Advisor at Booz Allen Hamilton, even put forward a "10V" model, that is, volume, variety, velocity, veracity, validity, value, variability, venue, vocabulary, and vagueness. In this section, we are more interested in the four Vs stated above: volume, velocity, variety, and value, which are listed below:

Volume means that the scale of big data can be very large, and usually such big data needs to be measured by a larger unit such as TB, PB, or even EB, which represent 10^{12} B, 10^{15} B, and 10^{18} B. Therefore, big data brings about challenges for big data collecting, cleaning, storage, management, processing, transferring, and visualization. Such operations can hardly be handled by traditional architectures.

Velocity means the processing velocity of big data. The operations must be rapid, especially for data collection and data analysis so as to maximally utilize the commercial value hidden in big data.

Variety means the various types of data. Data can be divided into structured data, semistructured data, and unstructured data. Typical semistructured and unstructured data include audio, video, webpages, and text. In general, each data item can contain many features, and data from different data sources may have their own formats, which causes problems in the analysis phase.

Value describes the high potential value of big data. For instance, it is interesting and inspiring in astronomy to discover surprising, rare, unexpected, and new objects or phenomena. Also, the discovery of a new distribution trend or law is of great value. Nowadays, data has become an important production factor that could be comparable to material assets and human capital. With the development of multimedia, social media, and IoT, enterprises will collect more information, leading to an exponential growth of data volume. Big data will have a huge and increasing potential in creating values for almost all fields.

In conclusion, there have been considerable discussions from both industry and academia on the definition of big data. In addition to developing a proper definition, it is also of great importance to extract its value, how to use data, and how to transform "a bunch of data" into "big data."

1.3 Development of big data

The development of big data is closely related to the development of databases. In the late 1970s, a database machine was proposed to store and analyze data. However, with the rapidly increasing data volume, the capacity of a single computer system became inadequate in processing larger datasets, and parallel systems were in urgent need. Therefore, in the 1980s, "sharing nothing" emerged to meet the requirements of the increasing data volume. The parallel system was based on the usage of a cluster where each machine had its own processor, memory, and disk. The first successful commercial parallel database system was call Teradata. In June 1982, one of the milestone events in databases occurred when Teradata delivered the first parallel database system with a storage capacity of 1 TB, which helped a large-scale retail company in North America expand its data warehouse. In the late 1990s, researchers in the database field recognized the strengths of parallel databases.

Challenges came up with the development of databases. For example, the Internet services have been popular since the end of the last century and are still booming. With larger volumes of data generated by the Internet, indexes and queries grew rapidly. Search engine companies are typical examples. To cope with the challenges in data management and analysis at the Internet scale, Google developed the GFS and MapReduce programming models. Moreover, users, sensors, and other kinds of data sources also generate great amounts of data. Some researchers believe that only a fundamental revolution in traditional computing architecture and large-scale data processing mechanisms can help cope with these problems.

Another milestone in big data occurred in 2011 when EMC/IDC released a research report titled, "Extracting Values from Chaos" in which researchers presented a new concept, "big data." They also introduced the potential of big data in the research that successfully triggered the great interest and attention in both industry and academia on big data.

Many major Internet companies started their big data projects over the last few years. For example, IBM has invested \$16 billion on 30 acquisitions related to big data since 2005. Other companies, including EMC, Oracle, Microsoft, Google, Amazon, and Facebook, also have plans in big data. Big data was also under the spotlight in academia. In 2008, a big data special issue was published in Nature, and in 2011, Science also launched a special issue on the core technology of "data processing" in big data. In 2012, the European Research Consortium for Informatics and Mathematics (ERCIM) News published a special issue on big data. At the beginning of 2012, a report titled "Big Data, Big Impact" was presented at the Davos Forum in Switzerland. In the report, big data was announced as a new kind of economic asset, just like currency or gold. Gartner issued Hype Cycles from 2012 to 2013, which divided big data computing, social analysis, and stored data analysis into 48 emerging technologies that require the most attention.

Many national governments also paid more attention to big data. Taking the US government as an example, in March 2012, the Obama administration authorized a $200 million investment to start the "Big Data Research and Development Plan," which became a second major scientific and technological development initiative after the "Information Highway" initiative in 1993. Then, in July 2012, the "Vigorous ICT Japan" project was issued by Japan's Ministry of Internal Affairs and Communications, which indicated that the big data development should be a national strategy and application technologies should be the focus. In the same month, the United Nations presented a "Big Data for Development" report that summarized how governments utilized big data to better serve and protect their citizens [1].

1.4 Big data in astronomy

Astronomy is probably one of the oldest observational sciences. Humans have been observing and trying to understand the mysterious luminous objects in the sky for more than 10,000 years, and this has been an important part of human culture. However, today astronomy has changed. In the past, a single astronomer or a small group might observe a couple of objects. Now such methods seem to be out of date, as large digital sky surveys predominate. In other words, modern astronomy has become a data-rich field.

Nowadays, large digital sky surveys have become dominant in astronomy thanks to the development and construction of large-scale telescopes. As a consequence, datasets captured by these large-scale telescopes are incredibly huge. According to the statistics, about 10 terabytes of image data are involved in a general sky survey, and the scale is still increasing. The data mainly include billions of detected sources, such as stars, galaxies, and quasars, with their measured attributes. New technologies in electronics have brought about new instruments that lead to the exponential increase of information in astronomy. Thus, datasets are orders of magnitude larger, more complex, and more homogeneous than in the past. In summary, astronomers have to confront the challenges of big data.

It is true that the data volumes now are several orders of magnitude larger than what astronomers and statisticians used to handle. These datasets might also be more complicated and higher dimensional. Of course, such huge datasets can be great opportunities for astronomers, allowing them to make world-changing discoveries. However, challenges come first, including management, storage, and the combining and analysis of data. Whereas some powerful techniques have already been tested or applied in these new astronomical applications, others will have to be created, in collaboration, by astronomers, statisticians, and computer scientists.

What do these images captured by large-scale telescopes look like? The current and forthcoming data (probably more than 100 terabytes) span the full range of wavelengths, radio through x-ray and beyond, and potentially provide a panchromatic and less-biased view of the sky. Actually, the universe looks different at different wave bands. It is known that some wave bands such as x-rays, radio, infrared, and ultraviolet cannot be observed by the human eye. To show these invisible wavelengths, detectors are specially designed to capture these wave bands. Detectors then produce false color images that refer to images using visible colors to show invisible lights. That means the colors used in false color images are not real, but just for bringing out important details. Also, astronomers observe the images at different wavelengths to discover important information about the nature of celestial objects. Such a panchromatic approach to the universe survey reveals a more complete physical picture.

In general, astronomical data begin with digital images over a certain portion of the sky at a certain wavelength. Scientists try to remove instrumental effects from the data to generate a quantifiable image of flux, that is, light energy per square centimeter per second, as a function of two spatial coordinates (projected on the sky). In some cases, datacubes rather than two-dimensional pictures are produced in which the other dimensions can be the wavelength, time, or other attributes. Now that dimension time is being considered in astronomical imaging, a lot of variable stars and transient phenomena are being revealed by repeated views of the night sky, including Type Ia supernovae (exploding stars used to quantify dark matter and dark energy in the universe) and planetary transits (when a planet passes directly between a larger body and the observer) [2]. Some of us may have witnessed the planetary transit when Venus passed across the disk of our sun. Such a phenomenon is valuable because it can help us discover planets orbiting more distant stars outside our solar system, and this has become a general method to detect new planets in astronomy. The Kepler space telescope is a good example. In March 2009, NASA launched the Kepler space telescope with the purpose of detecting exoplanets, which are planets outside our solar system. By the time it retired, Kepler had observed more than 500,000 stars and detected at least 2600 planets. Among the planets detected by Kepler, there was a super Earth planet named Kepler-452b that was believed to be suitable for human habitation, which caused a sensation.

Once raw data is collected, some source-finding algorithm is then run to identify individual or discrete astronomical sources including stars, galaxies, quasars etc. Also, the way their flux is distributed spatially is parameterized in wavelength and so forth. The number of independent measured parameters for each source then defines the dimensionality of a parameter space, and each source can be represented as a vector in this parameter space of observed attributes. Instances of parameters usually include fluxes, flux ratios (also known as colors), sizes, and measures of the image shapes and concentrations. Many modern digital sky surveys parameterize hundreds of attributes for each detected source. This parameter space representation then, in principle, contains all the information present in the original data in a condensed form. In other words, it ignores the "empty" pixels but is still related to the detected source. In addition, the original data are often transformed into a quantitative form that is suitable for statistical analysis.

In summary, such a systematic and data-driven approach will probably enable new science in addition to what can be done by individual researchers. It will enable meaningful, effective experiments within these vast parameter spaces [3].

1.5 Statistical challenges in astronomy

Today, astronomy is rich with data. Because of the particularity of astronomy research, many challenging problems arise. Nowadays, data collected by digital telescopes are extremely massive. Despite such huge data volume, it is still difficult to determine whether these observed groups of celestial bodies or phenomena can constitute a typical and unbiased sample of the vast underlying population of similar objects. Because there are countless stars, galaxies, molecular clouds, and other celestial bodies in the universe, those detected are only a tiny part, making it hard to infer the underlying distribution of the population. Besides, some faint transient signals will subside in no time after they emerge from the noise once or several times. Such faint transient signals are of great importance for quantifying the role of dark energy and other cosmology studies, especially in time-domain astronomy. However, such signals usually last for an extremely short while, and can be difficult to capture and record. In addition, measurement errors and flux limits may occur frequently when collecting data by telescopes, which leads to truncation. When this happens, some common statistical methods for big data analysis may fail. Furthermore, models used to fit astronomical spectra are usually nonlinear models with confidence limits on the best-fit parameters, which will make the analysis more complex.

Some other main challenging questions are:

- Is there any intrinsic physical relationship between two or more properties of a class of objects, especially when the collected data are affected by confounding variables or observational selection effects?
- How to quantify continuous structures observed in the sky such as the cosmic microwave background and intergalactic gaseous media?
- How do we interpret the vast range of temporally variable objects: periodic signals from rotating stars or orbiting extrasolar planets, stochastic signals from accreting neutron stars or black holes, explosive signals from magnetic reconnection flares or γ-ray bursts?

Such questions are very common today, and many researchers have made efforts on these challenges. New technologies are also expected to emerge to deal with these problems in more efficient ways [4].

2 Increasing data volumes of telescopes

Astrophysics and cosmology are rich with data. The advent of wide-area digital cameras on large aperture telescopes has led to ever more ambitious surveys of the sky. Data volumes of entire surveys a decade ago can now be acquired in a single night. Fig. 2.2 illustrates the data volume of some existing and upcoming telescopes including the Very Large Telescope (VLT), the Sloan Digital Sky Survey (SDSS), the Visible and Infrared Telescope for Astronomy (VISTA), the Large Synoptic Survey Telescope (LSST), and the Thirty Meter Telescope (TMT). We see that modern digital telescopes can easily generate TB-level data during a

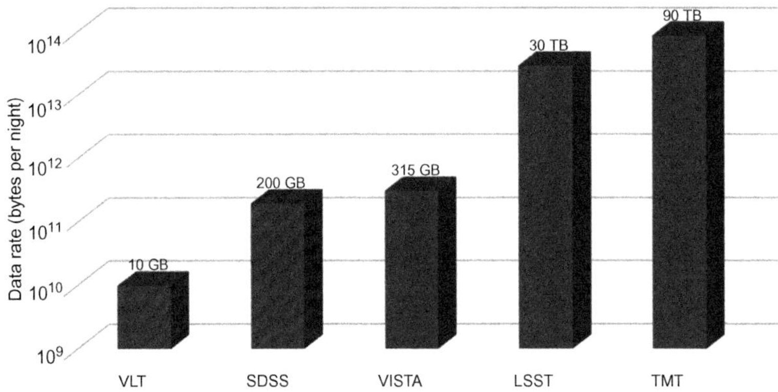

Fig. 2.2 Data volumes of the Very Large Telescope (VLT), the Sloan Digital Sky Survey (SDSS), the Visible and Infrared Telescope for Astronomy (VISTA), the Large Synoptic Survey Telescope (LSST), and the Thirty Meter Telescope (TMT).

single night. In the following, we will talk in more detail about the telescopes in Fig. 2.2, except VLT.

2.1 Sloan digital sky survey

It was the Sloan Digital Sky Survey (SDSS), designed in the 1990s by James Gunn and many colleagues, that brought astronomy into the big data era—and here "big" really does mean massive. The entire telescope was designed around a thorough visible-light survey of one-third of the celestial sphere, obtaining positions and brightness for a billion stars, galaxies, and quasars along with spectra of a million objects. With a modest cost of around $100 million, the SDSS telescope, still thriving today, produces about 200GB of data every night, sending that to a database that stands at around 50TB. The scientific impact of the SDSS has been phenomenal: more than 3000 papers on topics ranging from comets to cosmology, generating today more than 160,000 citations.

2.2 Visible and infrared survey telescope for astronomy

The VISTA (Visible and Infrared Survey Telescope for Astronomy) is a wide-field reflecting telescope with a 4.1-m mirror, located at the Paranal Observatory in Chile. It is operated by the European Southern Observatory and started science operations in December 2009. VISTA is a survey telescope working at infrared wavelengths, and is one of the largest telescopes in the world dedicated to surveying the sky at near-infrared wavelengths. The scientific goals of the VISTA surveys, include many of the most exciting problems in astrophysics today, range from the nature of dark energy to the threat of near-Earth asteroids. Its performance requirements were so challenging that it peaks at 55 megabytes/s data rate with a maximum of 1.4 terabytes of data per night.

2.3 Large synoptic survey telescope

One of the largest of the planned projects is the Large Synoptic Survey Telescope (LSST), rated the top priority new telescope by the US National Academy of Sciences with work scheduled to start around 2020. It will photograph half the sky every three nights using wide-field optics and a gigantic gigapixel charge-coupled device (CCD) camera. Particular scientific goals of the LSST include studying dark energy and dark matter, mapping small objects in our solar system, detecting transient optical events, and mapping the Milky Way. Millions of transients are expected to be detected by LSST per night, and this requires real-time detection to enable follow-up observations. Each night, up to 30 TB of images are expected to be produced, and to save these images, a 60 PB database will be used; the reduced object table will contain around 20 billion rows with dozens of columns [5].

2.4 Thirty meter telescope

The Thirty Meter Telescope (TMT) is an extremely large telescope (ELT) that was proposed by the US National Academy of Sciences in 2000; in that same year, the University of California and Caltech started to develop it. The telescope is designed for observations from near-ultraviolet to mid-infrared, that is, 0.31 to 28 μm wavelengths. Moreover, its adaptive optics system will contribute to correcting for image blur caused by the atmosphere of the Earth, helping it reach the potential of such a large mirror. According to Wikipedia, the TMT will have the highest elevation and will be the second-largest telescope among existing and planned ELTs once it is built. The TMT is expected to explore the nature of dark matter and dark energy, to understand the formation and growth of black holes, and to detect exoplanets. It is estimated that the TMT can give approximately nearly 90 TB data per night, which urgently requires more advanced solution to cope with the storage and analysis of such massive volumes of data.

3 Existing methods for the value chain of big data

In this section, we will pay more attention to the four phases of the value chain of big data, that is, data generation, data acquisition, data storage, and data analysis. Each will be introduced in the following:

3.1 Data generation

Data generation is the beginning of big data. Fig. 2.3 shows some current sources of big data, such as trading data, mobile data, user behavior, sensing data, Internet data, and other sources that are usually ignored. For example, nowadays Internet data has become a major source of big data where huge amounts of data in terms of searching entries, chatting records, and microblog messages are produced every day. Such data are closely related

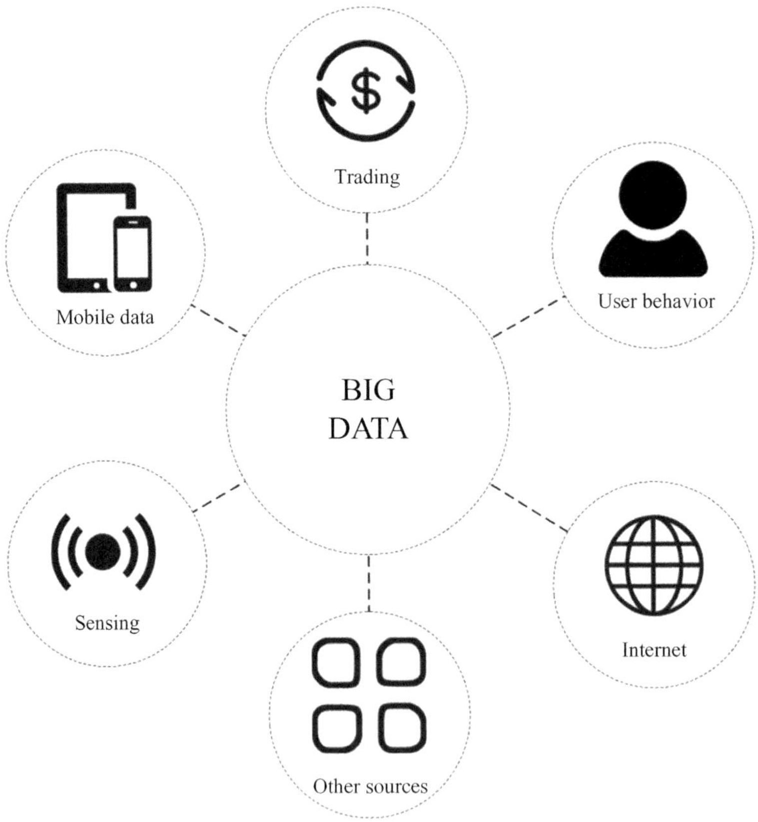

Fig. 2.3 Some sources of big data.

to people's daily lives, and may contain users' behavior. For individuals, the data seems valueless; however, useful information including user habits and hobbies can be determined and collected through the exploitation of such accumulated big data. Big data even makes it possible to predict users' behaviors and emotional moods.

Internet data is one of the most successful data sources utilized by many Internet companies to generate user portraits and provide personalized recommendation services. Other main sources of big data include the operation and trading information in enterprises, logistic and sensing information in the Internet of Things (IoT) networks, human interaction information, position information in the Internet world, etc. In addition, digital telescopes also generate massive data ranging from hundreds of GB to tens of TB or even larger, which is a rising source of big data.

3.2 Data acquisition

Data acquisition is the second phase of the big data system, including data collection, data transportation, and data preprocessing. The main task in this phase is to utilize an efficient transmission mechanism to transmit data to a proper storage management system to support different analytical applications after we collect the raw data.

Data collection is directly related to data generation. Some general methods for data collection include log files, sensing, methods for acquiring network data, and various packet capture technologies. However, the collected datasets often contain much redundant or useless data that unnecessarily increase storage space and get follow-up data analysis in lower efficiency. One example is that high redundancy is very common among datasets collected by sensor networks for environmental monitoring. Data compression technology is usually utilized to reduce the redundancy. Other data preprocessing operations are also helpful to ensure efficient data storage and exploitation.

Once the raw data have been collected, the data will be saved in a data storage system for further processing and analysis. In general, big data is mainly stored in a data center, which consists of multiple integrated server racks interconnected with an internal connection network. In some cases, to improve computing efficiency or facilitate hardware maintenance, data is expected to be transmitted from one data center to another. Therefore, data transmission consists of two types: inter-DCN transmissions and intra-DCN transmissions.

Inter-DCN transmissions occur between one data center and another, and are usually achieved with the existing physical network infrastructure. One major technology is IP-based wavelength division multiplexing network (WDM) architecture. So far, the single channel rate can reach 40 GB/s in the WDM optical transmission systems. For wireless systems, orthogonal frequency-division multiplexing (OFDM) is believed to be one of the main candidate technologies for future high-speed optical transmission. It also supports parallel transmission by segmenting a high-speed data flow to transform it into low-speed sub data flows that are able to be transmitted over multiple orthogonal subcarriers. Different from inter-DCN, intra-DCN transmissions are data transmissions that occur within data centers. Such transmissions rely on the communication within the data center. Nowadays, the internal connections of many data centers are designed in structures based on multicommodity network flows. In such structures, the servers in the data center are connected by switches that province Gbps-level transmissions. Recently, optical interconnection among the networks in data centers has also drawn great interest. Optical interconnection is a high-throughput, low-delay, and low-energy-consumption solution, and such networks in data centers provide a feasible solution that supports Tbps-level transmission bandwidth with lower energy consumption.

As for data preprocessing, there are still some common issues to resolve. Besides the redundancy mentioned above, the collected datasets vary with respect to noise, consistency, etc., because of the wide variety of data sources. Such problems will undoubtedly waste storage capacity to store meaningless data. Consequently, we shall sometimes preprocess data to integrate the data from different sources to make follow-up analysis more effective. In addition to lowering storage expenses, preprocessing data can also improve analysis accuracy. At present, common preprocessing techniques include integration and cleaning. The former means the combination of data from different sources and provides a uniform view of data while the latter means a process to identify inaccurate, incomplete, or unreasonable data, which will be then modified or removed to improve data quality.

3.3 Data storage

Big data has strict requirements on storage and management. Traditionally, data storage infrastructure is used to store, manage, look up, and analyze data with structured RDBMSs. Nowadays, data is already regarded as a new kind of asset, hence data storage devices are becoming increasingly more crucial. Big data storage is about the storage and management of massive datasets to reach the reliability and availability of data accessing. We will discuss important issues including massive storage systems, distributed storage systems, and big data storage mechanisms in this part. For one thing, the storage systems are expected to provide information storage service with reliable storage capacity, and for another, such systems must provide a powerful access interface for the query and analysis of a large amount of data.

Many storage systems have emerged to meet the demands of massive data. Existing massive storage technologies can be classified as direct attached storage (DAS) and network storage, and network storage can be further classified into network attached storage (NAS) and storage area network (SAN). DAS consists of various hard disks that are directly connected with servers. However, DAS is only useful when the scale of servers is small because each device takes a certain amount of I/O resource and is controlled by individual application software. Therefore, DAS is mainly used in personal computers and small-sized servers. In contrast, NAS is an auxiliary storage equipment of a network. In NAS, data are transmitted in the form of files through a hub or switch within the network. While NAS is network-oriented, SAN is especially developed for data storage with a scalable and bandwidth-intensive network.

In general, a data storage system can be divided into three parts: disk array, connection and network subsystems, and storage management software. Disk array is the foundation of a storage system and the fundamental guarantee for data storage. Connection and network subsystems provide connection among one or several disk arrays and servers. Storage management software handles data sharing, disaster recovery, and other storage management tasks of multiple servers.

To develop a large-scale distributed storage system for big data is also a challenge. To use a distributed system to store massive data, the following factors should be taken into consideration:

Consistency: a distributed storage system requires multiple servers to cooperatively store data. Because there are many servers, the probability of server failures will become larger. To cope with failures, data are generally divided into multiple pieces to be saved at different servers to ensure availability. However, server failures in parallel storage may cause inconsistency among different copies of the same data. In conclusion, consistency means that multiple copies of the same data should be identical.

Availability: as multiservers are used in distributed storage systems, server failures are inevitable. Availability refers to the entire system not being seriously affected in terms of reading and writing.

Partition tolerance: as mentioned above, there is a network that connect multiple servers in a distributed storage system. Also, network link/node failures may sometimes happen, which will lead to partition. A distributed storage system is partition-tolerant if it still works when the network is partitioned.

When talking about distributed storage systems, CAP theory should not be ignored. The theory, proved in 2002, indicated that a distributed system could not simultaneously meet the requirements of consistency, availability, and partition tolerance, which means at most two of them can be satisfied simultaneously. Now that consistency, availability, and partition tolerance cannot be achieved simultaneously, we can ignore one of the three requirements, according to different design goals and conditions. That means we can have CA systems, CP systems, and AP systems. CA systems do not support partition tolerance and cannot work when network failures happen. Therefore, it is a good choice to use CA systems in a single server. For example, some traditional small-scale relational databases can be supported by CA systems where consistency and availability are guaranteed. However, CA systems cannot be expanded to use multiservers. To enable large-scale storage, partition tolerance must be satisfied. Compared with CA systems, CP systems ensure partition tolerance. In general, CP systems are useful for the scenarios with moderate load but have strict requirements on data accuracy. BigTable and Hbase are two instances of CP systems. Also, AP systems ensure partition tolerance. However, AP systems only guarantee eventual consistency instead of strong consistency. Therefore, AP systems mainly apply to the scenarios with frequent queries but low requirements on accuracy. Such scenarios can be found in online social networking services (SNS) systems where there are many concurrent queries to the data but a certain number of data errors are acceptable.

As for storage mechanisms, existing mechanisms can be classified into three bottom-up levels: file systems, database systems, and programming models. File systems are the foundation of the applications at upper levels. Among existing file systems, GFS is one of the most successful ones, and is produced by Google. GFS is an expandable distributed file system developed for large-scale, distributed, and data-intensive applications. It consists of cheap commodity servers to satisfy fault-tolerance and provides high-performance services. In addition, GFS performs more frequent reading than writing in large-scale file applications. However, a single point of failure and poor performances for small files are the main limitations of GFS. Other file systems have also emerged to satisfy different restrictions. For example, HDFS and Kosmosfs are two file systems developed based on open source codes of GFS. Microsoft has its Cosmos to support its search and advertisement business. Alibaba developed TFS and FastDFS for trading systems, and Facebook has Haystack, which performs well at storing a large number of small-sized images. In summary, distributed file systems have become a mature field after years of exploration. So, we will continue focusing on the other two levels.

Over the last 30 years, database technology has been evolving. Many database systems are developed to handle datasets at different scales and support various applications. In the era of big data, traditional relational databases are losing efficiency when handling massive datasets. NoSQL databases (i.e., nontraditional relational databases) are becoming a more popular research topic for big data storage. Some main features of NoSQL databases include flexible modes, support for simple and easy copy, simple API, eventual consistency, and support of large volume data, making NoSQL databases the core technology for big data. There are three main NoSQL databases in this section: Key-value databases, column-oriented databases, and document-oriented databases, each based on certain data models.

Key-value databases store data corresponding to key-values. Each key in the database should be unique so that the queried values can be determined by the keys. Such databases have a simple structure and can perform with high expandability and shorter query response time compared to traditional databases. Some of main key-value databases include Dynamo by Amazon, Voldemort Redis, Tokyo Canbinet and Tokyo Tyrant, Memcached and Memcache DB, Riak and Scalaris, etc. Compared with key-value storage, document storage can support more complex data forms rather than just considering keys and values. Because documents do not follow strict modes, there is no need to conduct mode migration. In addition, key-value pairs can still be saved. MongoDB, SimpleDB, and CouchDB are three representatives of document databases. As for column-oriented databases, they are different from traditional relational databases as they store and process data according to columns other than rows. Both columns and rows in column-oriented databases are segmented in multiple nodes to achieve expandability. The column-oriented databases are mainly inspired by Google's BigTable, which is a distributed, structured data storage system designed to process the PB-level data among thousands of commercial servers.

Now, we continue to talk about programming models. Traditionally, we have some parallel models, including MPI and

OpenMP, to support parallel programs. However, such models become inadequate when handling massive datasets. Recently, some new parallel programming models have been proposed to effectively improve the performance of NoSQL and reduce the performance gap to relational databases. Such models have become the cornerstone for big data technology.

One of the most popular programming models in big data is MapReduce. It allows a large number of clusters of commercial servers to realize parallel computing and distribution. In MapReduce, there are only two functions, map and reduce, both of which need to be implemented by users. The map function is responsible for processing input key-value pairs and generating intermediate key-value pairs. Then, those intermediate key-value pairs are combined according to the keys and transmitted to the reduce function, which further compresses the value set into a smaller set. One advantage of MapReduce is that it releases programmers from complicated steps for developing parallel applications. Programmers only have to consider the two functions. The very first MapReduce framework performed badly in multiple datasets in a task, which has been mitigated by some recent improvements. However, the MapReduce framework only supports map and reduce functions that are nontransparent, making the framework unable to contain all the common operations. In consequence, developers have to spend extra time on implementing the basic functions that are typically difficult to maintain or reuse. In order to improve programming efficiency, some language systems have been developed such as Scope of Microsoft, Pig Latin of Yahoo, Hive of Facebook, and Sawzall of Google.

Inspired by the above programming models, other studies have also focused on programming modes for more complex computational tasks, such as, iterative computations, fault-tolerant memory computations, incremental computations, and flow control decision-making related to data.

3.4 Data analysis

Data analysis is the final and the most crucial phase among the four steps in big data. It mainly includes analytical methods for conventional data and big data, analytical architecture for big data, and software for the mining and analysis of big data. Through the analysis of datasets in different fields, we can gain different potential values. However, data analysis is extremely complex and can change very frequently. In this section, we will talk about the methods, architectures, and tools for big data analysis.

3.4.1 Traditional data analysis methods

Before talking about big data analysis, let us review those conventional methods for data analysis. Traditional data analysis is using statistical methods to analyze data; to concentrate, extract, and refine useful data hidden in a batch of original datasets; and to identify the inherent law of the subject matter so as to maximize the value of data. This plays an important role in making development plans for a country, forecasting customer demands for commerce, and understanding market trends for companies. Big data analysis is a special kind of data analysis with more massive volumes of data. Therefore, many traditional methods in data analysis may still work in big data analysis. Several representative traditional methods for data analysis are examined in the following and many of them are from statistics and computer science.

Factor analysis is aimed at explaining relations among many elements with only a few factors. In this method, several closely related features are grouped into a factor, and then a few such factors are used to reveal the most information of the original data.

Cluster analysis is a statistical method for classifying objects according to some features so that objects in the same category are expected to have high similarity with each other. This is an unsupervised study method.

Correlation analysis is a method for identifying the law of relations, such as correlation, correlative dependence, and mutual restriction, among recorded phenomena. Based on the results, accordingly conducting forecast and control can be planned by decision makers. In general, such relations can be classified into two types, that is, function and correlation. Function reflects the strict dependence relationship among phenomena and is also called a definitive dependence relationship. In comparison, correlation refers to some undetermined or inexact dependence relations. The numerical value of one variable may correspond to that of several other variables, and such numerical value presents a regular fluctuation surrounding their mean values.

Regression analysis is a mathematical tool for revealing correlations between one variable and some other variables. Regression analysis determines dependence relationships among variables hidden by randomness or noise, which may transfer complex and undetermined correlations among variables into simple and regular ones.

A/B testing is also called bucket testing. It is a method aimed at identifying a way to improve target variables by comparing the tested group. However, in the big data scenario, a larger number of tests have to be executed and examined.

Statistical analysis is based on the statistical theory, that is, a branch of applied mathematics. In statistical theory, randomness and uncertainty models are created based on probability theory, which provides a description and an inference for big data. At present, statistical analysis is widely used in many fields, including economics and medical care.

Data mining is to extract hidden, unknown, but potentially valuable information from massive, incomplete, noisy, and random data. The 10 most influential data mining algorithms were selected by the IEEE International Conference on Data Mining Series in 2006. They include C4.5, k-means, SVM, Apriori, EM, Naïve Bayes, and Cart. These algorithms mainly come from machine learning, covering classification, clustering, regression, statistical learning, and so forth.

3.4.2 Big data analytic methods

With the coming of the big data era, people have begun to pay more attention to methods that can extract key information from massive data rapidly so that enterprises and individuals can gain value from it. At present, the main processing methods of big data are listed as follows:

Bloom filter is a space-efficient probabilistic data structure using a series of hash functions. It provides quick queries about whether a certain element is in a set. However, false positives may happen, which means a query returns either "possibly in set" or "definitely not in set." Another drawback is that such structures only allow add operations rather than remove operations.

Hashing is a common method in computer science that essentially transforms data into shorter fixed-length numerical values or index values. Hashing has such advantages as rapid reading, writing, and high query speed, but it is hard to find a sound Hash function to make the performance stable.

Index effectively reduces the expense of disk reading and writing. It also helps improve insertion, deletion, modification, and query speeds in both traditional relational databases that store structured data and other technologies that manage semistructured and unstructured data. However, additional space is needed for storing index files, and such index files need maintaining dynamically after data are modified.

Triel is also called trie tree, digital tree, or prefix tree. It is a variant of Hash Tree mainly applied to rapid retrieval and word frequency statistics. The key idea of Triel is to utilize common prefixes of character strings to reduce comparison on character strings to the greatest extent, which improves query efficiency.

Parallel computing refers to simultaneously utilizing several computing resources to achieve a computation task compared to traditional serial computing. In parallel computing, problems are usually divided into small subtasks that can be easily handled by a single computer. Then these tasks are assigned to several separate processes (usually on different machines) to be independently executed. The results generated by each node are then combined to get the objective result. As mentioned above, MapReduce is one parallel programming model, and other models include MPI and Dryad.

3.4.3 Architecture for big data analysis

Considering the 4Vs of big data, different analytical architectures shall be designed to satisfy different requirements. According to timeliness requirements, big data analysis can be classified into real-time analysis and offline analysis.

Real-time analysis is mainly applied in electronic commerce and finance. As data constantly change, rapid data analysis is required and analytical results shall be returned with a very short delay. Parallel processing clusters using traditional relational databases and memory-based computing platforms are two of the main existing architectures of real-time analysis. For example, Greenplum from EMC and HANA from SAP are both real-time analysis architectures.

Offline analysis is mainly used for applications that do not require a short delay or response time. Such circumstances include machine learning, statistical analysis, and recommendation algorithms. Offline analysis generally works by importing data from a special platform through data acquisition tools. Many Internet companies utilize the offline analysis architecture based on Hadoop to reduce the expense of data format conversion and improve the efficiency of data acquisition. Examples include Taobao's open source tool Timetunnel, LinkedIn's open source tool Kafka, and Facebook's Scribe. Such tools are able to satisfy the demands of data acquisition and transmission with up to hundreds of MBs per second.

Big data analysis can also be classified into memory level analysis, business intelligence (BI) level analysis, and massive level analysis, which are listed in the following.

Memory-level analysis performs effectively when the memory of a cluster can hold the total data volume. With the development of technology, modern computers are equipped with larger memories than before. Nowadays, it is not rare that the memory of a server cluster can easily surpass hundreds of GBs or even several TBs, making it possible for memory-level analysis. Memory-level analysis is extremely suitable for real-time analysis. The development of the SSD (solid-state drive) has also been helpful in improving the capacity and performance of memory-level data analysis.

BI analysis is used when the data volume is larger than the memory capacity but may be imported into the BI analysis environment. At present, mainstream BI products are able to support data analysis at the level over TB.

Massive analysis is for the case when the data scale has completely exceeded the capacities of BI products and traditional relational databases. Currently, most massive analysis chooses HDFS of Hadoop to store data and uses MapReduce for data analysis. Because of the huge volume of data, most massive analysis belongs to offline analysis.

4 Current statistical methods for astronomical data analysis

We have talked about general methods for big data analysis. For astronomical data collected by various digital telescopes, astronomers have some statistical methods to perform the analysis. Many methods are based on the methods mentioned above. We will look at these statistical methods for astronomical data analysis in the following.

4.1 Nonparametric statistics

Nonparametric statistical inference is suitable for data that do not depend on assumptions on the underlying distribution of the datasets. In many cases, we just assume all distributions are normal, and that measurement uncertainties are constant and increase as $\sqrt{n}$ as the sample size increases. Actually, our knowledge about astronomical populations and processes is so limited that astronomers have no idea about whether the observed data is normally distributed or in other distributions. Therefore, nonparametric methods for statistical inference become particularly attractive to astronomers. The advantage of a nonparametric statistic is that it is a distribution-free method, which makes it extremely robust against some highly skewed distributions. However, many nonparametric methods require that the datasets should be univariate.

4.2 Data smoothing

Density estimation procedures can smooth sets of individual measurements into continuous curves or surfaces. Nonparametric density estimation can achieve this regardless of the underlying distribution of the datasets. In general, univariate data are collected into histograms that show frequencies of occurrences grouped into bins. However, histograms are rarely used for statistical inference for some disadvantages. A major disadvantage is the discrete points between bins cannot reflect the continuous behaviors of most physical quantities. To overcome the shortcomings, density estimation is operated for data smoothing.

4.3 Multivariate clustering and classification

The datasets collected in astronomy are usually in the form of table where each row represents a sample of a celestial object with columns as the features. Such datasets are known as multivariate datasets. When the population is homogeneous, methods from multivariate analysis such as principal components analysis and multiple regression can be applied. However, a mixture of astronomical classes is often recorded, and subclassifications are common. For example, spiral galaxy morphologies were classified into Sa, Sb, and Sc categories by Hubble. Supernovae can be divided into Types Ia, Ib, II, and so forth. Therefore, many methods in supervised learning and unsupervised learning are introduced. These methods include k-means partitioning, MLE clustering, support vector machines, k-nn, and naïve Bayes.

4.4 Nondetections and truncation

Truncation is very common in astronomical surveys due to the limitations of the telescope. For example, many faint objects or phenomena cannot be detected because of magnitude-limited or flux-limited surveys. Only the closest or brightest members of a celestial population can be recorded, which is different from those controlled studies in social sciences where accurately randomized and stratified subsamples can be conducted. In consequence, the data points usually look “left-censored.” To deal with truncation is difficult because less information is known about the full population, but some relevant methodology has been proposed for this problem. In 1971, the Kaplan-Meier (KM) estimator was formulated to estimate the parameters of the underlying distribution using censored data. Then Lynden-Bell-Woodroofe (LBW) estimator, which is similar to the KM estimator, was proposed for processing truncated datasets. Both estimators require independent truncation values of true values. In addition, a few extensions of the LBW estimator have been produced for more complex cases.

4.5 Spatial point processes

Spatial point datasets are multivariate datasets in which some of the variables are interpreted as spatial dimensions. This is common in astronomical surveys, such as Kuiper Belt Objects in the two-dimensional sky, galaxies in a three-dimensional redshift survey, and photons on four-dimensional x-ray images. Basic spatial point processes are stationary and isotropic, which means properties are invariant under translation and rotation. Complete spatial randomness (CSR) is a stationary Poisson spatial point process pattern that can generate random locations. More complex models were also proposed using nonstationary processes. At present, extensive methodology for interpreting such spatial point processes has been applied in various fields [6].

5 Platforms for big data processing

In this section, we will focus on some widely used platforms for big data processing. In general, the big data platforms can be classified into horizontal scaling and vertical scaling. Horizontal scaling is also known as “scale out.” It is formed by distributing the workload across many servers that may even be a commodity machine. Multiple independent servers are connected together to expand the processing capability. Vertical scaling is also called “scale up.” It involves more processors, larger memory, and faster hardware within usually a single server and a single instance of an operating system. In Table 2.1, we compare the strengths and weaknesses of both types of scaling.

Table 2.1

Scaling	Advantages	Drawbacks
Horizontal scaling	• Increases performance in small steps as needed • Financial investment to upgrade is relatively less • Able to scale out the system as much as needed	o Software has to manage all the data distribution and parallel processing complexities o Types of software that take advantage of horizontal scaling are limited
Vertical scaling	• Most of the software can easily take advantage of vertical scaling • Convenient to manage and install hardware within a single machine	o Substantial financial investment is needed o It is not possible to scale up vertically as much as needed. o System has to be more powerful to handle future workloads and initially the additional performance is not fully utilized

5.1 Horizontal scaling platforms

Some of the popular horizontal scale-out platforms include peer-to-peer networks and Apache Hadoop. Nowadays, researchers have already started to develop the next generation of horizontal scale-out tools to overcome the weaknesses of existing platforms. These platforms will be examined in more detail in the following.

Peer-to-peer networks involve millions of machines connected in a network. This network architecture is designed to be decentralized and distributed. It may be one of the oldest existing distributed computing platforms. Each node in the system is called a peer, and peers communicate and exchange messages by using the message passing interface scheme called MPI. Each node can store a certain amount of data and the scale out is practically without limitations, which means millions of nodes or even more can be involved. However, the major bottleneck in peer-to-peer networks arises in the communication among different nodes. Broadcasting messages in such systems is low cost, but the aggregation of data or analytic results is much more expensive. Moreover, the messages are sent over the network in the form of a spanning tree with an arbitrary node as the root where the broadcasting is initiated.

MPI is the standard software communication paradigm that is used in this network. It has been in use for many years and is well established and thoroughly debugged. One of the main features of MPI is the state preserving process. This means processes can be kept alive as long as the system operates and there is no need to load the same data again and again as other frameworks do, such as MapReduce. All the parameters can be saved locally. Therefore, MPI is suitable for iterative processing. Another important feature of MPI is its hierarchical master/slave paradigm. Under such a paradigm, a slave node can become the master for other processes, which is extremely useful and flexible for dynamic resource allocation, especially when the slave node has large amounts of data to process.

Actually, MPI is available for many programming languages, including C++ and Java. It provides methods to send and receive messages and data among processes. One of the most important methods in MPI is "broadcast," which broadcasts the data or messages over all the nodes. Another method called "barrier" is also frequently used to put up a barrier so that all the processes can synchronize and reach a certain state before working further.

Although it seems that MPI is perfect enough for developing algorithms for big data analysis, some drawbacks also arise. For example, MPI does not have any mechanism to deal with faults. Because peer-to-peer networks are completely unreliable hardware, when MPI is used on top of such architectures, a single node failure can lead to the breakdown of the entire system. Therefore, users must implement some fault tolerance mechanism within the program to avoid such tragedies. Other frameworks such as Hadoop are robust to fault tolerance and are becoming increasingly popular, meaning that MPI is not as widely used as before.

Apache Hadoop is an open source framework for storing and processing massive datasets using clusters of commodity hardware. Hadoop is developed to scale up to hundreds and even thousands of nodes with high fault tolerance. The components of a Hadoop stack are illustrated in Fig. 2.4. As the figure shows, the lowest layer of Hadoop is the Hadoop distributed file system (HDFS), which is a distributed file system that is widely used to

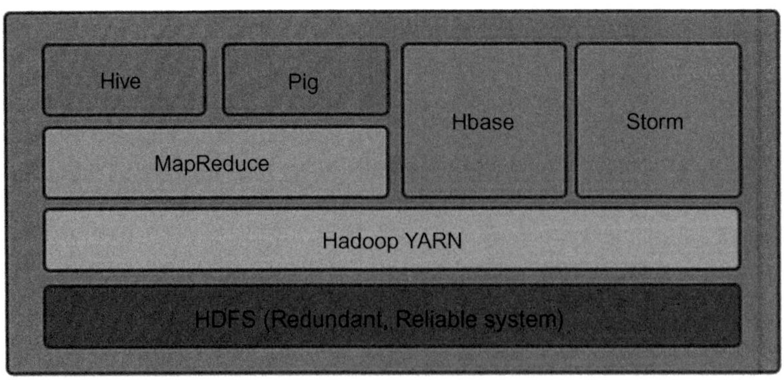

Fig. 2.4 Hadoop stack.

store data across clusters of commodity machines. The distributed file system supports high availability and excellent fault tolerance. Hadoop YARN is a layer responsible for resource management and job scheduling across the cluster. MapReduce is the programming model used in Hadoop. As mentioned above, MapReduce divides the entire task into two functions, that is, map and reduce. The map functions load the data from HDFS, process it, and then gain some intermediate results that are sent to the reduce function. Reduce functions aggregate the received intermediate results to generate the final output and write the output to HDFS. Typically, a Hadoop job involves operating several map functions and reduce functions across different nodes in the cluster.

Other platforms based on MapReduce have also been developed, and these are called MapReduce wrappers. These MapReduce wrappers can support better control over the MapReduce program and support aid in the source code development. Two popular wrappers are Apache Pig and Hive. Both can make code development easier by avoid making programmers deal with the complexities of MapReduce coding. In addition, programming environments such as DryadLINQ have emerged that allow end users more flexibility over MapReduce because they can have more control over the coding. DryadLINQ is developed by Microsoft, using LINQ (a parallel programming language) and a cluster execution environment named Dryad. Programmers can just use Visual Studio as the tool for better debugging and development, and even interoperate with other languages such as standard. NET.

Spark is regarded as the representative of a next-generation paradigm for big data processing. It is designed to overcome the disk I/O limitations on Hadoop and to improve the performance of earlier systems. What makes Spark unique and efficient is that Spark can support in-memory computations. Compared to Hadoop, which requires loading data from a disk, Spark allows data to be cached in memory to perform in a more efficient way. At present, Spark has become a general framework for massive dataset processing. It also supports many mainstream programming languages, including Java, Python, and Scala. Spark can run up to 100 times faster than Hadoop MapReduce for some certain tasks when the data is suitable to be cached in memory. Even when data resides on the disk, Spark operates up to 10 times as rapidly as Hadoop does. Spark is also compatible with Hadoop frameworks as it can run on Hadoop Yarn manager and load data from HDFS, making it simple to run Spark on different systems.

Berkeley data analytics stack (BDAS): BDAS was developed by researchers at the University of California at Berkeley. The BDAS

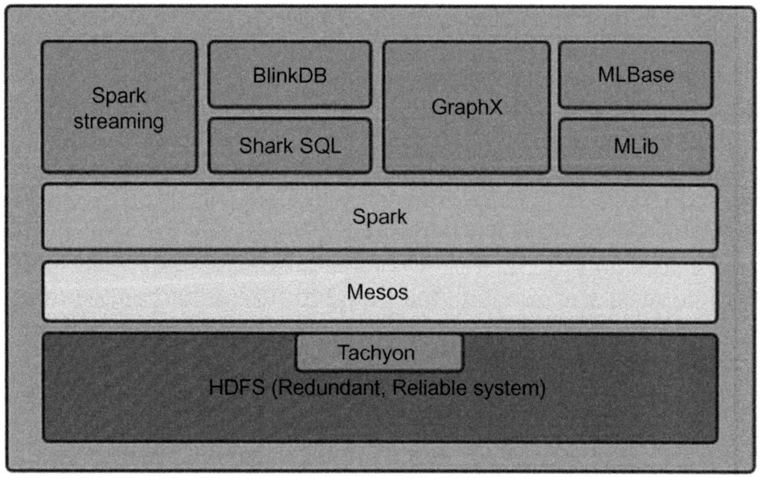

Fig. 2.5 Berkeley data analytics stack.

stack is shown in Fig. 2.5. As is seen, there is a layer called Tachyon that is on the upper level of HDFS. Actually, this component is based on HDFS, and it achieved higher performance than HDFS by utilizing memory in a more efficient way. Tachyon works by caching those files that are frequently read in memory so as to access the disk as little as possible. Thus, these cached files can be accessed at memory speed. Tachyon is also compatible with Hadoop MapReduce. In other words, MapReduce programs can run over Tachyon just as they directly run on HDFS. In addition, Tachyon also supports operations for raw tables. With Tachyon, users can load tables with hundreds of columns, and Tachyon can help identify the columns that are frequently used to cache them in memory for further use.

Another crucial layer in BDAS is Apache Mesos, shown in Fig. 2.5 above Tachyon. Mesos is used for cluster management that provides effective resource isolation and sharing across distributed applications or frameworks. It allows Hadoop, Spark, and other applications to share a dynamically shared pool of resources. Supported by Mesos, it is possible for tens of thousands of nodes to work together. Mesos provides APIs in programming languages such as Java, Python, and C++ to allow users to develop new parallel applications. In addition, Mesos includes the capability for multiresource scheduling, making it a powerful resource manager.

The layer above Mesos is Spark, which replaced Hadoop MapReduce in BDAS architectures. On the top of the stack are various Spark wrappers. For example, Spark Streaming is used for large-scale real-time stream processing. Blink DB can provide queries with bounded errors and bounded response times, even when the datasets are huge. GraphX is a distributed graph system based on Spark, and MLBase provides a distributed machine learning library based on Spark.

5.2 Vertical scaling platforms

Vertical scale-up paradigms are mainly related to hardware technologies. Typical vertical scale-up paradigms include high performance computing clusters (HPC), multicore processors, graphics processing units (GPU), and field programmable gate arrays (FPGA). We will talk about each in the following.

5.2.1 High performance computing (HPC) clusters

HPC clusters are also called blades or supercomputers. In general, HPC clusters have thousands of cores and a different variety of disk organization, cache, and communication mechanisms according to user requirements. One major drawback of HPC clusters is high expense because such clusters require high-end hardware to reach excellent performance. Because the hardware in HPC is top-quality, fault tolerance is perfectly guaranteed. Therefore, MPI can be used as an HPC communication scheme as the fault tolerance will not be a major drawback.

5.2.2 Multicore CPU

Multicore means that a computer has several processing cores. These cores usually have shared memory but use the same disk. With the development of technology, each chip can hold more cores, and each core can perform more operations than before. In general, each core has its own cache but all cores share the same memory, and modified shared invalid protocol (MSI) can be helpful to synchronize data in caches of each core. At present, multicore CPUs mainly conduct parallelism through multithreading. That means a task is usually divided into several threads to be executed by different cores simultaneously. Most mainstream programming languages, among which Java is the most popular choice, include libraries to generate threads to run multithreading programs.

5.2.3 Graphics processing unit (GPU)

Recently, GPU has received more attention because of its high performance when running machine learning algorithms. Since the CUDA framework was developed, some libraries have provided machine learning algorithms that can operate on a GPU so as to gain speed over using a CPU. However, one major drawback of GPU is its limited memory. Typically, the maximum GPU memory can reach tens of GBs, which is far from the requirements for handling big data. Once the data size surpasses the GPU memory capacity, the performance will be heavily affected because of frequent disk access.

5.2.4 Field programmable gate arrays (FPGA)

FPGA is a special kind of hardware that is custom built for specific applications. FPGAs are programmed using hardware descriptive language (HDL). To develop an FPGA application usually costs much more than other platforms because it is customized hardware. One example of FPGA is applied in the network security applications where FPGA plays a role as a hardware firewall and is much more efficient than software firewalls in filtering large amounts of network data [7].

References

[1] M. Chen, S. Mao, Y. Liu, Big data: a survey, Mob. Netw. Appl. 19 (2) (2014) 171–209.
[2] E.D. Feigelson, G.J. Babu, Big data in astronomy, Significance 9 (4) (2012) 22–25.
[3] G.J. Babu, S.G. Djorgovski, Some statistical and computational challenges, and opportunities in astronomy, Stat. Sci. 19 (2) (2004) 322–332.
[4] E.D. Feigelson, G.J. Babu, Statistical challenges in modern astronomy, 2004. arXiv preprint astro-ph/0401404.
[5] J. Kremer, K. Stensbo-Smidt, F. Gieseke, K.S. Pedersen, C. Igel, Big universe, big data: machine learning and image analysis for astronomy, IEEE Intell. Syst. 32 (2) (2017) 16–22.
[6] E.D. Feigelson, G.J. Babu, Statistical methods for astronomy, in: Planets, Stars and Stellar Systems: Volume 2: Astronomical Techniques, Software, and Data, Springer, 2013, pp. 445–480.
[7] D. Singh, C.K. Reddy, A survey on platforms for big data analytics, J. Big Data 2 (1) (2015) 8.

PART

B

Big data processing

3

Preprocessing pipeline on FPGA

Tian Huang[a,b], Yongxin Zhu[c,d], Yu Zheng[d]

[a]*Astrophysics Group, Cavendish Lab, Cambridge University, Cambridge, United Kingdom.* [b]*Institute of High Performance Computing, Agency for Science, Technology and Research (A*STAR), Singapore, Singapore.* [c]*Shanghai Advanced Research Institute, Chinese Academy of Sciences, Shanghai, China.* [d]*School of Microelectronics, Shanghai Jiao Tong University, Shanghai, China*

The complex characteristics and implicit regulations contained in astronomical radio data have aggravated the dilemma in astronomy data processing. In this case, data preprocessing has become an essential step to provide solutions by means of data compression, feature extraction, format transformation, information coding, and so on. Many of the methods enable scalable and parallel design approaches to achieve further acceleration, among which the FPGA pipeline is an efficient and adaptable technology for most datasets.

This chapter introduces ADC, filtering, time-frequency transforming, and correlator and data reduction on FPGA as the preprocessing pipeline. Moreover, the FPGA pipeline technology will be demonstrated in detail, and the specific design issues will be displayed as a reference. Through these discussions, readers can have a better understanding of the design concepts in data preprocessing.

1 FPGA interface for ADC

The digital preprocessing pipeline starts from ADCs. The FPGA is usually the first step of the preprocessing pipeline. The FPGA is capable of handling multiple high-speed data streams from ADCs and applies early stage preprocessing in real time at a relatively low clock speed without incurring much energy consumption. The power efficiency feature is critical to radio astronomy applications because power-demanding computing platforms are usually the source of radio frequency interference (RFI).

Big Data in Astronomy. https://doi.org/10.1016/B978-0-12-819084-5.00012-2

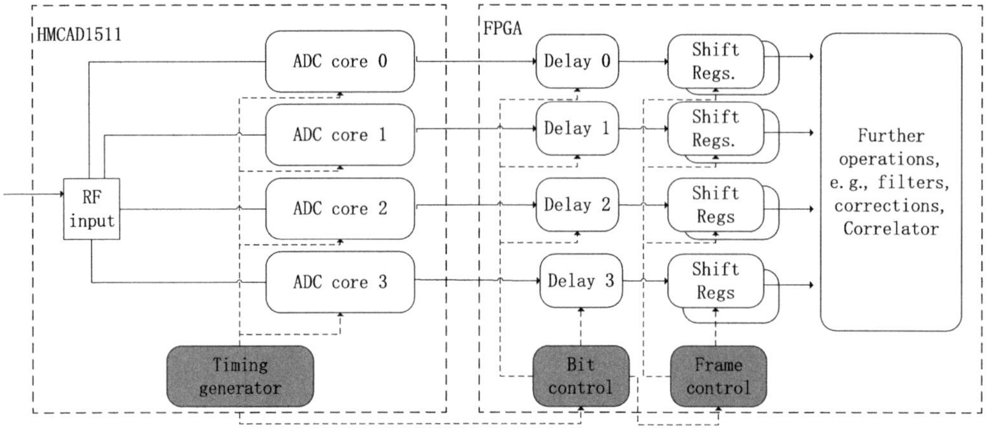

Fig. 3.1 Diagram of FPGA interface for HMCAD1511 ADC.

We take SNAP from CASPER as an example. The FPGA interface for HMCAD1511 [1,1a] on SNAP [2] is presented to show how the FPGA turns serialized high-speed ADC data streams into data bytes that are ready for subsequent calculations.

In Fig. 3.1, HMCAD1511 has four identical ADC converters (core). The four cores are used in an interleaved mode to quadruple the conversion (sample) speed. ADC cores convert the RF input into serialized digital signals and pass through delay and shift register units to the subsequent DSP operation.

1.1 ADC interleaving

Wideband frequency scanning applications require the system to detect the transmitted frequencies instantaneously. Such applications require the receiver chain to have a Nyquist band that accommodates all the possible frequencies [3]. For example, for a system requiring frequency detection from 100 MHz to 4 GHz, the sampling system will require a clock of 8 GHz. Such wideband applications are also required for the reconstruction of transmitted pulses with high time resolution for RADAR applications. They are essential for the reconstruction of pulses with very short rise times.

Time interleaving is a technique that allows the use of multiple identical analog-to-digital converters (ADCs) to process regular sample data series at a faster rate than the operating sample rate of each individual data converter [4]. In very simple terms, time interleaving (IL) consists of time multiplexing a parallel array of

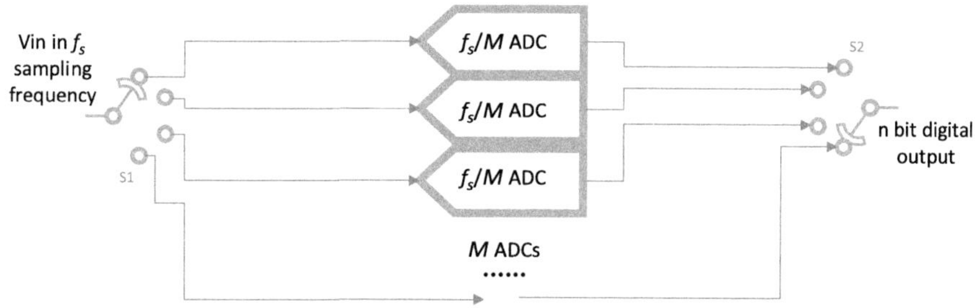

Fig. 3.2 An array of M time interleaved n-bit ADCs. The sample rate of each one is f_s/M; the resulting sample rate of the time interleaved ADCs is f_s.

M identical ADCs, as shown in Fig. 3.2, to achieve a higher net sample rate f_s (with sampling period $T_s = 1/f_s$) even though each ADC in the array is actually sampling (and converting) at the lower rate of fs/M. So, for example, by interleaving four 10-bit/100 MSPS ADCs, one could in principle realize a 10-bit/400 MSPS ADC.

1.2 Bit alignment

Due to the imperfection of manufacturing, there are always skews among different ADC cores and the electrical connections between the ADC and FPGA. Skew can lead to failure in conversion from serial data into parallel ones. Fig. 3.3 shows the waveform of the shift registers at the deserialization stage of FPGA. The double date rate (DDR) clock signal comes from the ADC output. The incoming data stream is sampled by the two shift registers alternately at the rising and falling edge of the clock.

T_1 and T_2 are the time between the rising/falling edge and the change of data, respectively. If the skew causes the rising/falling

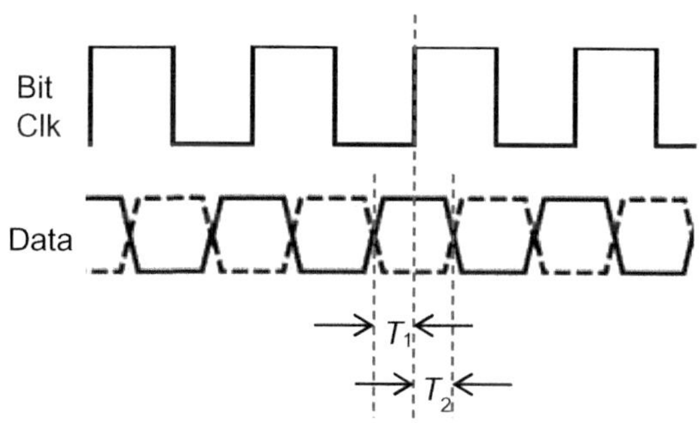

Fig. 3.3 The purpose of the delay is to make sure that the rising/falling edge of the clock signal happens in the middle of the data eye, that is, $T_1 = T_2$.

edge to happen at the change point of data, that is, $|T_1| < \varepsilon$ or $|T_2| < \varepsilon$, where ε is a very small amount of time, the data being sampled into the shift registers would be uncertain. To avoid this, the skew has to be compensated. Ideally, $T_1 = T_2$ would lead to the most reliable sampling condition, in which there are enough safe margins for jitters in the circuits.

To make sure that the shift registers behind have the best timing to capture each bit of the data stream, one delay unit is inserted before each shift register. By adjusting the delay unit, one can ensure that the rising/falling edge locates roughly in the middle of the data eye, and thus reduce the uncertainty of sampling. In the Xilinx FPGA 7 series, the corresponding programmable delay primitive is called IDELAY. IDELAY is a multitap, wraparound, delay primitive with a calibrated tap resolution. It allows incoming signals to be delayed on an individual input pin basis. The tap delay resolution is contiguously calibrated by the use of an IDELAYCTRL reference clock [5].

1.3 Stream deserialization

After going through the delay unit, the data is bit-aligned and ready to be deserialized. The process of deserialization can vary according to different ADC applications. For example, an ADC core can run in standalone and single data rate mode, which is very simple and straightforward. But this working mode is limited by the FPGA clock frequency and can hardly apply to data streams at the throughput above the scale of Gbps.

In order to facilitate the digital signal processing, ADCs in radio astronomy today usually work in interleaving mode and send data to DSP in DDR mode. Fig. 3.4 shows the process of deserialization from the point of view of data.

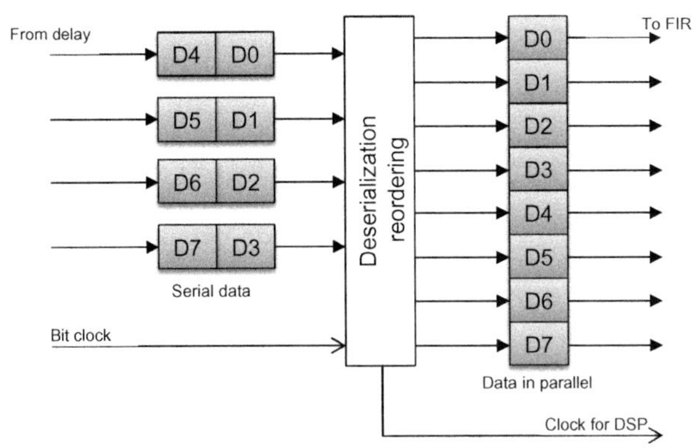

Fig. 3.4 8-bit deserialization in DDR and interleaving modes.

The input data on the left go into the deserialization unit sequentially and go out in parallel.

The input data are generated by four ADC cores coordinated in interleaving mode. For example, D0 and D4 are generated by the first ADC core, and D3 and D7 are generated by the last ADC core. In DDR mode, the ADC uses both the rising edge and the falling edge of the bit clock to synchronize the data transmission. For example, D0 is sampled at the rising edge whereas D4 is sampled at the falling edge. D0 arrives at FPGA ahead of D4.

In the Xilinx FPGA 7 series, the right candidate for deserialization is called ISERDES. It is a dedicated serial-to-parallel converter with specific clocking and logic features designed to facilitate the implementation of high-speed source-synchronous applications [5]. By using this built-in unit, one can avoid the additional timing complexities encountered when designing deserializers in the FPGA fabric [6].

The reordering functionality ensures that the data in parallel are sorted in a natural order [5]. Although the subsequent DSP logic does not necessarily require the input data to be in a natural order, the reordering functionality decouples the design between the interface and the following DSP logic and provides better flexibility in the design stage.

Deserialization and the ordering unit are the boundaries of different clock domains. On the left, the bit clock is synchronized with the input data, whereas the clock for DSP is synchronized with the data in parallel. The frequency of the clock for DSP is two times lower than the bit clock.

2 FIR filtering

The purpose of the Finite Impulse Response (FIR) filtering stage is to suppress the frequency components out of the band, which would otherwise cause in-band noises because of the leakage and scalloping loss [7] of the Discrete Fourier Transform (DFT) or Fast Fourier Transform (FFT) in the next stage.

2.1 Leakage

It is well known that astronomical data analysis is vulnerable to noise and loss, FIR filtering is essential before converting the signal from the time domain to the frequency domain. An example is given below to demonstrate the importance of FIR filtering.

Leakage happens because the impulse response window is finite. If DFT/FFT is carried out without suppressing the out-of-band frequency components, DFT/FFT is actually applying a

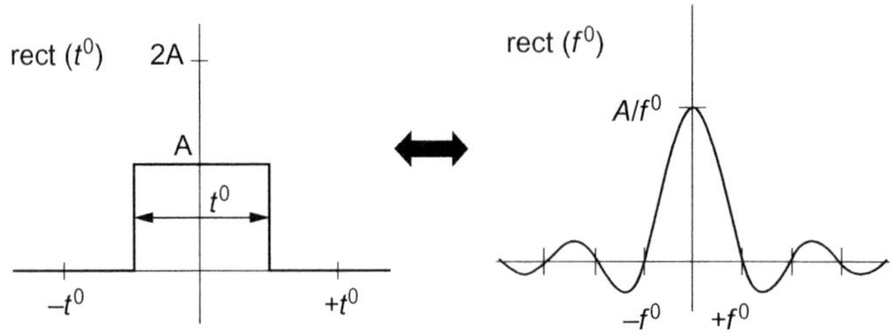

Fig. 3.5 A rectangular function in the time domain is equivalent to a sinc function in the frequency domain.

rectangular function over the signal. The width of the rectangular function is equal to the length of DFT/FFT. As show in Fig. 3.5, a rectangular function in the time domain is a sinc function in the frequency domain.

The leakage depends on the sampling frequency and the number of points in the transform; an input tone appears in more than one output frequency bin. As a result, out-of-band frequency components leak into in-band signals through side lobes, causing problems. If this tone is not strong enough, this effect can go unnoticed. But in the case of a strong radio frequency interference (RFI) signal, the leakage can drown out astronomical signals of interest in the nearby bins. This phenomenon is demonstrated in Fig. 3.6.

The DFT/FFT spectrum of a tone at 5.1 MHz would not only have a peak around 5.1 MHz, but also nonzero values in other frequency bins, which may cover with the frequency band that we are interested in.

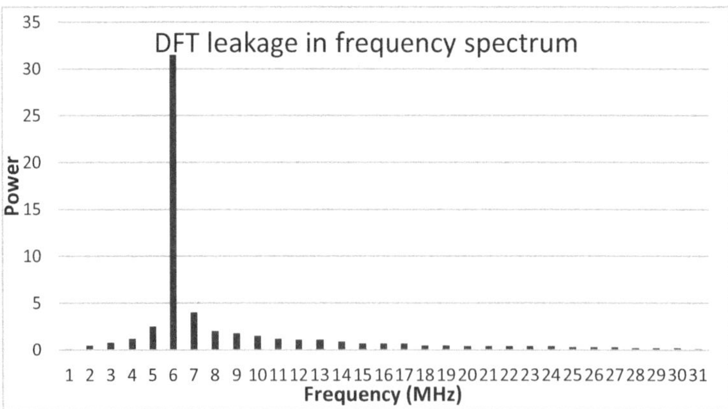

Fig. 3.6 Demonstration of DFT leakage—a tone at 5.1 MHz, sampled at 128 MHz, and Fourier-transformed with 64 points—appears to varying levels in all the output frequency bins.

2.2 Scalloping loss

The scalloping loss is the coherent gain of a frequency half a bin from a component on the DFT divided by the coherent gain of the window.

If one sweeps the frequency of a tone across the DFT spectrum, he or she will observe that the response characteristics in the pass band are not flat. Fig. 3.7 demonstrates the shape of the responses.

We sweep the frequency of a sine wave between 50 MHz and 80 MHz and plot the maximum value of the power spectrum. The X axis is the frequency of the sine wave, whereas the Y axis is the peak value of the corresponding power spectrum. The length of the DFT is 8192 points. The sampling frequency is 256 MHz. The frequency component that falls approximately in the middle of two frequency bins would suffer a loss of approximately 3.8 dBm, which is shown as the blue line, if there is no pre-windowing (equivalent to a rectangle window) applied before DFT/FFT. The graph resembles a scallop (the decorative loops in curtains or tablecloths).

Suppose that we apply the Hann window of the same number of points before computing the DFT. The DFT after the Hann window is shown as an orange line. Eq. (3.1) shows the loss given an arbitrary window $a(k)$:

$$SL = \frac{\left| \sum_{k=0}^{N-1} a(k) e^{-\frac{j\pi k}{N}} \right|}{\sum_{k=0}^{N-1} a(k)} \tag{3.1}$$

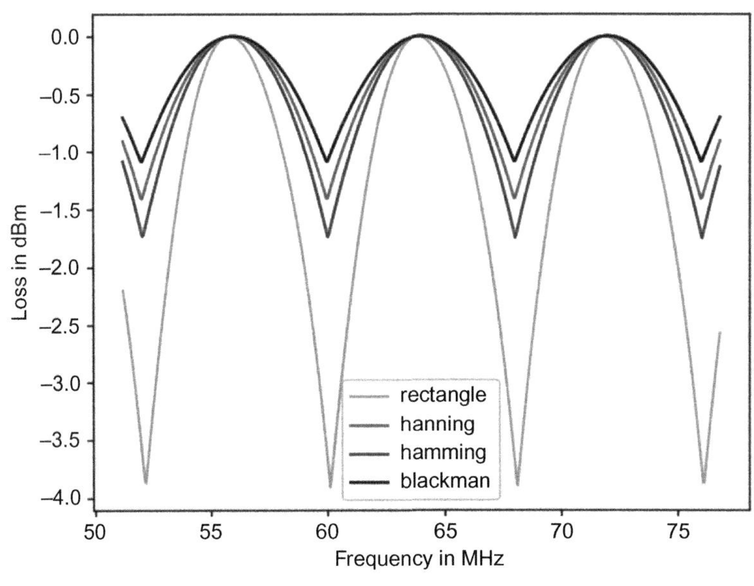

Fig. 3.7 Scalloping loss of different filtering windows.

The DFT after the Hann window in this graph is adjusted for the coherent gain of the window. The scalloping loss with the Hann window is approximately 1.3 dB, much smaller than its counterpart from the rectangle window. Thus, the scalloping loss is a measure of the shape of the main lobe of the DFT of the window.

There are other options as well such as the Hamming and Blackman window. The Hanning window is one of the commonly used windows.

2.3 Polyphase filter

In order to mitigate the leakage and scalloping loss problem, before DFT/FFT, we have to apply a window with frequency domain characteristics that are flat in the pass band and filter band while also suppressing signals in the filter band down to an unnoticeable level. The so-called "window method" for FIR filter design is also based on the convolution theorem for Fourier transforms.

The polyphase filter bank (PFB) is a type of FIR filter for mitigating the aforementioned drawbacks of the straightforward DFT. The PFB not only produces a flat response across the channel (suppress the scalloping loss), but also provides excellent suppression of filter-band signals (suppresses the leakage). Spectrometers and correlators are typical beneficiaries of the PFB technique. If we make a peer-to-peer comparison of the single-bin frequency response of a PFB with a direct FFT, we can find that the width of the polyphase window is usually much bigger than that of the FFT, for example, the width of the polyphase window is eight times the width of the FFT window [7].

Here is a brief introduction of how a PFB works from the point of view of the data stream. The structure of the PFB is shown in Fig. 3.8.

In a PFB, an input block of data of size $N \times P = M$ is read, and multiplied point by point with a window function. The whole block of data is severely decimated by the action of a commutator, which distributes the time samples among the branches, or "partitions," with a cycling period N.

Each partition is multiplied by a window function noted as $h(n)$. Functions such as sinc would be good choices for $h(n)$ because the shape of its pair in the frequency domain is an approximate of the rectangular function. The multiplication step can be seen as applying weight on the data. The weighted data are then summed together and output to subsequent operations. In the context of a spectrometer, the output is then passed to a regular DFT routine to get an N-point transform that exhibits less leakage. The FIR filtering structure can be represented by Eq. (3.2):

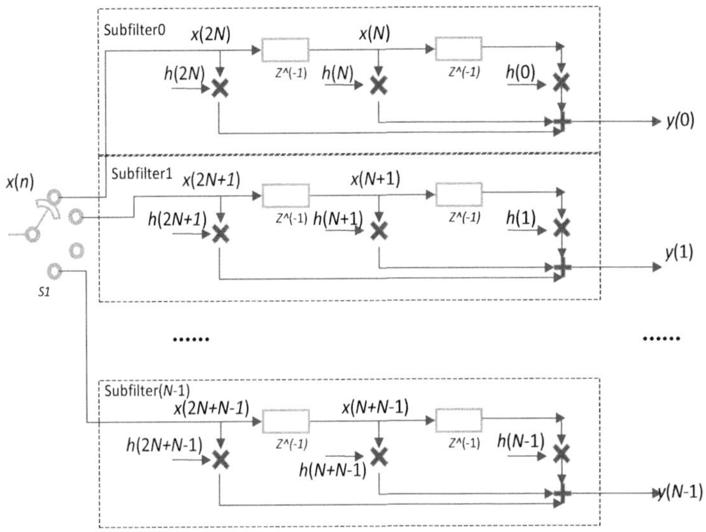

Fig. 3.8 The FIR filter structure realization of a PFB with $P = 3$ taps and N subfilters. The commutator at the left rotates in the clockwise direction, and makes one complete rotation in the duration of a one-unit delay. The output of this structure is $y(n)$, which is the input to an N-point DFT.

$$y(n) = \sum_{p=0}^{P-1} x(n+pN)h(n+pN) \tag{3.2}$$

where the $h(n+pN)$ is called P-tap "polyphase subfilters." The N such subfilters together with the following DFT stage are collectively called a PFB. As the entire data stream of length M is divided into N partitions, each subfilter is a decimated-by-N version of $h(n)$. Each subfilter is essentially an all-pass filter. The only difference between the subfilters is their phase response, which is why this structure is called a "polyphase" filter bank.

Polyphase filtering and PFBs correct deficiencies at a modest computational overhead. On FPGAs, a PFB typically consumes about 1.5 times more resources than a direct FFT. In many cases, the data quality advantages outweigh this increase in cost. PFBs have become an important tool in radio astronomy as a way of excising radio frequency interference because they make possible the elimination of only the specific channels in which the interference occurs.

3 Time-frequency domain transposing

After FIR filtering, N real samples in the time domain are transformed to provide N values of complex amplitude in the frequency domain. The negative frequencies are redundant, and only $N = 2$ spectral points need be retained.

The clock rate of the preprocessing line inside FPGA is usually below 250 MHz. In wideband spectrometer applications, the Nyquist sampling frequency is usually several times higher than the maximum clock rate of FPGA. There are a few techniques that realize real-time FFT on FPGA.

3.1 Real-valued FFT

In radio astronomy, the input data for the FFT are purely real; complex-valued FFT algorithms can still be used. One simple approach creates a complex sequence from the real sequence, that is, real data for the real components and zeros for the imaginary components. The complex FFT can then be applied directly. However, the output is conjugate-symmetric, that is, the negative frequency terms are the complex conjugates of the corresponding positive-frequency terms. The negative-frequency terms are redundant, so this method is not efficient.

A complex-valued FFT algorithm can be used to process two real-valued sequences at the same time. Suppose $x_1(n)$ and $x_2(n)$ are real-valued sequences of length N, and $x(n)$ is a complex-valued sequence defined as:

$$x(n) = x_1(n) + jx_2(n), 0 \le n \le N-1 \tag{3.3}$$

The FFT of the two N-length sequences $x_1(n)$ and $x_2(n)$ can be found by performing a single N-length FFT on the complex-valued sequence and some additional computation. These additional computations are referred to as the split operation, and are shown below:

$$X_1(k) = \frac{1}{2}[X(k) + X^*(N-k)], k = 0, 1, \ldots, N-1 \tag{3.4}$$

$$X_2(k) = \frac{1}{2j}[X(k) - X^*(N-k)], k = 0, 1, \ldots, N-1 \tag{3.5}$$

By using this technique, two input time series can share a single complex-value FFT computing unit. This technique can also be applied along with other techniques such as demultiplexing to further improve the efficiency of the preprocessing pipeline on FPGA.

3.2 Demultiplexing

As the working frequency of an FPGA is generally slower than that of a high-speed ADC used with broadband correlators, the input data stream in the time domain is usually demultiplexed before going through the DFT stage. The Fourier transform of the demultiplexed time series can later be combined to form the frequency spectrum of the original time series.

Let's take a radix-2 decimation-in-time (DIT) DFT as an example. The DFT is defined by the formula:

$$X_k = \sum_{n}^{N-1} x_n e^{-\frac{2\pi i}{N}nk}, k \in [0, N-1] \tag{3.6}$$

We can use radix-2 DIT to compute the DFTs of the odd-indexed inputs $(x_{2m+1} = x_1, x_3, \ldots, x_{N-1},)$ and the even-indexed inputs $(x_{2m} = x_0, x_2, \ldots, x_{N-2},)$, respectively. Then, we combine these two results with some additional computation to produce the DFT of the whole sequence.

$$X_k = \sum_{m=0}^{N/2-1} x_{2m} e^{-\frac{2\pi i}{N}(2m)k} + \sum_{m=0}^{N/2-1} x_{2m+1} e^{-\frac{2\pi i}{N}(2m+1)k} \tag{3.7}$$

By factoring a common multiplier $e^{-\frac{2\pi i}{N}k}$ out of the second term, we have:

$$X_k = \sum_{m=0}^{\frac{N}{2}-1} x_{2m} e^{-\frac{2\pi i}{\frac{N}{2}}mk} + e^{-\frac{2\pi i}{N}k} \sum_{m=0}^{N/2-1} x_{2m+1} e^{-\frac{2\pi i}{\frac{N}{2}}mk} \tag{3.8}$$

$$X_k = E_k + e^{-\frac{2\pi i}{N}k} O_k \tag{3.9}$$

Because the complex exponential is periodic, $X_{k+N/2}$ can also be obtained from E_k and O_k.

$$X_{k+N/2} = E_k + e^{-\frac{2\pi i}{N}k} O_k \tag{3.10}$$

Therefore, the DFT of length N can be recursively splitted into two DFTs of size $N/2$, which is the key idea of the commonly used FFT algorithm. We can make use of this technique and apply FFT over the output from the FIR filtering stage.

The output of the FIR filtering stage is decimated in time, as shown in Table 3.1. It coincides with the input of a radix-2 FFT algorithm. Actually, the FIR filtering stage is designed such that the output can be directly feed into an FFT algorithm. Fig. 3.9 shows a diagram of a radix-2 FFT of length N.

The radix-2 FFT of length N is composed of two sub-FFTs of length $N/2$. The inputs of the sub-FFT are decimated or demultiplexed in time from the original time series, which are in the same shape as the output of the FIR filtering stage. The demultiplexed

Table 3.1 N partitions of a continuous series of data.

$x(0)$,	$x(N)$,	$x(2N)$,
$x(1)$,	$x(N+1)$,	$x(2N+1)$,
$x(2)$,	$x(N+2)$,	$x(2N+2)$,
...		
$x(N-1)$,	$x(2N-1)$,	$x(3N-1)$,

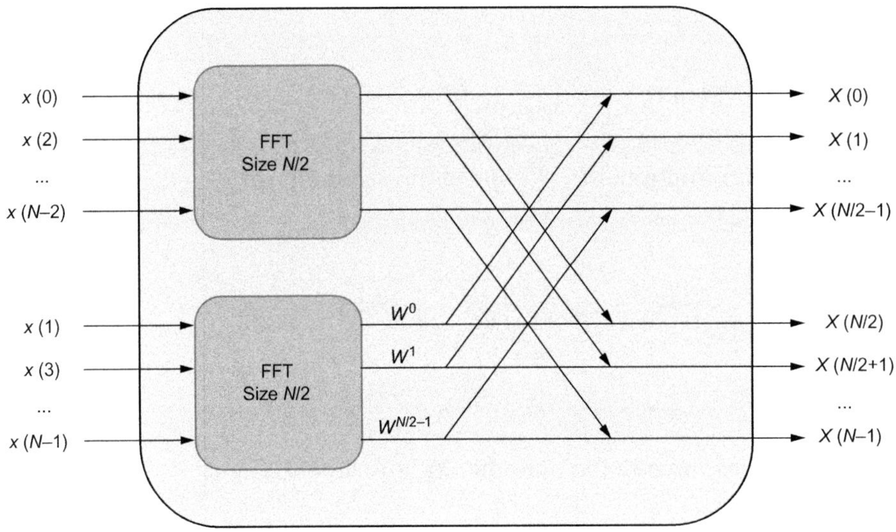

Fig. 3.9 An example of a demultiplexed radix-2 FFT.

and filtered data out of the FIR filtering stage can be directly fed into the FFT algorithm without any reordering.

The output of the two sub-FFT algorithms is combined to form the DFT of the original time series. The outputs are grouped into even-indexed and odd-indexed outputs. We can recover the ordering by using a reorder buffer [8].

The input time series above are demultiplexed in the time domain. The FIR filtering stage together with the FFT stage are collectively called the PFB, which is commonly used in radio astronomy and digital communication systems. The data stream can also be demultiplexed in the frequency domain. It is then necessary only to cross-correlate corresponding frequency channels between each antenna, so the number of cross-correlators per signal pair is again equal to n for each lag [9,10].

4 Correlators based on FPGA

Correlation is a highly parallel signal processing application that is used heavily in radio astronomy for imaging and other astronomical measurements. Correlators typically operate on huge datasets and often require real-time processing as the storage of raw data is impractical, resulting in substantial computational requirements.

For low-frequency radio astronomy, software implementation on general purpose hardware is a reliable solution for correlation. The computational demands are proportional to BMN^2, where B is the signal bandwidth, M is the number of independent beams, and N is the number of antennas. Astronomers require more powerful telescopes to achieve higher levels of sensitivity and resolution as well as increased agility and field of view; this poses more demands on the digital signal processing unit.

In modern radio telescope systems, the use of field programmable gate arrays (FPGAs) has become common practice for real-time radar signal processing. These FPGA chips are programmed using open-source signal processing libraries, which are designed to be flexible, scalable, and chip-independent. Some projects, such as Submillimeter Array (SMA) telescope and the Atacama Large Millimeter Array (ALMA), have already been practicing FPGA-powered correlators.

4.1 FPGA-based correlator for SMA radio telescope

The Submillimeter Array (SMA) consists of eight 6 m (20 ft) diameter radio telescopes arranged as an interferometer for submillimeter wavelength observations. The SMA Wideband Astronomical ROACH2 Machine (SWARM) [11] is a 32 GHz bandwidth VLBI capable correlator and phased array that was designed and deployed at the Smithsonian Astrophysical Observatory's Submillimeter Array (SMA) in 2017. The total bandwidth of 32 GHz sets the most fundamental and demanding requirement for SWARM in that the instantaneous processed bandwidth matches the aggregate bandwidth of the receivers. The preprocessing pipeline is shown in Fig. 3.10.

Unlike in many other CASPER correlators, the SMA Wideband Astronomical ROACH2 Machine (SWARM) correlator engines

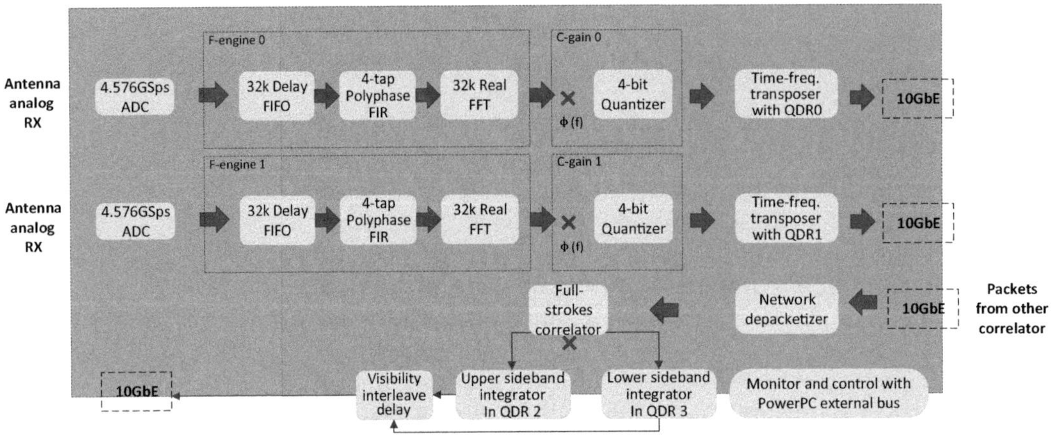

Fig. 3.10 Correlator in preprocessing pipeline on FPGA.

(X-engines) are colocated with the F-engines (conversion of a sequence of discrete time domain samples to the Fourier domain), which means that they use the same processing boards as the F-engines. While this has presented challenges in terms of clocking the FPGA design at high clock rates, the approach was intended to reduce the total number of ROACH2 boards (thus reducing cost) as well as using fewer Ethernet switch ports for the corner turn.

Additionally, the corner-turn switch ports are all used full-duplex at very nearly 10 Gbps in both directions. The SWARM X-engines compute all cross-correlation products regardless of whether the SWARM board is in dual-polarization mode, or in single-polarization mode. So, although we consider SWARM to be an 8-element full Stokes correlator, it could also be thought of as a 16-element single Stoke correlator. Additionally, autocorrelations are produced that have proven useful for calibrating the data. In total, the X-engines produce 120 complex-valued cross-correlations and 16 real-valued autocorrelations; each pair of real-valued autocorrelations can be crammed into a single complex number, thus reducing the total output components to 128.

4.2 FPGA-based correlator of ALMA radio telescope

The Atacama Large Millimeter/Submillimeter Array (ALMA) is an astronomical interferometer of 66 radio telescopes in the Atacama Desert of northern Chile. The ALMA correlator, a powerful supercomputer, has been fully installed and tested at its remote, high-altitude site in the Andes of northern Chile.

The ALMA observatory has two arrays. The main array has 64 antennas, each of which has a diameter of 12 m. The smaller one, which is called the Atacama Compact Array (ACA), has 12 7 m antennas. There is a correlator for each of these arrays. The main correlator is called an FXF (hybrid) type, in which the input signals are digitized at full bandwidth, divided into frequency channels using digital filters, and resampled at the appropriate lower rate before cross-correlation between all antenna pairs. Finally, the cross-correlated data are Fourier transformed to the frequency domain. The correlator for the ACA is an FX type. It will first do the Fourier transform of the data and then the cross-correlation.

The input stage of the correlator is the tunable filter bank (TFB) card. Sixteen FPGAs are assembled on one TFB card. The TFB performs a frequency division of the digitized input wideband (2 GHz) into 32 subchannels 62.5 MHz (or 31.25 MHz) wide [12]. The station card implements a bulk delay for geometric delay compensation and provides signal routing between the TFB and the correlator subsystem. The correlator portion of the system in each of the four quadrants consists of 32 correlator "planes" with each plane implementing a 64-by-64 correlation matrix. Fig. 3.11 shows the configuration in a single plane. The LTA (long-term accumulator) subsystem seen on the right takes short 1 ms or 16 ms integrations from the correlator array and provides longer-term integration.

The input sample format is in 3-bit, sampled at 4G sample/s per baseband channel, whereas the correlation sample format is in 2-bit or 4-bit, as specified in the observation schedule. General purpose processing units such as CPU or GPU do not have hardware support for these irregular data formats, nor can they handle the astronomical amount of the data throughput efficiently. The ALMA correlator demonstrates the power of FPGA, which can host computation units of customized bit width, sophisticated demultiplexing, and pipelines that well exploit the parallelism of the computation and handle large-scale data throughput in a power-efficient manner.

5 General architectures for data reduction design and implementation

5.1 Software design analysis

Because the FPGA pipeline is applied on the basis of a software algorithm, exploring the software algorithm design can provide meaningful guidance in hardware implementation. In order to

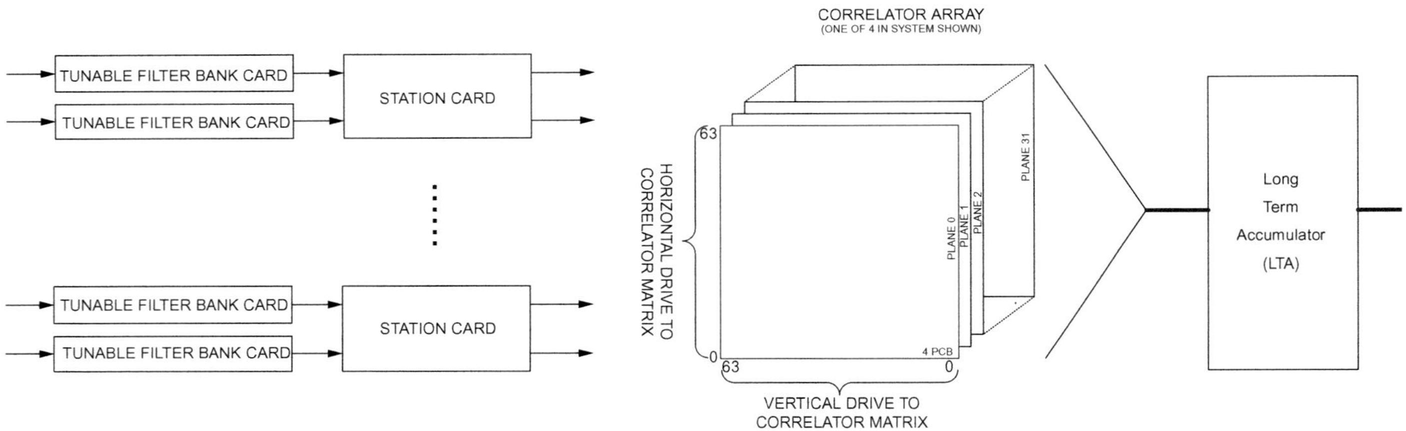

Fig. 3.11 ALMA correlator, simplified for the scope of this chapter.

design a customized preprocessing algorithm to fit the target astronomical dataset, there are some general design flows, including data analysis, preprocessing method configuration, and objective compression selection.

Astronomical radio data contain multiple dimensions, and the data operations within units in these dimensions tend to be separated to some degree. The independence in data processing units can be utilized to design parallel algorithms. The general preprocessing methods include but are not limited to data merging and compression, feature extraction, sampling, and quantization.

For data merging and compression operations, the operation usually takes place within some adjacent units. In Escoffier et al. [12], the merge is executed within the opposite polarization. Or, as described in Zheng et al. [13], the merge is operated in a sliding window with polynomial computation. This fact encourages us to reuse the operation units within a merge operation by traversing along an extensive dimension and dividing pipeline stages. By this means, full use of the pipeline technology can be made.

Feature extraction is a regular method in image preprocessing but is also applicable for astronomical radio data because astronomical radio data contain multiple sample dimensions in space, and the space resolution can be analogical. Most feature extractions are achieved through computational granularity enabling the pipeline architecture. Combining different levels of computational granularity usually achieves lower latency for astronomical data processing.

The data sample, quantization, and coding are indispensable steps in transforming the ordinary data into a standard digital format for subsequent processing and transmission. This procedure is accompanied with some basic data transformation and encoding such as DFT, VLBI, and calibration. These operations can perform in parallel in software design, and there exists the feasibility to implement the operations with pipeline technology.

5.2 Hardware implementation hierarchy

The complete preprocessing system of FPGA-based hardware implementation is typically composed of many major components.

The data transmission layer can provide an interface for data transmission between a server and an FPFA platform. In a server, the CPU is dominant for data management, and the layer is responsible for supporting the data communication and transmission between the CPU and FPGA. The performance of the data transmission layer is evaluated by throughput, bandwidth, and stability indicators. The mainstream PCIe-based CPU-FPGA platforms use direct memory access (DMA) for an FPGA to access the data from a CPU. First, the FPGA needs a memory controller IP to read the data from the CPU's DRAM to its own DRAM through PCIe. Then, the FPGA performs specified acceleration in its accelerator function units (AFUs).

Despite data transmission, the kernel function module plays a major role. The module contains many arithmetic units to execute the preprocessing steps, and the logic units are distributed to form many processing elements (PE) to fully utilize the onboard resource. The processing elements can perform partition to the data matrix and divide them into groups, and the work groups can run in parallel while each partition has an internal pipeline design and can be executed with fewer latencies.

Onboard storage is another key issue in FPGA hardware implementation. The onboard memory resource in FPGA is mainly composed of DDR RAM, Block RAM, Flip-Flops, and Lookup Tables (LUTs), but the resource is limited and the data processed in a period have a capacity bound. In this case, the memory resource must be properly distributed to achieve the temporary variable caching, ultimate result storage, and available input and output queues to satisfy the data transmission condition.

5.3 A case study of a typical compression design

This section tries to give a specific example to show the superiority of pipeline design. The example is an astronomical radio data compression scheme that makes use of the preprocessing steps with FPGA pipeline technology, and it is proved to achieve improved performance compared with the software experiment results.

5.3.1 *Workflow of data reduction based on FPGA*

The major component of the compression scheme is a preprocessing hardware design to achieve the function of bit splicing and unit merging. It is implemented with OpenCL framework where the program resolves into multiple modules, respectively running in host mode or kernel mode. The distribution is due to the following considerations. The effective bit-width is determined according to the maximum range within the row partition, which is calculated sequentially. So, the module is arranged in host mode. The judgment determines that parts of the array rows need to be merged and the others just need to be maintained. Besides, the selected rows need to be combined to reconstruct the array order.

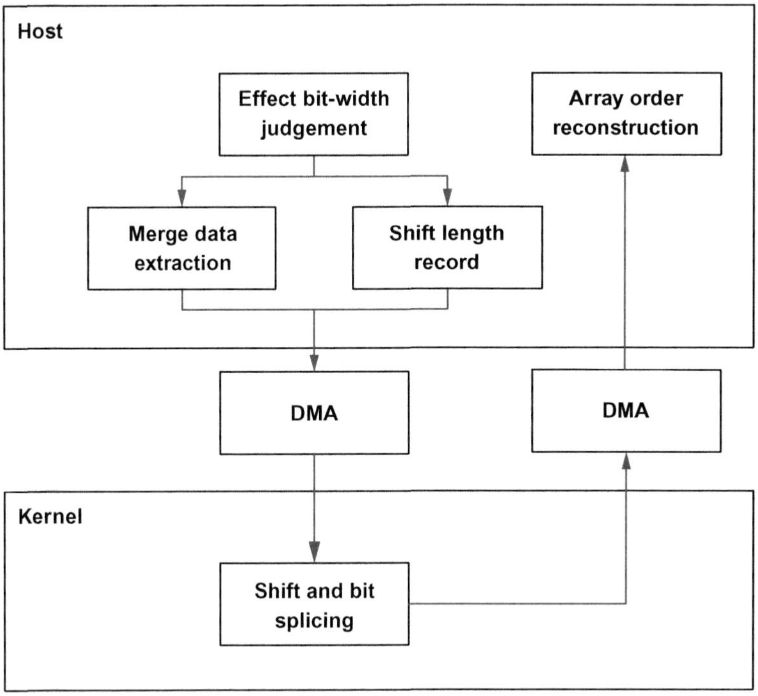

Fig. 3.12 FPGA-based data reduction workflow [13].

So, only the selected rows are extracted and passed to the FPGA kernel. The kernel is responsible for the shift and bit splicing module because the function and operation in this module can be realized by parallel and pipeline technology. After the kernel operation, the merged row data are fetched and the array order reconstruction will be performed in host mode. The complete design flow is shown in Fig. 3.12.

5.3.2 Data communication interface

In the OpenCL design flow, the data communication between host and kernel is essential and can lead to one of the bottlenecks in global performance. In order to take advantage of the throughput of the FPGA DDR, the design applies a uniform data transmission format and wraps the data into frames, which can effectively improve the DMA data transmission performance.

5.3.3 Design issues in pipeline

OpenCL is able to implement parallelization by means of work group division and concurrent operations. In order to coordinate the mechanism with FPGA onboard computing resources to achieve load balance, an important approach is to repeatedly optimize the workload division and work group scale configuration. Moreover, the traverse loop in the algorithm is optimized to minimize loop-carried dependencies, and the work items are effectively organized and synthesized to avoid repeated calculation.

6 Conclusion

In this chapter, we described major steps of the preprocessing pipeline on FPGA with highlights in ADC interleaving, deserialization, FIR filtering, time frequency domain transposing, correlator and data reduction. Due to the constraints on increasing performance and power, it seems FPGA plays more roles in preprocessing that were previously taken on by software on the CPU or GPU.

References

[1] Analog Device, HMCAD1511: High Speed Multi-Mode 8-Bit 30 MSPS to 1 GSPS A/D Converter, https://www.analog.com/media/en/technical-documentation/data-sheets/hmcad1511.pdf.

[1a] CASPER community, ADC16_interface. https://github.com/casper-astro/mlib_devel/tree/master/jasper_library/hdl_sources/adc16_interface.

[2] CASPER Community, Smart Network ADC Processor, https://casper.ssl.berkeley.edu/wiki/SNAP.

[3] P. Shetty, P. Poshala, Wide Bandwidth Receiver Implementation by Interleaving Two Gigasampling ADCs, TI, 2015. http://www.ti.com/lit/an/snaa286/snaa286.pdf.

[4] G. Manganaro, D.H. Robertson, Interleaving ADCs: Unraveling the Mysteries, Analog Device, 2015. https://www.analog.com/en/analog-dialogue/articles/interleaving-adcs.html.

[5] M. Defossez, Serial LVDS High-Speed ADC Interface, Xilinx, 2012. https://www.xilinx.com/support/documentation/application_notes/xapp524-serial-lvds-adc-interface.pdf.

[6] Xilinx, 7 Series FPGAs SelectIO Resources, UG471, https://www.xilinx.com/support/documentation/user_guides/ug471_7Series_SelectIO.pdf, 2018.

[7] J. Chennamangalam, The Polyphase Filter Bank Technique, https://casper.ssl.berkeley.edu/wiki/The_Polyphase_Filter_Bank_Technique, 2016.

[8] R.P. Escoffier, The MMA Correlator, MMA Memo 166, National Radio Astronomy Observatory, 1997.

[9] B.R. Carlson, P.E. Dewdney, Efficient wideband digital correlation, Electron. Lett. 36 (2000) 987–988.

[10] R. Perley, P. Napier, J. Jackson, B. Butler, B. Carlson, D. Fort, P. Dewdney, B. Clark, R. Hayward, S. Durand, M. Revnell, M. McKinnon, The expanded very large array, Proc. IEEE 97 (2009) 1448–1462.

[11] R.A. Primiani, K.H. Young, A. Young, N. Patel, R.W. Wilson, L. Vertatschitsch, B. B. Chitwood, R. Srinivasan, D. MacMahon, J. Weintrou, SWARM: a 32 GHz correlator and VLBI Beamformer for the submillimeter array, J. Astron. Instrum. 5 (4) (2016).

[12] R.P. Escoffier, G. Comoretto, J.C. Webber, A. Baudry, C.M. Broadwell, J.H. Greenberg, R.R. Treacy, P. Cais, B. Quertier, P. Camino, A. Bos, A. W. Gunst, The ALMA correlator, Astron. Astrophys. 462 (2007) 801–810.
[13] Y. Zheng, Y. Zhu, Y. Song, T. Nan, W. Li, A lossless astronomical data compression scheme with FPGA acceleration, in: Proceedings of IEEE 32nd System on Chip Conference, 2019.

4

Real-time stream processing in radio astronomy

Danny C. Price
Centre for Astrophysics and Supercomputing, Swinburne University, Hawthorn, VIC, Australia; Department of Astronomy, University of California at Berkeley, Berkeley, CA, United States

1 Introduction

A major challenge in modern radio astronomy is dealing with the massive data volumes generated by wide-bandwidth receivers. For example, in the upcoming Square Kilometre Array SKA1-mid telescope, each of 133 antennas is expected to generate 100 Gb/s, for 13.3 Tb/s of data that must be processed in real time [1]. These data will be augmented with an extra 64 streams of 40 Gb/s data from the existing MeerKAT telescope, bringing the total data rate to 15.8 Tb/s; within the correlator the aggregate data rate will exceed 57 Tb/s! Overall, the SKA1 is expected to produce over five times the global internet traffic as of 2015 [2]. Existing telescopes are already producing massive data volumes: for example, the Australian SKA Pathfinder, ASKAP, has a prebeamformer data rate of 103 Tb/s [3], and the Canadian Hydrogen Intensity Mapping Experiment (CHIME) digitizes and processes an impressive 13.1 Tb/s [4]. These telescopes are shown in Fig. 4.1.

Such massive data rates are difficult to redistribute and cannot be recorded as final data products. Data rates are often too great for a single device to cope, and so processing must be split across multiple devices working in parallel. These devices must work in unison to process incoming data in real time, reduce the data volume to a manageable size, and output a science-ready data product.

The aim of this chapter is to give a broad overview of how digital systems for radio telescopes are commonly implemented, with a focus on real-time stream processing over multiple compute devices. Accompanying chapters delve deeper into specific

Big Data in Astronomy. https://doi.org/10.1016/B978-0-12-819084-5.00013-4

Fig. 4.1 The Canadian Hydrogen Intensity Mapping Experiment (CHIME, *left*), the Australian Square Kilometre Array Pathfinder (ASKAP, *center*), and artist's conception of the Square Kilometre Array telescope (SKA, *right*). Image Credits: Z22/Wikimedia, Ant Schinckel, SKA Project Development Office.

aspects: Chapter 3 gives an overview of preprocessing pipelines on field-programmable gate arrays (FPGAs), and Chapter 5 discusses how signals from a telescope are digitized, split into channels, and prepared for data transport.

In this chapter, we use cross-correlators as a primary example of a stream processing pipeline. We refer readers unfamiliar with correlators to the freely available [5], which gives a comprehensive overview of interferometry in radio astronomy.

This chapter is structured as follows. First, we introduce the concept of stream processing (Section 2), then summarize system architectures and technologies that are commonly used (Section 3). Section 4 gives an overview of Ethernet, which has become ubiquitous for data transport between compute nodes. We then discuss data preprocessing ("first-stage" signal processing, Section 5), to prepare the data streams for redistribution. Section 6 details logistical concerns of data redistribution, and the common problem of "corner-turning" for data transposition. Section 7 discusses the "second-stage" data processing to form final science data products and strategies for data recording. We conclude with a discussion of promising technologies for future systems.

2 Stream processing

A telescope's output can be considered as a stream of data; unlike files, a data stream does not have a clearly defined start or endpoint. It is illustrative to think of a telescope's output as a stream of *frames*, where each frame is an array of data at a given time step. A frame may have multiple dimensions, depending on the number of inputs and transformations applied to the data. For example, a frame may have a shape

$$(N_{\text{ant}}; N_{\text{beam}}; N_{\text{pol}}; N_{\text{chan}}) \tag{4.1}$$

where N_{ant} is the number of antenna elements in the telescope, N_{beam} is the number of beams on the sky, N_{pol} is how many polarizations the telescope samples have, and N_{chan} is the number of frequency channels that the data are divided into. For simplicity, we assume all data in the array are of the same datatype (e.g., 8-bit signed integer).

From this view, the purpose of a telescope's digital signal processing (DSP) system is to apply a series of transformations from one or more frame *sources* to produce a data product that is written to one or more outputs, or *sinks*. One may design a basic set of operations, or *blocks*, as:

- *Stream splitting/merging*. A frame can be split upon an axis to produce multiple data streams with smaller frames. Alternatively, multiple data streams may be merged into a single frame. The rate at which frames are output (i.e., the frame rate) remains the same, but the size of the frame will change, which will affect the output data rate.
- *Intraframe transformation*. Using only data within the frame, a transform is applied to modify the data. The shape and datatype of the frame may change, but the rate at which frames are processed remains constant.
- *Interframe transformation*. Data from multiple frames are combined; for example, averaged in time to lower frame rate, or buffered up to create an additional frame axis (e.g., a "subframe" axis). Unlike intraframe transforms, the output frame rate may change.

Stream processing systems may be thought of as a *pipeline* through which data flow like water from the source to the sink.[a] The time it takes from start to finish is known as the pipeline latency. Importantly, transformations in a pipeline are run concurrently, not applied in serial. If a processing block does not finish processing a frame before the next one arrives, a *bottleneck* will arise, which will slow the entire pipeline. A pipeline may also have multiple data streams that run in parallel, as long as they are independent from one another at that stage of the pipeline.

A simple example diagram representing a pipelined implementation of a cross-correlator is given in Fig. 4.2. In the diagram, three antennas act as data sources, which output their data streams in parallel. We have labeled their axes "pol" for polarization, and "count" for data from the analog-to-digital converter (ADC). After applying a Fast Fourier Transform (FFT), which adds a "channel" axis, the three parallel streams are then merged into

[a]For those with a background in graph theory, a pipeline is a directed acyclic graph where the vertices are processing blocks.

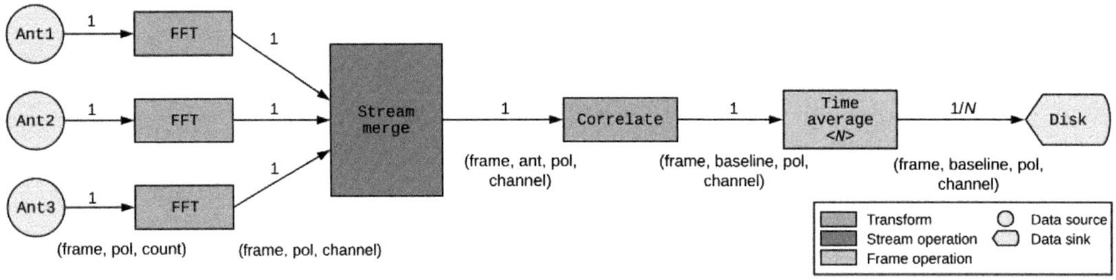

Fig. 4.2 Simple example of a pipeline to cross-correlate three antennas, using data streams. Each antenna outputs a data stream, and a Fast Fourier Transform (FFT) is applied to each to form channels. The three streams are merged together, and then cross-correlation is applied. Multiple frames are then averaged together, before being written to disk. The number above the arrow represents the frame rate; frame dimensions are shown in brackets below the streams.

one, which effectively adds an axis to the data frame (labeled "ant" for antenna). The data frame is then fed to a correlation transform, which changes the frame shape to add a "baseline" axis. Time averaging is then applied, which requires multiple frames to be added together. The output is the written to disk (a data sink) as a file.

3 Heterogeneous signal processing

Unless all DSP is conducted on a single processing board, data must be redistributed to other parts of the digital system. A common design approach is to have a DSP "frontend," on which first-stage signal processing and reduction is done, that connects via high-speed links to a DSP "backend" that applies second-stage signal processing. The backend ingests data from one or more frontend streams, and reduces the data into final science data products. This division of labor allows different DSP platforms to be used as desired an approach known as heterogeneous signal processing.

It is common, particularly for larger systems, for both first-stage and second-stage systems to consist of multiple DSP boards or servers (or more generally, nodes). These nodes need to be interconnected to form a single system. The choice of how to connect boards is dependent upon what data need to get where. For example, a correlator might consist of N antennas, each of which is connected to frontend board that channelizes the data. At early stages, the data streams are independent, so frontend boards might be located at the antennas. Cross-correlation between antenna data streams could then be performed by a second-stage board that ingests, merges, and processes the N data streams in a central facility. As cross-correlations are computed on a per-channel basis, one could parallelize the computation across several boards, each processing a subset of channels from all antennas.

A diagram showing a heterogeneous approach to implementing an FX correlator, where signals are channelized before cross-correlation, is shown in Fig. 4.3. In this example, N processing nodes are used to channelize data from N antennas. Their output data streams are sent via an Ethernet switch to L correlator nodes running in parallel. In this example, the problem of data redistribution (Section 6) is solved by use of a commercial off-the-shelf network switch. This approach is detailed further in Refs. [6, 7].

While DSP nodes may run independently from one another, for radio astronomy it is vital that inputs are kept synchronized or else phase coherence will be lost. A common approach is to distribute a single sampling clock to the ADCs on each frontend node. A pulse-per-second, derived from global positioning system (GPS), may also be distributed to the frontend boards, to demarcate an accurate absolute start time. As long as data frames are accurately time tagged, data processing after the ADC may be

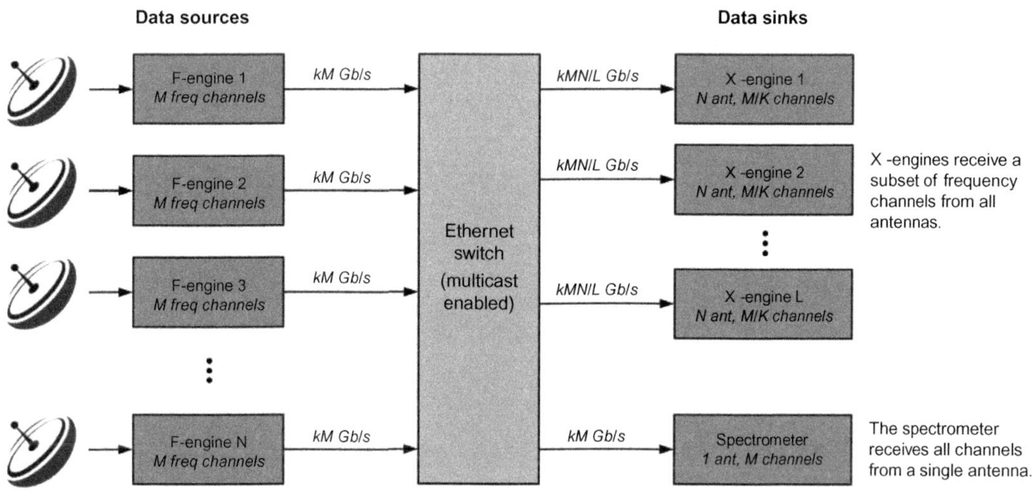

Fig. 4.3 Diagram showing a common approach to building an FX correlator. The outputs of N antennas are digitized and channelized into M channels using a DSP frontend (e.g., FPGA boards). These channels are split into L sub-bands, which are sent over Ethernet to and cross-correlated by L backends running in parallel (e.g., GPU servers). Diagram modified from Fig. 1 of J. Hickish, Z. Abdurashidova, Z. Ali, K.D. Buch, S. C. Chaudhari, H. Chen, M. Dexter, R.S. Domagalski, J. Ford, G. Foster, D. George, J. Greenberg, L. Greenhill, A. Isaacson, H. Jiang, G. Jones, F. Kapp, H. Kriel, R. Lacasse, A. Lutomirski, D. MacMahon, J. Manley, A. Martens, R. McCullough, M.V. Muley, W. New, A. Parsons, D.C. Price, R.A. Primiani, J. Ray, A. Siemion, V. van Tonder, L. Vertatschitsch, M. Wagner, J. Weintroub, D. Werthimer, A decade of developing radio-astronomy instrumentation using CASPER open-source technology. J. Astron. Instrum. 5(4) (2016) 1641001–12.

asynchronous, although frames may need to be buffered in memory as they propagate through asynchronous parts of a pipeline.

3.1 Common architectures

The world's radio telescopes are powered by four dominant platforms for digital signal processing: Central Processing Units (CPUs), Graphics Processing Units (GPUs), Field-Programmable Gate Arrays (FPGAs), and Application-Specific Integrated Circuits (ASICs). As their strengths and weaknesses are key to choosing which platform to use, we provide a brief overview here.

ASICs. As the name suggests, ASICs are digital circuits designed for a specific purpose. ASICs were once common within radio astronomy, and power the Karl Jansky Very Large Array (JVLA) and Atacama Large Millimeter Array (ALMA, Fig. 4.4) correlators [8, 9]. However, the one-time cost to fabricate a custom chip on nanometer-scale processes is significant, and economies of scale can be hard to realize for most experiments. For the most part, ASICs have been replaced with off-the-shelf FPGAs and GPUs, which offer a faster development cycle thanks to repurposable codebases and shared tool flows. Indeed, FPGAs are often used for prototyping ASIC designs. Nevertheless, for large-scale projects such as the SKA, ASICs offer significant power savings that

Fig. 4.4 One quadrant of the ASIC-powered ALMA correlator. The ALMA correlator processes up to 16-GHz bandwidth from 64 antennas. Image Credit: ALMA (ESO/NAOJ/NRAO), S. Argandona.

could reduce operational costs [10]. The functionality of ASICs is generally defined using a hardware description language (HDL).

CPUs. The ubiquitous CPU is the most flexible of the four DSP platforms, as it is designed specifically for general-purpose computation. The two dominant suppliers of CPUs for server-class computers are Intel and Advanced Micro Devices (AMD). Both Intel and AMD chips are based on the x86-64 instruction set architecture, so codes are intercompatible between the two. For high-performance computing, the Intel Xeon and AMD Ryzen Pro branded processors are marketed, which support larger amounts of RAM and have advanced features not found on desktop-class chips. An interesting alternative is the IBM POWER9 CPU, which is notable in its support for NVLINK, an 80-GB/s bus designed for faster memory copy to NVIDIA GPU cards.

In mobile and low-power embedded platforms, ARM (previously advanced RISC machine) has emerged as the dominant architecture. From a radio astronomy standpoint, ARM processors are commonly found as systems-on-chip (SoC) that directly interface with FPGAs, but they are generally not used for heavy DSP work.

GPUs. A recent trend is for CPUs to be joined by GPUs as "coprocessors," where computations are offloaded to the GPU. The types of computation required for realistic, high frame-rate graphics are a good match for many astronomy algorithms, where a single instruction (e.g., a multiplication operation) can be applied to multiple data in parallel.[b] GPU processors are, to date, not used without a host system.

The use of GPUs for nongraphics processing is known as general-purpose GPU programming, or GP-GPU. As with CPUs, there are two dominant players: NVIDIA and AMD. NVIDIA GPUs are programmed using the compute-unified device architecture, or CUDA programming model, whereas AMD requires the use of open computing language (OpenCL), an open-source standard maintained by the Khronos Group. While OpenCL is also supported by NVIDIA GPUs, achieving maximum performance on either vendor's offerings requires targeting of specific instructions and tuning of kernels to match the underlying architectures. For example, the CHIME correlator leverages a fused multiply-add instruction "mad24" only available on AMD GPUs [12], whereas the xGPU correlator code leverages a dot-product accumulate "dp4a" instruction only available on some NVIDIA GPUs [13]. As such, writing highly performant code compatible with both

[b]The term single instruction, multiple data (SIMD) was popularized by Ref. [11] as part of his seminal taxonomy of computer architectures.

vendors is burdensome, and generally code is written in CUDA if targeting NVIDIA GPUs, or OpenCL if targeting AMD GPUs.

FPGAs are user-reconfigurable integrated circuits, consisting of an array of programmable logic blocks. These blocks are connected together by a reconfigurable mesh, such that different functionality is achieved by connecting different components together to form a digital circuit. The two dominant FPGA vendors are Xilinx Inc. and Intel (formerly Altera). As with GPUs, differences in implementation of interfaces and logic elements mean that firmware is designed to target a specific chip. While FPGAs are also used as coprocessors alongside CPUs in high-performance computing, for radio astronomy applications they are more commonly found on discrete boards with peripheral interfaces such as ADCs and high-speed Ethernet, to perform first-stage DSP tasks such as channelization and packetization (see Chapter 5).

FPGAs are well suited to interface with ADCs and other peripherals due to high-speed (>1 Gb/s) transceivers, over which a SerDes (Serializer/Deserializer) link can be setup. To route data from an ADC to an FPGA, the parts are placed onto a circuit board and the relevant pins of the ADC are connected to the relevant pins of the FPGA; alternatively, the FPGA pins are connected to a mezzanine connector that allows a variety of daughter boards to be connected. Similarly, the FPGA transceivers are used to drive high-speed Ethernet interfaces, over which signals can be transported in Ethernet packets.

FPGAs are most commonly programmed using an HDL, such as VHDL or Verilog, but also support OpenCL. Higher-level tools, such as MATLAB Simulink and National Instruments LabView, are also available. The Collaboration for Astronomy Signal Processing and Electronics Research (CASPER) have pro-vided the radio astronomy community with open-source FPGA-based hardware and development libraries tools for over a decade [7], promoting design reuse and lowering entry barriers. Several other FPGA boards for radio astronomy exist, including the ASTRON Uniboard [14], the CSIRO Redback and Gemini boards [15, 16], and the ICE board from McGill Cosmology Instrumentation Laboratory [17].

4 Ethernet interconnect

Ethernet is an industry-standard collection of computer networking technologies used in virtually all computer networks. Computer networking is a vast topic, well outside the scope of this chapter; nevertheless, a general understanding is important to fully appreciate implementation concerns of Ethernet-based data redistribution, which has become commonplace. To that end, a gentle introduction is provided here to guide our discussion. We refer the reader to [18, 19] for a comprehensive overview; the latter is now freely available online at http://www.tcpipguide.com. Our goal for this section to is to give the reader an idea of the steps required to design a well-scoped and fully functional data redistribution network for a small-to-medium size DSP system.

4.1 High-speed Ethernet technologies

Ethernet interconnect uses either passive copper links, or fiber optic links driven by active transceivers. For short distances (<10 m), more economical copper cables can be used, but longer distances require fiber optic links, which can span several kilometers.

Ethernet defines a variety of connector types and speeds. As of writing, 10 and 40 GbE are widespread and economical, with 100 GbE becoming increasingly available. The SFP+ (small form-factor pluggable transceiver) is most widespread for 10 GbE, whereas and QSFP+ (quad SFP) is used for 40 GbE. A 40-GbE link can be split into four 10-GbE links, meaning a 32-port 40-GbE switch can operate as a 128-port 10-GbE switch if supported by the switch firmware.

Two common multi-interface devices are used in Ethernet networks: switches and routers. In general, only switches are required for data redistribution networks. Switches work by identifying what devices are connected, then directing incoming packets to their destination via the most suitable port. Routers act more like a gateway between two networks, sending packets from one network to another.

4.2 TCP/IP

TCP/IP refers to a family of transport protocols, including TCP (Transport Control Protocol), UDP (User Datagram Protocol), and IP (Internet Protocol). UDP and TCP are the prevailing protocols used to transport data across a network, and sit on top of IP. IP defines how data are relayed across a network: data are packaged up as discrete "datagrams" that can be lost, duplicated, or delivered out of order. This nonguaranteed strategy contrasts starkly with simpler electrical signaling schemes, as would be used in backplanes to connect transceivers of neighboring boards. It introduces an extra level of complexity and processing overhead, but offers advantages for large networks, as it offers a means to dynamically share a link with multiple hosts.

Within TCP/IP, all network interfaces are assigned an IP address, a numerical label akin to a telephone number, by the network administrator. Every interface also has to have a MAC address, a 48-bit identifier that is generally set by the manufacturer so as to be globally unique. MAC addresses are usually expressed as six hexadecimal pairs, e.g., 00:00:00:00:00:00. IP (v4) addresses are 32 bits,[c] expressed in a dot decimal notation of four integers between 0 and 255, separated by dots, e.g., 192.168.0.1. In a private network, you can decide yourself how to allocate IP addresses, but interfaces connected to the broader internet must be assigned IP addresses by the internet provider.

Along with the IP address, every interface must have a subnet mask defined. A subnet mask is used to subdivide the 32-bit IP address range into subnetworks in which traffic is contained; see Chapter 9 of Ref. [19].

An example IP address table. One of the first tasks to bringing up a Ethernet-based data redistribution network is to allocate IP addresses to all network interfaces. To put IP addressing into context, consider an example FX correlator, with four F-engines running on FPGAs, connected to three X-engines servers with GPUs. Note that we are following the approach shown in Fig. 4.3, shown as a stream processing diagram in Fig. 4.6.

An example IP allocation table for the putative correlator is shown in Table 4.1. We have chosen to use the 192.168.10.0 network, and set a subnet mask 255.255.255.0.

4.3 TCP and UDP

As mentioned earlier, TCP and UDP are the prevailing protocols used for network data communications. Both protocols send data as a series of datagrams, or “packets” with roughly kB sizes. The main difference between the two is that TCP checks if data are transferred successfully, and if not, will resend missing data. For large, complex networks like the internet, this makes TCP a reliable method to transfer data; however, it has significant overhead due to the requirement that the receiver must send acknowledgments and requests to resend missing data back to the transmitter. Data must also be buffered so that they can be retransmitted if required. In contrast, UDP is an unguaranteed, one-directional

[c] The latest version IPv6, uses 128-bit addresses expressed with a hexadecimal notation, e.g., 2001:0db8:0000:0000:0000:8a2e:0370:7334; this length is necessary to service the global internet, but not for private internal networks where the shorter IPv4 addresses may still be used.

Table 4.1 Example IP allocation for a small correlator.

Device	IP address	Subnet mask	MAC address
F-engine 1	192.168.10.101	255.255.255.0	02:00:00:00:00:01
F-engine 2	192.168.10.102	255.255.255.0	02:00:00:00:00:02
F-engine 3	192.168.10.103	255.255.255.0	02:00:00:00:00:03
F-engine 4	192.168.10.104	255.255.255.0	02:00:00:00:00:04
X-engine 1	192.168.10.201	255.255.255.0	02:01:00:00:00:01
X-engine 2	192.168.10.202	255.255.255.0	02:01:00:00:00:02
X-engine 3	192.168.10.203	255.255.255.0	02:01:00:00:00:03

stream of packets that the receiver does not need to acknowledge with a response.

UDP is therefore easier to implement, and faster overall transmission speeds can be achieved due to the lower overhead. While packet loss during transmission is a potential issue, this is rarely encountered in well-scoped data redistribution networks: packet loss is a symptom of long (i.e., high latency) links with contention due to congestion, and/or hardware failure.

4.4 UDP datagram structure

Units of data are passed over UDP as packets, or more correctly *UDP datagrams*.[d] UDP datagrams are themselves embedded into *IP datagrams*, which are themselves embedded into an *Ethernet frame* (not to be confused with the use of frame elsewhere in this chapter), rather like nested matryoshka dolls (see Fig. 4.5).

At each layer, there is a header and a data payload. Headers are several bytes in size (IPv4 header is 20 bytes, and UDP header is 8 bytes), while payloads can be thousands of bytes—the maximum size of an Ethernet frame is 1500 B, or up to 9000 B if "jumbo frames" are enabled in the Ethernet hardware. Due to the headers, small datagrams have appreciable overhead, so larger datagrams are preferred for high-throughput applications. In addition, the number of packets per second decreases as packets get larger,

[d]The term "packet" is used generically to refer to any type of IP message. TCP calls its messages segments, while UDP calls its messages datagrams.

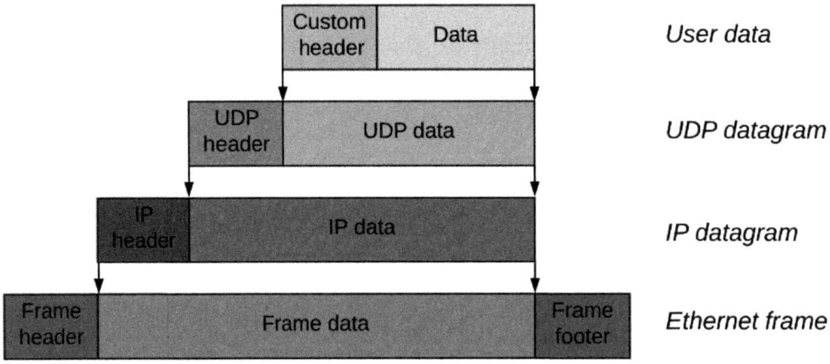

Fig. 4.5 The layers within an Ethernet frame for a UDP datagram. Data units are nested within the data payload of each lower layer. Within the UDP data payload, a user may define their own data structure, such as a header plus data payload.

which is beneficial when capturing large data volumes (see Section 7.2).

While this is admittedly complex, as long as we are happy to adhere to the TCP/IP standard, methods for handling the lower levels come "baked in" to all Ethernet devices. On FPGAs, Ethernet IP cores are provided by vendors and third parties. As such, the astronomer/engineer only needs to focus on how to structure their data within the UDP datagram's payload: the yellow part of Fig. 4.5.

A common choice is to set aside a few bytes as the start of the UDP data payload and define a custom header that describes the data well enough to reconstruct the data stream. At its simplest, the header would just be a counter so that packets can be reordered correctly in case they are received out of order. If there are multiple nodes transmitting data, a few bytes of header could be allocated for a unique board identifier (e.g., an 8-bit number between 0 and 255). If a single frame from the data stream cannot fit within the maximum packet size of 8972 B—for example, if there are 32,768 channels of 32-bit datatype—then extra information is needed to identify what chunk of the frame is being transmitted, so the frame can be faithfully reconstructed. While specifications for streaming protocols over UDP exist, such as SPEAD,[e] it is common for the UDP payload data structure to be custom to a given experiment.

[e]Streaming Protocol for the Exchange of Astronomical Data, https://spead2.readthedocs.io/.

4.5 Multicast

IP Multicast is a useful technique that allows incoming data sources to be distributed to multiple data sinks. Instead of sending to the IP address of a single node, data are sent to a multicast IP address (between 224.0.0.0 and 239.255.255.255). Any nodes that want to receive these data "subscribe" to the multicast data by joining the multicast group (defined by the multicast IP address chosen). For this to work, the network must have a device acting as a multicast router, which uses Internet group management protocol (IGMP) to establish group membership. Network switches use IGMP snooping to maintain a list of which links it should retransmit multicast packets over (alternatively, switches can broadcast multicast packets over all ports, which is generally undesirable).

Multicast is used in the MeerKAT telescope as a method to allow user-supplied equipment access to data streams from the first-stage FPGA boards. They also use multicast as part of their corner-turn operation: sub-bands are sent to different multicast groups.

5 First-stage data processing

Now that we have introduced common technologies, we now discuss how these technologies may be used to implement stream processing systems. The first step in a radio astronomy processing pipeline is to convert incoming signals from analog to digital, using an analog-to-digital converter (ADC). This step is detailed in Chapters 3 and 5, along with common first-stage data processing approaches; here, we provide a short overview to highlight the challenges faced downstream.

5.1 Data rates

ADCs act as a data sources in a stream processing pipeline. The data rate from the ADC depends upon the sampling rate, F_s, and the number of bits per sample, N_{bits}:

$$\text{Data rate (b/s)} = F_s \times N_{\text{bits}} \tag{4.2}$$

For example, a 1 Gsample/s ADC with 8-bit resolution outputs an 8 Gb/s data stream. The Nyquist Shannon sampling theorem dictates that the ADC sampling rate must be twice the analog bandwidth B of the incoming signal, $F_s = 2\text{B}$. For current-generation radio telescopes, receiver bandwidths may span several gigahertz.

In order to capture the two orthogonal polarization states ($N_{\text{pol}} = 2$), two ADCs are required per telescope receiver. Some telescopes, such as ASKAP, have multiple receivers per antenna:

the ASKAP phased array feed consists of hundreds of receivers that are digitally combined to form beams on the sky [3]. Similarly, receivers may split their bandwidth over several sub-bands. The aggregate ingest data rate for an array of radio telescopes is thus given by

$$\text{Data rate (b/s)} = (F_s \times N_{\text{bits}}) \times N_{\text{ant}} \times N_{\text{pol}} \times N_{\text{rxb}} \times N_{\text{sub}}, \tag{4.3}$$

where N_{rxb} is the number of receiver elements per antenna, and N_{sub} is the number of sub-bands.

Table 4.2 shows aggregate data rates for a selection of radio telescopes where total ingest exceeds 1 Tb/s. These rates may be exceeded at parts in the pipeline, due to increased datatype size (e.g., bit growth within an FFT), or an increase in frame size (e.g., after cross-correlation across antenna pairs). Nevertheless, data rates must eventually be reduced by many orders of magnitude in order to form science data products that can be written to disk.

5.2 Channelization

After digitization, it is common for input signals to be split into channels using a FFT or polyphase filter bank (PFB; see [5, 23] Section 8.8, for an overview). Channelization is motivated by two reasons. Firstly, many processing tasks, including cross-correlation and beamforming, can be parallelized across frequency channel. Secondly, as shown in Fig. 4.3, a data frame can be split across channels and distributed across multiple processing boards.

The channelizer part of a stream processing pipeline is commonly referred to as an "F-engine." This term is used broadly to include gain correction, requantization of signals to lower bit-depth, and preparation of the data stream for signal transport. F-engines are explored in more detail in Chapter 5.

Table 4.2 Aggregate data rates after digitization for selected radio telescopes.

Telescope	F_s (MHz)	N_{bits}	Nant	N_{rxb}	N_{pol}	N_{sub}	DR (Tb/s)	Ref.
ASKAP	608	12	36	188	2	1	98.8	[3]
CHIME	800	8	1024	1	2	1	13.1	[4, 17]
ALMA	4000	3	64	1	2	4	6.1	[9]
MeerKAT	1712	10	64	1	2	1	2.2	[20]
SMA	4000	8	8	1	1	4	2.0	[21]
JVLA	2048	8	26	1	2	2	1.7	[22]

6 Data redistribution

Getting data from multiple source nodes and routing them to multiple sinks nodes is a nontrivial task for large installations. In this section, we discuss common approaches to data redistribution, with a focus on the use of Ethernet for multinode stream processing.

6.1 The corner-turn problem

The problem of data redistribution is exemplified by correlators with a large number antenna inputs. In a so-called FX correlator, the data stream from each antenna is channelized into M channels by a filter bank, referred to as an "F-engine." The F-engine is fed into an "X-engine" that correlates the signals from all antenna pairs together. In the F-engine, each antenna is independent, so a separate F-engine can be run on each of N antenna inputs, to produce M channels. The X-engine computes the cross-correlation for every pair of antennas, but each channel can be processed independently. The X-engine can therefore be split into L separate X-engines, each processing M/L channels.

A stream processing diagram for a simple FX correlator is given in Fig. 4.6, with $N=4$ F-engines connected to $L=3$ X-engines. The

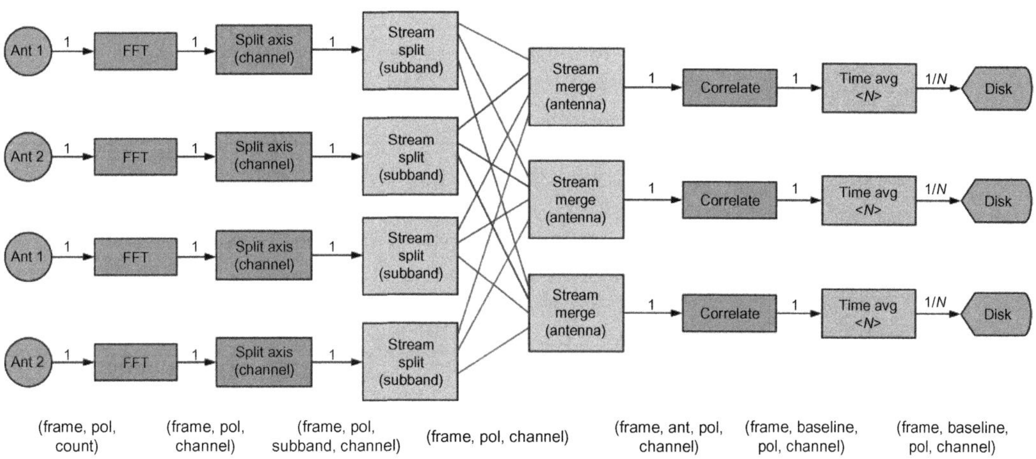

Fig. 4.6 A stream processing pipeline showing a four-element cross-correlator with a corner-turn operation. Here, after the FFT operation the frame is split into three sub-bands. Three separate correlator streams run in parallel, each processing one sub-band. This stream splitting then merging is used to group a fraction channels from every antenna input, and is known as a corner turn; as the number of antennas increases the corner-turn operation can become difficult.

data streams are split into sub-bands and then merged by antenna; this process is known as a *corner-turn*, and is conceptually similar to an $N \times L$ matrix transpose.

As N and L increase, the corner-turn becomes increasingly difficult, by virtue of the $N \times L$ stream split/merge operations required. Adding to the complexity are the massive volumes of data that are being transported, and limitations on the data rate each link can handle. Choosing an appropriate and scalable data redistribution scheme is an important design choice in any radio astronomy instrument.

6.2 Data backplanes

One method to redistribute data is to use a custom backplane into which processing boards are plugged. The backplane implements a network of links, generally using differential signaling over copper lines. This method is only appropriate for FPGAs and ASICs, as their transceivers can drive the differential lines of the backplane to be routed to other FPGAs or ASICs within the system. Backplanes also provide a convenient way to route power and clock signals to the boards. Nevertheless, there are limitations to how large the redistribution network can grow, as the number of transceivers on any FPGA is finite; plus, there are physical routing considerations as more boards are added to the backplane. Only devices specifically designed for the backplane may be connected, which limits their portability and compatibility with other technologies.

The CHIME backplane. An example of a modern backplane is that used in the CHIME telescope [4, 17, 24]. CHIME is the largest radio interferometer build to date, based on the metric of number of inputs squared times bandwidth (N^2B), processing data from $N = 2048$ inputs, each with B = 400 MHz. CHIME uses the "ICE" FPGA-based signal processing and networking system, in which up to 16 FPGA motherboards are plugged into 9 U (~40 cm) rack-mountable "crates." Each crate has a custom backplane that provides high-speed, full-mesh connectivity between the 16 boards; 64 × 10 Gb/s Ethernet links are also provided to interconnect crates (4 links per FPGA board). The backplane is constructed from 24 substrate layers, and interfaces the FPGA boards using high-density Impact-series Molex connectors that can pass up to 25 Gb/s.

A total of 6.6 Tb/s of data flows through the CHIME corner-turn system before being fed into 256 GPU-based correlator nodes. The corner-turn is done in several stages. The first is conducted on the FPGA boards, each of which processes 16 antenna streams. The second stage uses the backplane full-mesh network, to gather sub-bands from 256 antenna inputs per crate. Thirdly, multiple crates are connected using the 10 Gb/s Ethernet links, and each FPGA board ends up with 32 channels from 512 inputs; two more corner-turn stages are then conducted on the GPU servers.

6.3 Packetized Ethernet interconnect

An alternative method is to connect all boards via a high-speed Ethernet network. This approach is becoming increasingly common, as GPU-based instruments gain traction and Ethernet switches and interface cards become cheaper and faster.

The simplest method is to connect all interfaces, i.e., all data sources and all data sinks, to a single switch. Some care is needed to ensure that instantaneous packet rates are not too "bursty," so as to momentarily exceed the output port's data rate; many switches have hardware ow control to alleviate this issue, but ow control buffers on ports are often small.

The MeerKAT data network. The MeerKAT telescope [20] implements an impressive data redistribution network [25] that allows multiple racks of second-stage "user-supplied equipment" (USE) to access the incoming data streams from 64 antennas. USE subscribes to incoming data streams using multicast join requests; data from the first-stage signal processors are divided into sub-bands and each sub-band is sent to a separate multicast IP.

Essentially, the MeerKAT data redistribution acts as if one giant switch, with 648 available ports. This virtual switch is implemented using a folded Clos topology (see example in Fig. 4.7), where there are 36 "leaf" switches connected to 18 "spine" switches, all switches are the same hardware. Each leaf switch is connected to every spine switch, but not to other leaf switches. Each USE rack may have a leaf switch physically located within it, for convenience.

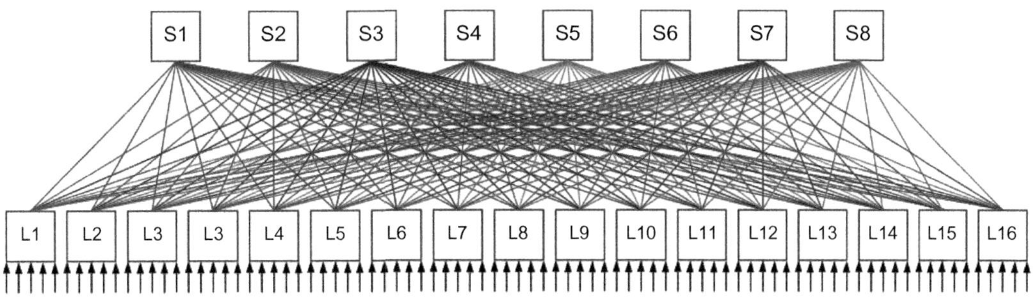

Fig. 4.7 A network with a folded Clos topology, with eight spine nodes connected to 16 leaf nodes. Equipment only connects to leaf switches (shown with black arrows).

As there are multiple paths a packet could take, a packet routing scheme must be implemented to avoid loops and saturation. MeerKAT uses the Open Shortest Path First (OSPF) IP protocol, but manual routing tables can also be implemented.

7 Second-stage processing

Given the widespread use of CPU/GPU servers for second-stage processing, in this section we discuss factors relevant to achieving high performance on CPU/GPU servers, and detail common approaches.

7.1 Performance modeling

To keep hardware costs down, each node within a second-stage processor should capture and process as large a bandwidth as possible. The limit will depend on whether the processing pipeline is compute bound or bandwidth (I/O) bound. In a compute-bound task, the amount of data that can be processed is limited primarily by the time taken for computational operations to be performed. Conversely, in an I/O-bound task, the movement of data from source to sink is the primary performance limitation. Both bounds are encountered in radio astronomy pipelines, depending on the telescope specifications and signal processing requirements.

The roofline performance model [26] offers an intuitive way to determine whether blocks in a pipeline are likely to be compute or I/O bound. For a given block, the achievable number of operations per second (ops/s) is, to first order, given by:

$$\text{Achievable ops/s} = \min(\text{peak performance, peak bandwidth} \times \text{operational intensity}), \tag{4.4}$$

where the underlying hardware specifications set the peak computational performance, and relevant I/O bandwidths (memory, PCI bus, network interfaces, etc.) set the peak bandwidth. The term "operational intensity" refers to the number of computational operations required per byte of I/O traffic.

Achieving the peak performance of the computer hardware can be a difficult task. Some useful avenues or attack, starting with efficient packet capture, are outlined.

7.2 Packet capture

High-speed packet capture is arguably the most difficult part of bringing up a data processing pipeline on a compute server. It invariably requires programming in a low-level language such as C, as scripting languages such as Python are not performant.

The easiest way to explain why is to take a dive into some basic C code to capture a packet. In Unix/BSD, packets are captured by creating a *socket*, which is a communication endpoint that acts as a le descriptor. Let's take a look at the synopsis from the Linux programmer's manual[f]:

```
include <sys/types.h> /*See NOTES */
#include <sys/socket.h>
int socket(int domain, int type, int protocol);
```

Creating a socket requires three things to be set: a `domain`, a `type`, and a `protocol`. The domain specifies which communication protocol we wish to use; for TCP/IP we choose `AF_INET`. For UDP, we set the type argument to `SOCK_DGRAM` (or `SOCK_STREAM` for TCP), and the protocol argument should be set to 0 (not required for UDP). So a UDP socket is created with the call:

```
int socket(AF_INET, SOCK_DGRAM, 0);
```

Once created, (and skipping several lines of setup), the `recvfrom`[g] call is used to receive a packet from the socket:

```
ssize_t recvfrom(int sockfd, void *buf, size_t len, int
                 flags, struct sockaddr *src_addr,
                 socklen_t *addrlen);
```

where `sockfd` is a socket you've created, `*buf` is a pointer to a memory buffer into which the packet payload should be copied, `len` is the size of the buffer, `flags` are optional flags, `sockaddr` holds the IP address and port to receive data from, and `addrlen` is just the size of the `sockaddr` struct (i.e. `sizeof(struct sockaddr)`). The recvfrom call is handled by the operating system kernel. When the network interface card (NIC) receives a packet, its driver will issue a hardware interrupt request (IRQ) to the kernel, to let it know data has arrived.

Receiving a stream of UDP data thus requires writing a loop and calling `recvfrom` multiple times, to get the packet from the NIC into kernel space, and then up into user space where your packet capture code runs. Once the packet is in user space, one still needs to parse the information in the UDP packet to make sure they have been received in the correct order, and then place their data in the right spot in memory to recreate the data frames. All this must be done before the next packet is received, or multiple processing threads have to be used to keep up.

[f] http://man7.org/linux/man-pages/man2/socket.2.html.
[g] https://linux.die.net/man/2/recvfrom.

At high packet rates, keeping up can become difficult. For example, consider a 10 Gb/s stream of packets of size 1000 B. The number of packets received per second is 1.25 M, which means 1.25 M calls to `recvfrom` are made, each with corresponding interrupts. The sheer number of kernel calls can quickly become a bottleneck, and the system may not keep up with the data rate.

7.3 Kernel bypass packet capture

When performance becomes an issue (likely for >10 Gb/s streams), the first x one might attempt is increasing the size of the packets from the data source. Moving from a 1000 B packet to an 8000 B packet would decrease the packet rate by a factor of 8. Another approach would be to use what is called a "raw socket," which bypasses the TCP/UDP layer and gives access to lower-level, unextracted packets, with packetmmap[h] to grab multiple packets with one call. We advise against this approach, in favor of the use of accelerator frameworks designed specifically for high-speed packet capture, such as DPDK,[i] and ibverbs,[j] which allow for data to be copied to memory without any kernel calls, known as *kernel bypass*. Bypassing the kernel to copy data is known as a zero copy, and when performed on data incoming from another device, is known as remote direct memory access (RDMA). Both DPDK and ibverbs require RDMA-capable network cards and drivers; for example, Mellanox NICs require their proprietary VMA[k] library to be installed in order to use ibverbs.

Example implementations. Table 4.3 shows data ingest rates for a selection of second-stage signal processing instruments that capture above 20 Gb/s per server. Note that computational bounds and technical constraints (e.g., power and cooling) are important considerations, so higher ingest rates do not directly reflect higher computational efficiency. The ibverbs approach is used by packet capture code running for the Parkes Ultra-Wideband Low receiver, to capture 24.8 Gb/s per server [30], and also in the SPEAD2 packet capture library[l] used by the MeerKAT telescope. In the TRAPUM backend for transient studies with MeerKAT [28], up to 56 Gb/s is captured per server, using a pair of 40 GbE NICs. The Hydrogen Epoch of Reionization Array (HERA, [27]) uses ibverbs to capture 2.34 Gb/s per server, again using a

[h] https://sites.google.com/site/packetmmap/.
[i] Data Plane Development Kit, https://www.dpdk.org/.
[j] http://rdmaconsortium.org/.
[k] https://www.mellanox.com/page/software_vma.
[l] https://spead2.readthedocs.io/.

Table 4.3 Data ingest rates (per server) for selected second-stage signal processing instruments.

Instrument	Ingest rate (Gb/s/server)	Packet size (B)	Capture method	Data pipeline	Ref.
HERA	68.0	4608	ibverbs	HASHPIPE	A
TRAPUM	56.0	1024	ibverbs/SPEAD2	PSRDADA	B
CHIME	25.6	8592	DPDK	kotekan	C
Parkes UWL	24.8	8592	ibverbs	PSRDADA	D
LEDA	21.4	7008	socket	PSRDADA	E

A [27], J. Hickish (personal commun.), B [28], E. Barr (personal commun.), C [29]; [4], D [30], E [31].

pair of 40 GbE NICs. The CHIME X-engine uses DPDK to capture 25.6 Gb/s per server [29]. The LEDA correlator [32] ingests 21.4 Gb/s per server, without the use of an accelerator library.

7.4 Ring buffers

Once captured from the NIC, data is captured into a buffer, from which a processing pipeline can grab data to process.

A *ring buffer*, or circular buffer, is a data structure used to emulate a wrap-around in memory. Ring buffers are contiguous blocks of memory that act as if they are circular, with no beginning or end, instead of 1D. In an asynchronous pipeline, when a write process and read process are sharing a contiguous block of regular memory, the writing process has to either stop when it reaches the end of the block, allocate more memory, or go back to the start address. In a ring buffer, a write process will automatically wrap around to the start address block of memory and will continues to write seamlessly.

A ring buffer will generally be an integer multiple of the size of the data frames being buffered, plus a “ghost region” to help synchronize the beginning and end of the buffer. Ring buffers can be used to ensure that a memory address being read by a process is not overwritten during the read by the writing process. However, if the read process is too slow, the writer will need to choose whether to stop writing until the reader catches up, or to continue writing and risk skipping data frames.

Ring buffers are as a standard data structure for buffer management in stream processing frameworks. An alternative approach, ping-pong buffering, can be considered a specific implementation of a ring buffer with exactly two fixed-length elements. In general, ring buffers are not required between processing elements in an FPGA-based pipeline, as operations are guaranteed to occur synchronously.

7.5 CPU/GPU pipeline frameworks

A number of open-source CPU/GPU pipeline frameworks have been deployed in radio astronomy systems, including:

PSRDADA,[m] where DADA stands for Distributed Acquisition and Data Analysis, was designed for recording and stream processing pulsar data. PSRDADA implements ring buffers in shared memory, and provides a C API for reading and writing from the buffers, along with monitor and control scripts. Pipelines are created by spawning multiple processes that communicate through ring buffers. The Swinburne Pulsar Instrumentation Package, SPIP[n] extends PSRDADA an object-oriented approach. PSRDADA is used in Ref. [33].

HASHPIPE,[o] the High Availability Shared Pipeline Engine, provides a C API for designing pipelines, where processing blocks are run in separate threads. As with PSRDADA, processing blocks are joined with ring buffers. HASHPIPE a derivative of an earlier pipeline called GUPPI, the Green Bank Ultimate Pulsar Processing Instrument [34].

Kotekan[p] [29] is a C/C++ pipeline used in the CHIME telescope, which launches processing blocks in threads. Ring buffers are implemented for both CPU and AMD GPUs.

Bifrost[q] [35] is written in C++ and Python, where Python is used as a high-level wrapper to ring buffers and kernels implemented using C++ and CUDA. Bifrost comes with a collection of configurable GPU-accelerated processing blocks that are combined into a performant pipeline using the high-level Python interface. At runtime, ring buffers are automatically created, and blocks are connected through these ring buffers as a directed graph. Bifrost is used to process data from the 256-antenna Long Wavelength Array station at Sevilleta

[m] http://psrdada.sourceforge.net/.
[n] https://github.com/ajameson/spip.
[o] https://casper.ssl.berkeley.edu/wiki/HASHPIPE.
[p] http://lwlab.dunlap.utoronto.ca/kotekan/.
[q] http://ledatelescope.github.io/bifrost/.

(LWA-SV), which implements a beamformer, correlator and direct imaging correlator [36].

Other efforts include PELICAN[r] [37], which is designed for static, quasi-realtime purposes. Outside radio astronomy, the gstreamer[s] framework has been used in gravitational wave detection [38]. The LOFAR telescope implements a pipeline called Cobalt [39] that uses similar design methodologies.

7.6 Disk I/O

Writing to disk drive (or more generally, a "data sink") is the final stage of most processing pipelines. The write speed of a drive is in general much slower than RAM, so this step can be a bottleneck. The two prevailing technologies are magnetic storage hard disk drives (HDDs) and solid-state drives (SSDs) that use flash memory. HDDs have spinning ferromagnetic disk platters, with maximum write speeds in the range 80–160 MB/s, for contiguous writes (for a current-generation 7200 RPM HDD). As of writing, capacities of up to 16 GB are commercially available.

SSDs have no moving components and offer higher write speeds (1–3 GB/s), but are more expensive per TB of storage space, and have smaller overall storage capacity. The flash memory in SSDs can only be written a certain number of times before failure, meaning sustained writing to SSDs will result in a short (months) lifetime. When an SSD fails, large portions of data will be lost. In contrast, HDDs are more susceptible to damage from physical shock, but unlike SSDs early warning signs are often given before failure, and total data loss is less common.

Data loss due to drive failure can be mitigated by using RAID (Redundant Array of Inexpensive Disks) configurations to introduce data redundancy. In a RAID configuration, multiple physical drives are combined into one virtual drive, and data are distributed across disks. Data read and write speeds can also be increased significantly, depending on the RAID configuration and number of drives.

On multiserver installations, distributed filesystems like Lustre and Ceph[t] are often used, so that all servers can access shared data storage via Ethernet. In this case, data may not be local to the compute resource (known as data locality), care must be taken that access patterns and read/write speeds can be sustained.

[r] https://github.com/pelican/pelican.

[s] https://gstreamer.freedesktop.org/.

[t] https://ceph.io/.

7.7 Performance tuning

Getting optimal performance out of a compute server requires knowledge of the underlying hardware and is a vast topic. Here, we briefly detail a few key points to consider when implementing a real-time stream processing pipeline that may not be commonly encountered in general CPU/GPU code optimization.

NUMA awareness. The components of a compute server are connected via buses, each with a finite bandwidth. Significant bottlenecks can arise on these buses if poor data transfer patterns are used. An important consideration for computers with multiple CPUs is that each CPU has its own system bus, and is the center of what is known as a nonuniform memory access (NUMA) node. Things like NICs, GPUs, and RAM are associated with a particular NUMA node, depending on which system bus they are located on. It is possible, but undesirable, for a NIC to be located on a different NUMA node than a GPU card. This is undesirable as memory access across NUMA nodes is slower than access within the node.

The bandwidth between NUMA nodes is determined by the interconnect technology. Intel chipsets use UltraPath Interconnect (UPI, previously Quick-Path), and AMD chipsets use Infinity Fabric (previously HyperTransport).

PCIe. Another bottleneck concern is the PCIe (Peripheral Component Interconnect Express) bus. All peripherals, such as NIC and GPU cards, are connected via the PCIe bus, which has a limited number of lanes. As of writing, a PCIe v3.0 slot for a graphics card has 16 lanes (x16), each with 1 GB/s bandwidth. Most NICs are housed on x8 slots, which means a maximum of 8 GB/s can be offloaded (64 Gb/s). The latest revision, PCIe 4.0, is not yet widespread but will double the bandwidth per lane.

Memory bandwidth. The memory bus can also be a bottleneck. For DDR4 memory (double data rate v4), the bandwidth is 256 times the clock speed; for DDR4-2666 this is 68 GB/s.

IRQ core binding. Interrupts (IRQs) are by default shared between CPU cores, but they are problematic for real-time stream processing. One can manually assign a CPU affinity to individual IRQs to stop interrupts being sent to CPU active CPU cores. Several pipelines provide tools to aid in IRQ bindings.

Kernel tuning. The default parameters used in Linux kernel are not optimal for real-time processing and high-speed data capture [31]. A specific suggestion is to increase the permitted buffer size for TCP in /etc/sysctl.conf file (net.core.wmem.max and net.core.rmem.max). We recreate the suggestions from [31] in Table 4.4, but caution that these may not be optimal for a given system.

Direct I/O. In modern operating systems, the kernel may try to cache a write request in memory. Consecutive read requests can

Table 4.4 Modifications to /etc/sysctl.conf, used in Ref. [31] to improve data transport performance.

Parameter	Value
kernel.shmmax	68,719,476,736
kernel.shmall	4,294,967,296
net.core.netdev max backlog	250,000
net.core.wmem max	536,870,912
net.core.rmem max	536,870,912
net.core.rmem default	16,777,216
net.core.wmem default	16,777,216
net.core.optmem max	16,777,216
net.ipv4.tcp mem	16,777,216 16,777,216 16,777,216
net.ipv4.tcp rmem	4096 87,380 16,777,216
net.ipv4.tcp wmem	4096 87,380 16,777,216
net.ipv4.tcp timestamps	0
net.ipv4.tcp sack	0
net.ipv4.tcp low latency	1

then be much faster, as the data are already loaded into memory. Direct I/O (using the O_DIRECT flag in C), forces the data to be written directly to the disk (in 512B chunks). For sustained high-bandwidth disk writes, direct I/O is preferable to avoid unnecessary caching.

8 Discussion

High-performance stream processing systems are fundamental to the operation of radio telescopes, and define the science data products that the telescope can provide. Building a high-performance system remains a challenging engineering task, but the shift toward industry-standard Ethernet is a welcome simplification.

This chapter has covered a lot of ground, with the aim of giving a broad overview of the technologies and approaches used for real-time stream processing in radio astronomy. In Sections 2 and 3 we introduced stream processing concepts and heterogeneous signal processing systems. Section 4 gave an overview of

Ethernet networking, which is used extensively in modern systems. We briefly introduced first-stage processing in Section 5, and discussed data redistribution in Section 6. Second-stage processing, most commonly performed on server-class computers, was presented in Section 7. This chapter concludes with a future outlook and discussion of promising next-generation technologies.

8.1 Future outlook

As of writing, Intel has $a > 95\%$ share of the high-end server market, based on the June 2019 Top500 supercomputer list (https://www.top500.org/), and NVIDIA has a >90% share of accelerator cards. However, AMD is expected to gain a significant market share, due to beating Intel to 7-nm process and other advances; this is well evidenced by the Frontier supercomputer, due for completion in 2021, which will be powered by AMD chips and is expected to be the World's fastest supercomputer.[u]

In theory, OpenCL supports CPUs, GPUs, and FPGAs, and FPGA accelerator PCIe cards are becoming more common. In practice, differences in architecture mean that code still needs to be written to target a given architecture for best results. Nevertheless, we may see OpenCL become more commonplace; the Heterogeneous Systems Architecture (HSA) foundation[v] provides a specification with similar goals.

A new type of memory, marketed by Intel as Optane, blurs the line between volatile RAM and flash storage, and products exist in both DDR and SSD formats. This, and similar technologies like Micron's 3D XPoint, may prove useful for buffering large volumes of data, avoiding disk I/O bottlenecks. We may also see adoption of RDMA access to storage devices, by approaches such as the NVMe over fabric (NVMe-oF) specification.[w]

We expect to see 25 GbE/100 GbE become commonplace in the coming years. Adoption of IP version 6 will also continue to increase, which brings the option of larger packets. IPv6 defines an optional jumbo payload, allowing single packets of several gigabytes, which would alleviate issues capturing high packet rates and allow larger data stream frames to be transmitted in a single packet. So-called smart NICs, which have an FPGA on the NIC itself, may also become more prevalent.

[u]https://www.amd.com/en/products/frontier.

[v]http://www.hsafoundation.com/.

[w]https://www.nvmexpress.org/.

With the huge momentum of deep learning and AI, we also expect to see machine-learning methods move into first- or second-stage processing, to help form novel science data products. Work to interface frameworks such as tensorflow[x] with pipelines is an area worthy of investigation.

Acknowledgments

D. Price thanks A. Jameson, B. Barsdell, and D. Macmahon for their valuable insights into CPU/GPU processing over the years.

References

[1] P. Dewdney, W. Turner, R. Braun, J. Santander-Vela, M. Waterson, G.-H. Tan, Ska1 System Baseline v2 Description. Technical Report 2, SKA Organization, Jodrell Bank Observatory, Cheshire, UK, 2015.

[2] J. Diamond, Square Kilometre Array Prospectus. Technical Report, SKA Organization, Jodrell Bank Observatory, Cheshire, UK, 2016.

[3] A.E. Schinckel, J.D. Bunton, T.J. Cornwell, I. Feain, S.G. Hay, The Australian SKA pathfinder, in: Ground-based and Airborne Tele-scopes IV, Volume 8444 of Society of Photo-Optical Instrumentation Engineers (SPIE) Conference Series, 2012, p. 84442A.

[4] CHIME/FRB Collaboration, M. Amiri, K. Bandura, P. Berger, M. Bhardwaj, M.M. Boyce, P.J. Boyle, C. Brar, M. Burhanpurkar, P. Chawla, J. Chowdhury, J.F. Cliche, M.D. Cranmer, D. Cubranic, M. Deng, N. Den-man, M. Dobbs, M. Fandino, E. Fonseca, B.M. Gaensler, U. Giri, A.J. Gilbert, D.C. Good, S. Guliani, M. Halpern, G. Hinshaw, C. Hofer, A. Josephy, V.M. Kaspi, T.L. Landecker, D. Lang, H. Liao, K.W. Masui, J. Mena-Parra, A. Naidu, L.B. Newburgh, C. Ng, C. Patel, U.L. Pen, T. Pinsonneault-Marotte, Z. Pleunis, M. Raei Ravandi, S.M. Ransom, A. Renard, P. Scholz, K. Sigurdson, S.R. Siegel, K.M. Smith, I.H. Stairs, S.P. Tendulkar, K. Vanderlinde, D.V. Wiebe, The CHIME fast radio burst project: system overview., Astrophys J. 863 (1) (2018) 48.

[5] A.R. Thompson, J.M. Moran, G.W. Swenson Jr., Interferometry and Synthesis in Radio Astronomy, third ed., 2017.

[6] A. Parsons, D. Backer, A. Siemion, H. Chen, D. Werthimer, P. Droz, T. Filiba, J. Manley, P. McMahon, A. Parsa, D. MacMahon, M. Wright, A scalable correlator architecture based on modular FPGA hardware, reuseable gateware, and data packetization, PASP 120 (873) (2008) 1207.

[7] J. Hickish, Z. Abdurashidova, Z. Ali, K.D. Buch, S.C. Chaudhari, H. Chen, M. Dexter, R.S. Domagalski, J. Ford, G. Foster, D. George, J. Greenberg, L. Greenhill, A. Isaacson, H. Jiang, G. Jones, F. Kapp, H. Kriel, R. Lacasse, A. Lutomirski, D. MacMahon, J. Manley, A. Martens, R. McCullough, M.V. Muley, W. New, A. Parsons, D.C. Price, R.A. Primiani, J. Ray, A. Siemion, V. van Tonder, L. Vertatschitsch, M. Wagner, J. Weintroub, D. Werthimer, A decade of developing radio-astronomy instrumentation using CASPER open-source technology, J. Astron. Instrum. 5 (4) (2016) 1641001–1641012.

[x] http://www.tensorflow.org.

[8] R. Perley, P. Napier, J. Jackson, B. Butler, B. Carlson, D. Fort, P. Dewdney, B. Clark, R. Hayward, S. Durand, M. Revnell, M. McKinnon, The expanded very large array, Proc. IEEE 97 (8) (2009) 1448–1462.
[9] A. Baudry, J. Webber, The alma 64-antenna correlator: main technical features and science modes, in: 2011 XXXth URSI General Assembly and Scientific symposium, 2011, pp. 1–4.
[10] L.R. D'Addario, D. Wang, An integrated circuit for radio astronomy correlators supporting large arrays of antennas, J. Astron. Instrum. 5 (2) (2016) 1650002–1650945.
[11] M.J. Flynn, Some computer organizations and their effectiveness, IEEE Trans. Comput. C-21 (9) (1972) 948–960.
[12] P. Klages, K. Bandura, N. Denman, A. Recnik, J. Sievers, K. Vanderlinde, GPU kernels for high-speed 4-bit astrophysical data processing, in: 2015 IEEE 26th International Conference on Application-specific Systems, Architectures and Processors (ASAP), 2015. arXiv e-prints, page arXiv:1503.06203.
[13] M.A. Clark, P.C. LaPlante, L.J. Greenhill, Accelerating radio astronomy cross-correlation with graphics processing units, Int. J. High Perform. Comput. Appl. 27 (2013) 178–192.
[14] A. Szomoru, The UniBoard, in: 10th European VLBI Network Symposium and EVN Users Meeting: VLBI and the New Generation of Radio Arrays, Volume 10, 2010, p. 98.
[15] G.A. Hampson, A. Brown, J.D. Bunton, S. Neuhold, R. Chekkala, T. Bateman, J. Tuthill, Askap redback-3 | an agile digital signal processing platform, in: 2014 XXXIth URSI General Assembly and Scientific Symposium (URSI GASS), 2014, pp. 1–4.
[16] E. Kooistra, G.A. Hampson, A.W. Gunst, J.D. Bunton, G.W. Schoonderbeek, A. Brown, Gemini FPGA hardware platform for the SKA low correlator and beamformer, in: 2017 XXXIInd General Assembly and Scientific Symposium of the International Union of Radio Science (URSI GASS), 2017, pp. 1–4.
[17] K. Bandura, A.N. Bender, J.F. Cliche, T. de Haan, M.A. Dobbs, A.J. Gilbert, S. Grin, G. Hsyu, D. Ittah, J.M. Parra, J. Montgomery, T. Pinsonneault-Marotte, S. Siegel, G. Smecher, Q.Y. Tang, K. Vanderlinde, N. Whitehorn, ICE: a scalable, low-cost FPGA-based telescope signal processing and networking system, J. Astron. Instrum. 5 (4) (2016) 1641005.
[18] W.R. Stevens, TCP/IP Illustrated, Vol. 1: The Protocols, first ed., Addison-Wesley Professional, 1994.
[19] C.M. Kozierok, The TCP/IP Guide: A Comprehensive, Illustrated Internet Protocols Reference, first ed., No Starch Press, 2005.
[20] J.L. Jonas, MeerKAT-the south African Array with composite dishes and wide-band single pixel feeds, IEEE Proc. 97 (8) (2009) 1522–1530.
[21] R.A. Primiani, K.H. Young, A. Young, N. Patel, R.W. Wilson, L. Vertatschitsch, B.B. Chitwood, R. Srinivasan, D. MacMahon, J. Weintroub, SWARM: a 32 GHz correlator and VLBI Beamformer for the sub-millimeter array, J. Astron. Instrum. 5 (4) (2016) 1641006–1641810.
[22] R.A. Perley, C.J. Chandler, B.J. Butler, J.M. Wrobel, The expanded very large array: a new telescope for new science, Astrophys. J. Lett. 739 (1) (2011) L1.
[23] D.C. Price, Spectrometers and Polyphase Filterbanks in Radio Astronomy, (2016). arXiv e-prints, page arXiv:1607.03579.
[24] K. Bandura, J.F. Cliche, M.A. Dobbs, A.J. Gilbert, D. Ittah, J. Mena Parra, G. Smecher, ICE-based custom full-mesh network for the CHIME high bandwidth radio astronomy correlator, J. Astron. Instrum. 5 (4) (2016) 1641004.
[25] M.J. Slabber, J. Manley, J. Mwangama, N. Ventura, MeerKAT data distribution network, in: Proceedings of the SPIE, volume 10707 of Society of Photo-Optical Instrumentation Engineers (SPIE) Conference Series, 2018, p. 107070H.

[26] S. Williams, A. Waterman, D. Patterson, Roofline: an insightful visual performance model for multicore architectures, Commun. ACM 52 (4) (2009) 65–76.
[27] D.R. DeBoer, A.R. Parsons, J.E. Aguirre, P. Alexander, Z.S. Ali, A.P. Beards-ley, G. Bernardi, J.D. Bowman, R.F. Bradley, C.L. Carilli, C. Cheng, E. de Lera Acedo, J.S. Dillon, A. Ewall-Wice, G. Fadana, N. Fagnoni, R. Fritz, S.R. Furlanetto, B. Glendenning, B. Greig, J. Grobbelaar, B.J. Hazelton, J.N. Hewitt, J. Hickish, D.C. Jacobs, A. Julius, M. Kariseb, S.A. Kohn, T. Lekalake, A. Liu, A. Loots, D. MacMahon, L. Malan, C. Malgas, M. Maree, Z. Martinot, N. Mathison, E. Matsetela, A. Mesinger, M.F. Morales, A.R. Neben, N. Patra, S. Pieterse, J.C. Pober, N. Razavi-Ghods, J. Ringuette, J. Robnett, K. Rosie, R. Sell, C. Smith, A. Syce, M. Tegmark, N. Thyagarajan, P.K.G. Williams, H. Zheng, Hydrogen epoch of reionization array (HERA), PASP 129 (974) (2017).
[28] B. Stappers, M. Kramer, An update on TRAPUM, in: MeerKAT Science: On the Pathway to the SKA, 2016, p. 9.
[29] A. Recnik, K. Bandura, N. Denman, A.D. Hincks, G. Hinshaw, P. Klages, U.-L. Pen, K. Vanderlinde, An Efficient Real-time Data Pipeline for the CHIME Pathfinder Radio Telescope X-Engine, (2015). arXiv e-prints, page arXiv:1503.06189.
[30] G. Hobbs, R.N. Manchester, A. Dunning, A. Jameson, P. Roberts, D. George, J.A. Green, J. Tuthill, L. Toomey, J.F. Kaczmarek, S. Mader, M. Marquarding, A. Ahmed, S.W. Amy, M. Bailes, R. Beresford, N.D.R. Bhat, D.C.J. Bock, M. Bourne, M. Bowen, M. Brothers, A.D. Cameron, E. Carretti, N. Carter, S. Castillo, R. Chekkala, W. Cheng, Y. Chung, D.A. Craig, S. Dai, J.R. Dawson, J. Dempsey, P. Doherty, B. Dong, P.G. Edwards, T. Ergesh, X.Y. Gao, J.L. Han, D.B. Hayman, B.T. Inder-muehle, K. Jeganathan, S. Johnston, H. Kanoniuk, M. Kesteven, M. Kramer, M. Leach, V.J. Mcintyre, V.A. Moss, S. Oslowski, C.J. Phillips, N.C. Pope, B. Preisig, D.C. Price, K. Reeves, L. Reilly, J.E. Reynolds, T. Robishaw, P. Roush, T. Ruckley, E.M. Sadler, J. Sarkissian, S. Severs, R.M. Shannon, K.W. Smart, M. Smith, S.L.S.C. Sobey, L. Staveley-Smith, A.K. Tzioumis, W. van Straten, N. Wang, L. Wen, M.T. Whiting, An ultra-wide bandwidth (704 to 4032 MHz) receiver for the Parkes radio telescope, PASA (2019). arXiv e-prints, page arXiv:1911.00656.
[31] J. Kocz, L.J. Greenhill, B.R. Barsdell, D. Price, G. Bernardi, S. Bourke, M.A. Clark, J. Craig, M. Dexter, J. Dowell, T. Eftekhari, S. Ellingson, G. Hallinan, J. Hartman, A. Jameson, D. MacMahon, G. Taylor, F. Schinzel, D. Werthimer, Digital signal processing using stream high performance computing: a 512-input broadband correlator for radio astronomy, J. Astron. Instrum. 4 (2015) 1550003.
[32] J. Kocz, L.J. Greenhill, B.R. Barsdell, G. Bernardi, A. Jameson, M.A. Clark, J. Craig, D. Price, G.B. Taylor, F. Schinzel, D. Werthimer, A scalable hybrid FPGA/GPU FX correlator, J. Astron. Instrum. 3 (1) (2014) 1450002–1450330.
[33] M. Bailes, A. Jameson, C. Flynn, T. Bateman, E.D. Barr, S. Bhandari, J.D. Bunton, M. Caleb, D. Campbell-Wilson, W. Farah, B. Gaensler, A.J. Green, R.W. Hunstead, F. Jankowski, E.F. Keane, V.V. Krishnan, T. Murphy, M. O'Neill, S. Oslowski, A. Parthasarathy, V. Ravi, P. Rosado, D. Temby, The UTMOST: a hybrid digital signal processor transforms the Molonglo observatory synthesis telescope, Publ. Astron. Soc. Aust. 34 (2017).
[34] S.M. Ransom, P. Demorest, J. Ford, R. McCullough, J. Ray, R. DuPlain, P. Brandt, GUPPI: Green Bank ultimate pulsar processing instrument, in: American Astronomical Society Meeting Abstracts #214, Volume 214 of American Astronomical Society Meeting Abstracts, 2009. 605.08.
[35] M.D. Cranmer, B.R. Barsdell, D.C. Price, J. Dowell, H. Garsden, V. Dike, T. Eftekhari, A.M. Hegedus, J. Malins, K.S. Obenberger, F. Schinzel, K. Stovall, G.B. Taylor, L.J. Greenhill, Bifrost: a Python/C++ framework for high-throughput stream processing in astronomy, J. Astron. Instrum. 6 (4) (2017) 1750007.

[36] N. Thyagarajan, A.P. Beardsley, J.D. Bowman, M.F. Morales, A generic and efficient E-field parallel imaging correlator for next-generation radio telescopes, MNRAS 467 (1) (2017) 715–730.

[37] B. Mort, F. Dulwich, C. Williams, S. Salvini, Pelican: Pipeline for Extensible, Lightweight Imaging and CAlibratioN, Astrophysics Source Code Library, 2015 ascl:1507.003.

[38] C. Messick, K. Blackburn, P. Brady, P. Brockill, K. Cannon, R. Cariou, S. Caudill, S.J. Chamberlin, J.D.E. Creighton, R. Everett, C. Hanna, D. Keppel, R.N. Lang, T.G.F. Li, D. Meacher, A. Nielsen, C. Pankow, S. Privitera, H. Qi, S. Sachdev, L. Sadeghian, L. Singer, E.G. Thomas, L. Wade, M. Wade, A. Weinstein, K. Wiesner, Analysis frame-work for the prompt discovery of compact binary mergers in gravitational-wave data, Phys. Rev. D 95 (4) (2017).

[39] P.C. Broekema, J.J.D. Mol, R. Nijboer, A.S. van Amesfoort, M.A. Brentjens, G.M. Loose, W.F.A. Klijn, J.W. Romein, Cobalt: a GPU-based correlator and beamformer for LOFAR, Astron. Comput. 23 (2018) 180.

5

Digitization, channelization, and packeting

Dongliang Liu, Shenghua Yu
National Astronomical Observatories, Chinese Academy of Sciences, Beijing, China

Digital receivers are a main subject that forms modern radio astronomical technology. Because of advancements in analog-to-digital converters (ADCs) and increases in digital signal processing capacity, massive amounts of astronomical data can be processed fast enough to match the increasing scientific observation requirements [1]. So in this chapter, we will introduce the core processing procedures in digital receivers to bridge analog RF to science processing data: digitization, channelization, and packeting, as shown in Fig. 5.1. Many other arithmetic operations can also be applied between or after these core procedures such as pulsar timing, dedispersion, beamforming, correlation, etc. [2–5], which will be discussed in further chapters.

1 Digitization

In a digital receiver, the analog RF input is downconverted into an IF, which is then digitized with high-speed ADCs with a large number of quantization levels; digital signal processing is then used to produce the desired science data [6]. Before the received signal is digitized by an ADC, the signal is amplified by an amplifier chain containing several amplifiers with different gains, noise figures, and third-order intermodulation points [7]. One of the purposes of using amplifiers in front of the ADC is to match the input signal to the ADC, and In general, adding amplifiers will also improve the sensitivity of the receiver. The output from the ADC is digital, which can then benefit from the major advantages related to digital signal processing. Once a signal is digitized, the following processing will all be digital. Digital signal processing is more

Big Data in Astronomy. https://doi.org/10.1016/B978-0-12-819084-5.00007-9

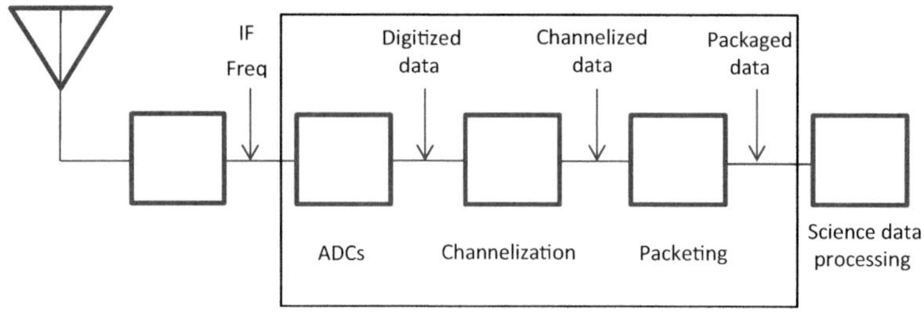

Fig. 5.1 Function of a digital receiver.

robust because there is no temperature drifting, gain variation, or DC level shifting as in analog circuits, therefore less calibration is required. The frequency resolution can also be very fine if high-resolution spectrum estimation techniques can be applied [8].

A classic digitization process can be described as follows: a sine wave in the time domain t can be represented by

$$x(t) = A\sin(2\pi ft)$$

where A is the amplitude and f is the frequency. A set of digitized data can be generated from this signal by sampling some discrete points on the sine wave. The device that performs this operation is called an analog-to-digital converter (ADC). The digitizing operation can be represented by a switch. It samples the input signal at a uniform rate. Of course, the input signal can also be sampled at a nonuniform rate, but only the uniform digitized signal will be discussed here because it is necessary for DFT operation [9–12].

In Fig. 5.2, the input is continuous but the ADC output is a pulse train with an envelope following the input. It is obvious that

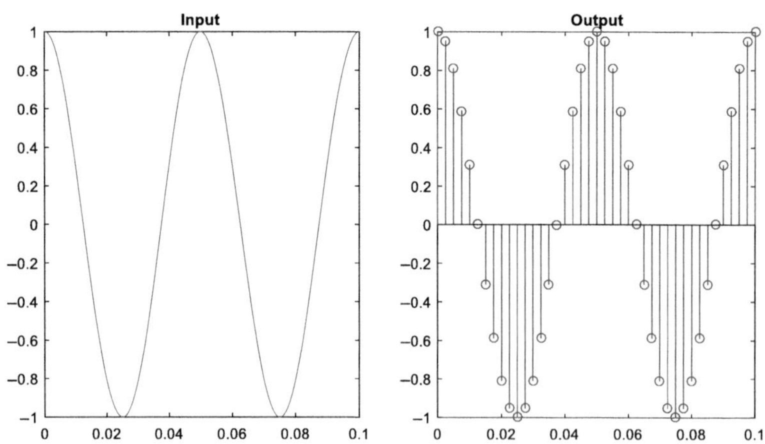

Fig. 5.2 A digitized waveform.

not all the information in the signal is transferred to the output. The information between the sampling points is lost. Thus, the output is no longer exactly equal to the input signal. Mathematically, the digitizing operation can be represented by multiplying the input signal with a comb function in the time domain, which can be written as

$$x(nt) = x(t) \sum_{n=-\infty}^{\infty} \delta(t - nt)$$

where n is an integer, t is the sampling interval, and $x(nt)$ represents the digitized signal. The digitized signal has an output only at integer multiples of t, as shown in Fig. 5.2.

For the hardware part, to realize the functionality of the ADC so as to quantize an input signal at a certain instant in time, the most common way is to place a track (sample) and hold circuit in front of a quantizer for the signal to be held constant at that instant. One of the most important elements of a data conversion system is the track and hold. Some types of extremely fast samplers have their sample command AC coupled and for a short period of time the signal is "sampled" and then held [13–19]. If the input signal changes rapidly and the digitizing process is slow, the accuracy of the output data will be poor. A track and hold circuit can create a very narrow aperture interval, which will take the input at the desired instant and keep the voltage constant for a relatively long period of time, such that the digitizing circuit can operate properly. A track and hold circuit is also used to reduce the aperture time of the sampling system. In general, it is necessary for the input signal to the ADC to be constant during the conversion process to avoid error. A successive approximation ADC uses an N-step algorithm when forming the conversion and if the signal varies during the conversion process, the wrong approximation would take place. Even flash converters can benefit from being driven from a track and hold circuit because the time delays of all the comparators are not identical [20, 21].

A simple track and hold circuit usually consists of a tracking switch and a holding capacitor [22]. The two amplifiers are used for impedance matching. When the switch is closed, the sample and hold is in the track mode and the voltage on the capacitor follows the input voltage. When the switch is opened, the voltage on the capacitor should stay constant; this is the hold mode. In the track mode, the operation can be divided into two time periods: accession and tracking. When the switch is closed, the voltage on the capacitor starts to change rapidly from the previously held value toward the input voltage until it finally reaches it. This period is called the accession time [23–29]. The voltage on the capacitor follows the input voltage; this period is called the tracking time. When the switch is open, the voltage on the capacitor stays at a constant value and is at the hold mode. After the hold mode, the voltage on the capacitor usually oscillates slightly due to a transient effect; this will be the settling time. After the settling time, the voltage will drop slightly due to charge leakage through the finite input impedance of the amplifier. The aperture time is the elapsed time between initiation and completion of the mode transition from track to hold. The aperture uncertainty is the variance of the aperture time. This time can be very short, usually in the nanoseconds to picoseconds for high-frequency ADCs [30–35]. The sampling time offset is the time interval between the sample to hold mode transition command and the actual initiation of the transition. The sampling time uncertainty is the variance of the sampling time offset [36].

The quantizer (comparator) is another common hardware element in a digitizing system. It is used to sense a level and then generate a digital signal, either a "1" or a "0," to report the result of that comparison to the rest of the system. The quantizer can usually be implemented by two components, a high-gain amplifier and the latching type approach. The fastest ADC is composed of an individual latch comparator for each quantizing level. The design of this type of converter would not be practical by any other technique, as only the latching comparator offers sufficient simplicity to allow for the necessary amount of integration. When a high-gain amplifier is used as a quantizer, many low-gain stages are cascaded to achieve high-gain bandwidth product. This is in contrast to the way a wideband operational amplifier would be designed. A design objective for a wideband operational amplifier would be to achieve high gain in a single stage to avoid accumulating an excessive amount of phase shift. It is important to achieve a phase characteristic approaching single pole response. Phase shift through a comparator is usually not important, although a high bandwidth and low propagation delay are desirable [37–40].

After these major components being assembled, the ADC can convert a continuous input voltage into discrete output levels, which can be represented by binary coded words, this is usually described as the bits number. In radio astronomical applications, the usual number of bits is from 4 to 12 [7]. The smallest discrete step size is called the quantization level. The conversion usually occurs at uniformly spaced time intervals. This is often referred to as the sampling time. In order to fulfill the Nyquist sampling criterion, it is often assumed that the input frequency to an ADC is half the sampling frequency. The Nyquist sampling frequency, named after electronic engineer Harry Nyquist, is half the sampling rate of a discrete signal processing system. It is sometimes known as the folding frequency of a sampling system.

The Nyquist-Shannon sampling theorem provides a sufficient condition for the sampling and reconstruction of a band-limited signal. When reconstruction is done via the Shannon interpolation formula, the Nyquist sampling frequency is also a necessary condition to avoid aliasing, in the sense that if samples are taken at a slower rate than twice the band limit, then there are some signals that will not be correctly reconstructed. However, if further restrictions are imposed on the signal, then the Nyquist frequency may no longer be a necessary condition. There is also another concept called the Nyquist rate. This should not be confused with the Nyquist frequency, which is the minimum sampling rate that satisfies the Nyquist sampling criterion for a given signal or family of signals. The Nyquist rate is twice the maximum component frequency of the function being sampled. For example, the Nyquist rate for the sinusoid at 0.8 fs is 1.6 fs, which means that at the fs rate, it is being undersampled. Thus, the Nyquist rate is a property of a continuous-time signal, whereas the Nyquist frequency is a property of a discrete-time system. When the function domain is time, sample rates are usually expressed in samples per second, and the unit of the Nyquist frequency is cycles per second (hertz). When the function domain is distance, as in an image sampling system, the sample rate might be dots per inch and the corresponding Nyquist frequency would be in cycles/inch [41–43].

As to the over and sub-samplings, we do not always follow the Nyquist-Shannon sampling theorem, which assumes that the input frequency to an ADC is half the sampling frequency. In order to avoid aliasing ambiguity, it is required that the input bandwidth (not necessarily the input frequency) of the ADC be less than half the sampling frequency. It is often desirable to have an input frequency higher than the maximum sampling frequency because the input bandwidth does not necessarily start from a direct current (DC) level. For example, if the maximum sampling frequency of the ADC is 200 MHz, the unambiguous bandwidth is 100 MHz, and the input frequency spectrum does not necessarily extend from DC to 100 MHz. It can be from 120 to 220 MHz, and this selection keeps the input bandwidth below an octave. An input bandwidth of less than one octave will eliminate the second harmonic, which can be generated by an analog front end or by the ADC nonlinear transfer characteristic. Octave bandwidth means the highest frequency is double the lowest frequency, such as from 1 to 2 GHz. Another nontrivial example of exploiting extra assumptions about the signal is given by the recent field of compressed sensing, which allows for full reconstruction with a sub-Nyquist sampling rate [44–46]. Specifically, this applies to signals that are sparse (or compressible) in some domain. As an example, compressed sensing deals with signals that may have a low overall bandwidth (the effective bandwidth EB), but the frequency locations are unknown, rather than all together in a single band, so that the passband technique does not apply. In other words, the frequency spectrum is sparse. Traditionally, the necessary sampling rate is thus 2B. Using compressed sensing techniques, the signal could be perfectly reconstructed if it is sampled at a rate slightly lower than 2 EB. With this approach, reconstruction is no longer given by a formula, but instead by the solution to a linear optimization program [47–50].

As to the quantization noise or error, the ADCs discussed in the above sections are assumed to be ideal. However, the performance of almost all the ADCs is not perfect, so this will introduce noises into the system. For example, the quantization steps may not be uniform. Some steps are wider and some are narrower. This quantity can be expressed in logarithmic form as 1.76 + 6.02b-dB; this is why the dynamic range of an ADC is often referred to as 6-dB per bit. In an extreme case, a certain quantization level is so narrow that it may never generate an output, and this is referred to as missing bits. The sampling window is not always stable and the window will jitter, which will have an adverse effect. There is also noise in the ADC circuit. For some ADCs, even without an input signal, the LSB may toggle in a random manner. Common sense tells us that noise will affect the sensitivity of a receiver. In many wideband receivers, the noise level is kept as low as possible. The noise level is not the only concern, as the dynamic range is also of concern [50–53]. High sensitivity (low noise) usually means low dynamic range. This is true in a digital receiver as well.

Finally, in radio telescope engineering, sometimes ADCs are used in parallel (other than interleaved) to increase sampling speed. If the input frequency of the ADC is high, these ADCs can be used to increase the bandwidth of the system. If the input frequency of the ADC is limited to half the sampling frequency, a parallel approach cannot be adopted because the ADC cannot receive high-frequency input signals. Here, we will introduce two schemes as the digitization solutions for the SKA-wideband single pixel feed (WBSPF) we have implemented. According to the two schemes, the specific functions of the backend system will then be designed and the performance and system size will then be estimated.

(1) High-dynamic range sampling scheme. The 10-bit sampling can cover the bandwidth of 1.25 GHz, 2.5 GHz, and 5 GHz (interwoven sampling) according to different software configurations. This scheme can completely cover band 2 (0.95–1.76 GHz) and partially cover band B (4.8–24 GHz). This index is higher than ska1-mid, which plans to cover the 2.5 GHz bandwidth at one time in band 5 with a resolution (dynamic range) of 8 bit. The AD sampling and transmission system uses a modular and configurable signal transmission system platform, which is planned to be placed at the feed source, and uses two channels of 20–40 G optical fiber to transmit digital signals to the CSP machine room. Its electromagnetic shielding, heat dissipation, and other technologies are relatively mature. Higher coverage bandwidth and resolution (dynamic range) are beneficial to the use of experiments for different scientific targets.

(2) Low dynamic range and large bandwidth sampling scheme. Sampling by 4 bit, the bandwidth of 7.5 GHz and 15 GHz (interwoven sampling) can be covered according to different software configurations. This scheme can cover most bands of band B (4.8–24 GHz) at one time. Ska1-mid cannot cover such a large amount of bandwidth at one time. The AD sampling transmission system also uses a modular and configurable signal transmission system platform.

2 Channelization

In the digital receiver system scheme, the idea of the channelized receiver is very simple so as to use the filter bank to sort signals with different frequencies [54]. If two signals appear in one channel, this input condition is beyond the capability of the receiver and may generate erroneous frequency information. Therefore, channelization can be viewed as one of the most important operations in building digital receivers. The equivalent analog operation is the filter bank. Therefore, digital channelization can be considered digital filter bank processing. It can also be considered an N-port network with one input and N-1 outputs. An input signal will appear at a certain output according to its frequency. By measuring the outputs from the filter bank, the frequency of the input signal can then be determined.

The straightforward approach for building a filter bank is to build individual filters, each one with a specific center frequency and bandwidth. Theoretically, each filter can be designed independently with a different bandwidth or shape. It is desirable to build a receiver with uniform frequency resolution to reduce the complexity, which means that the filters have the same shape and bandwidth [55–57]. Because the fast Fourier transform (FFT) converts the time domain data to frequency domain data and each element of the FFT results represents the signal's component at a certain frequency, we can consider FFT as a kind of filter bank. It is easier to build such a filter bank through FFT techniques than by using individual filter design because there is less computation. The filter response and four adjacent filters of FFT using rectangular windows are shown in Fig. 5.3.

So, the common method of performing channelization is by employing FFT. To build a receiver using FFT, the length and the overlap of the FFT are very important parameters. These parameters are related to the minimum pulse width (PW) and the frequency resolution, which determines the sensitivity of the receiver [58–61]. The frequency information can be obtained from the outputs of the digital filters. In order to obtain the input frequency, the filter outputs must be further processed. For example, if the FFT operation is performed on N points of data, the resulting N frequency outputs will occur at the same time. If the sampling frequency is fs, which is the input data rate, the corresponding input sampling time is ts = 1/fs. The output sampling rate is very important in building a receiver because the outputs from the FFT are usually further processed to obtain finer frequency resolution. The output rate also depends on the data overlapping rate. For example, if the input data are 100% overlapped, the output rate is also fs. If the input data are shifted by M samples between every FFT operation, the output sampling time is Mts, which corresponds to an output sampling rate of fs/M. For this processor, the input rate is the output sampling rate from the FFT. This output sampling rate determines the bandwidth of this processor. If a rectangular window containing N points is used to process the input signal with zero overlapping, the output sampling rate is fs/N. The corresponding output bandwidth is about fs/N. For a rectangular window in the time domain, the corresponding output in the frequency domain is a sinc function. This shape represents the response of one filter output. This filter shape is not desirable because the side lobes are very high. If a signal of frequency falls in channels B and C, it will also enter channel A through its first side lobe. This phenomenon limits the instantaneous dynamic range of the receiver. Because the output bandwidth is limited to fs/N, channel A cannot properly process a signal because it is outside its bandwidth. If channel A processes this signal, it may assign an erroneous frequency. As for the solutions, first, a higher output sampling rate will help this situation. Also, in order to lower the side lobes of the filter, a weighting function can be applied in the time domain. A weighting function widens the main lobe but suppresses the

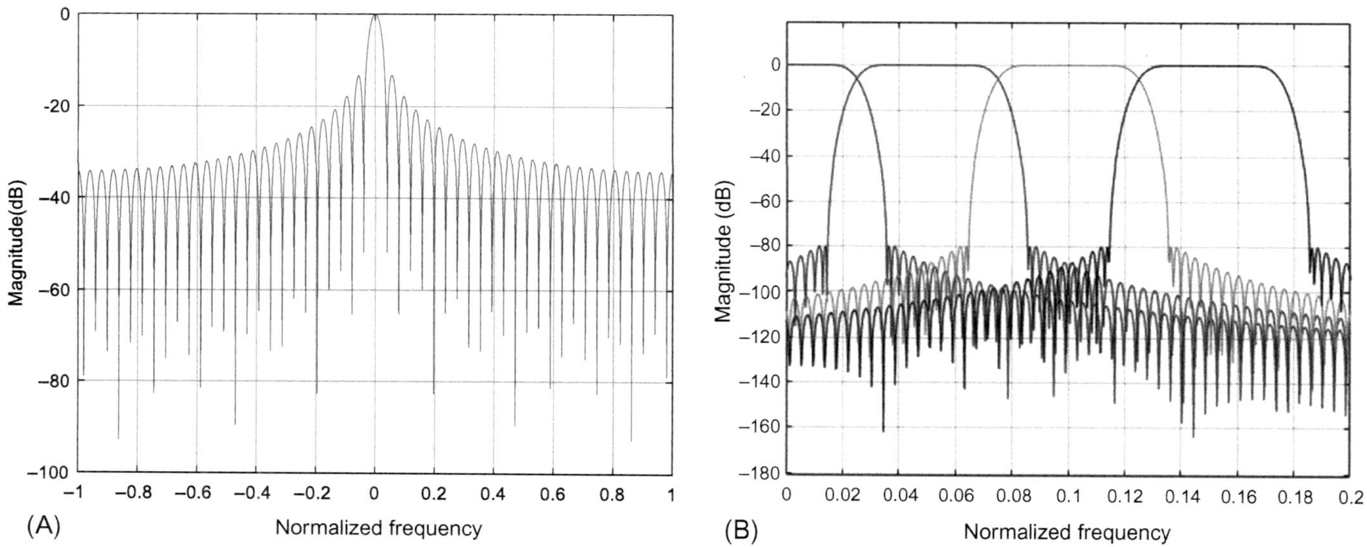

Fig. 5.3 FFT of a rectangular window: (A) detailed filter response and (B) four adjacent filters.

side lobes. Because the main lobe is wide, fewer channels are needed to cover the desired input bandwidth. In order to further process the signal, the output sampling rate must be increased to match the bandwidth. The need for increasing the output sampling rate can also be explained in the time domain. One can see that only the data near the center of the weighting function are given weights near unity. The data close to the edges of the window are heavily attenuated. If a zero overlapping approach is used, the contribution from these data in the FFT operation will be nonuniform, which is undesirable. If the windows are overlapped in the time domain, which is equivalent to increasing the output sampling rate, this deficiency can be remedied.

As for the output part of FFT, if 512 data points are used for the FFT and every output is kept, there are 256 independent outputs in the frequency domain. If 64 points are used for the FFT operation and every output is kept, there are 32 independent outputs in the frequency domain. The filter shape is independent of the length of the FFT, but the bandwidth is inversely proportional to the data length used in the FFT. If the FFT uses 256 data points but only one out of eight of the outputs is kept, there are a total of 32 outputs. Among these outputs, 16 carry redundant information. Therefore, only 16 outputs will be displayed in output. Each filter output is represented by a sinc function. Only a few side lobes are shown and the highest two are only a few dB below the main lobe. This filter bank has many holes (high insertion loss region). If an input signal falls in one of the holes, the receiver will miss it entirely. The shape of this filter is definitely unacceptable. So, to widen the individual filters and suppress the side lobes simultaneously, a window (or weighting) function can be applied to the input data. There are many different window functions. The one we have implemented here is the Hanning Window because it can provide the desired frequency response. The coefficients of the window can be obtained from the function of MATLAB. The Hanning Window function has the corresponding frequency domain response with very low passband ripples and side lobes lower than 70 dB. This is a desirable filter shape. One can see from the time domain response that the window function passes with moderate attenuation on fewer than 50 of the 256 samples. The rest of the input data are highly attenuated. The corresponding effect in the frequency domain is the wide bandwidth of each individual filter shape in the uniform filter bank [62–66].

The selection of the filter shape is one of the major tasks in designing a channelization system. It is desirable to keep the probability of one signal falling into two adjacent channels small; if a signal falls into one channel, the logic circuit of that channel will process the signal. If a signal falls into two adjacent channels, both circuits of the adjacent channels will process the signal. It is easier to measure one signal with one encoder circuit. If two channels measure one signal, the results must be compared to determine whether there are one or two signals. It also requires filters with a very sharp skirt. A sharp filter will have a relatively long transient time when the leading and trailing edges of the signal pass through it. During the transient period, both the amplitude and the frequency of the output signal change, as mentioned before. If the duration of the transient time is short, the steady state of the pulse might be too short to provide an estimation of the signal frequency. The long transient time eliminates the selection of very sharp filters. The transient effect must be evaluated in selecting filters. Another issue to be considered is the bandwidth of the processing circuit following the filter. Usually, the frequency response of the processing circuit is repetitive, such as in an FFT operation where the unique frequency is from 0 to fs/2. As discussed above, when the signal is near the edge of the band, it can be assigned to a wrong frequency. If the 3-dB band is used for further processing, there is usually an ambiguity at the edge of the circuit because of the presence of noise in the signal. This effect can put a signal on the wrong side of the filter and cause a frequency error equal to the bandwidth of the filter. In order to avoid this problem, the minimum bandwidth of the processing circuit should be equal to the 60-dB bandwidth rather than the 3-dB one. With this bandwidth, if a signal falls in between channels, the processing bandwidth is wide enough to avoid the ambiguity. For example, the channelized receiver considered earlier is used to demonstrate filter selection. Assume that the window function is the same, then the approach to obtain this filter is as follows: The sampling frequency is 4000 MHz, thus the unambiguous bandwidth is 2000 MHz because only 32 independent channels can be obtained from the 64-point FFT; the equivalent filter bank has 32 outputs. The bandwidth of each channel is 62.5 MHz (2000/32), which can be considered the 3-dB bandwidth. It is desirable to have at least 60-dB attenuation at a bandwidth of 125 MHz (62.5 × 2), which is double the 3-dB bandwidth.

Although the basic approach toward channelization discussed above is very flexible, it is not suitable for high-speed operation because of the limitation of the calculation speed. However, the same operation can be accomplished in hardware with much higher operation speeds. Now let us consider in more detail the process to generate the $y(n)$ values. The $y(n)$ values must be generated from input data shifting with time. One can see that each of these values can be generated from the convolution output of a filter with the input signal. For example, the 256-point window function in the time domain can be written as:

$$h(n) = \mathrm{h}(255)\delta(n) + h(254)\delta(n-1) + h(253)\delta(n-2) + \ldots$$
$$+ h(0)\delta(n-255)$$

where the δ function indicates the $h(n)$ value occurs at time n. The impulse sequence of the filter is written in an inverse way. This impulse function can generate the results through convolution with the input signal. Because the window function shown in Fig. 5.3 is symmetric in the time domain, this inverse is just a subscript change. This function can be decimated by 32, which results in 32 individual filters, each having eight taps. This filter decimation is often referred to as the poly phase filter. Each of the 32 filters has the response as indicated below:

$$h_0(n) = \mathrm{h}(255)\delta(n) + h(254)\delta(n-1) + h(253)\delta(n-2) + \ldots$$
$$+ h(0)\delta(n-255)$$
$$h_1(n) = \mathrm{h}(255)\delta(n) + h(254)\delta(n-1) + h(253)\delta(n-2) + \ldots$$
$$+ h(0)\delta(n-255)$$
$$\ldots$$
$$h_{31}(n) = \mathrm{h}(255)\delta(n) + h(254)\delta(n-1) + h(253)\delta(n-2) + \ldots$$
$$+ h(0)\delta(n-255)$$

These filters must be convolved with the proper input data to obtain the result. In order to obtain the correct data format, the input data must also be decimated by 32. When the decimated input signal and the decimated filter are convolved and reach the steady state, the output is equal to the result. So, when the 32-point FFT is performed, the input $y(n)$ values to the FFT are

$$y(0) = \mathrm{x}(32)h(0) + x(64)h(32) + \ldots + x(256)h(224)$$
$$y(1) = \mathrm{x}(33)h(1) + x(65)h(33) + \ldots + x(257)h(225)$$
$$\ldots$$
$$y(31) = \mathrm{x}(63)h(31) + x(95)h(63) + \ldots + x(287)h(255)$$

In this equation, the first data point is $x(32)$, thus the input is shifted 32 points. The hardware to accomplish this goal is shown in Fig. 5.4, which is also named the poly phase filter bank. In this figure, there are 32 filters and each filter has eight taps. Two cycles of input data are shown and each cycle contains 32 data points. The outputs are shown as $y(n)$ and they are used as the input of

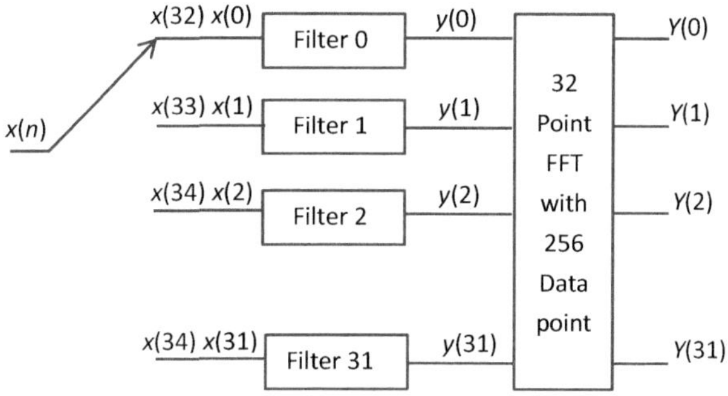

Fig. 5.4 Channelization approach with poly phase filter bank.

the FFT. The final results in the frequency domain are represented by $Y(k)$. In this case, the inputs are decimated by 32, and the final frequency domain also has 32 outputs. The input data are shifted 32 points, which is also the output frequency bin number. This gives a common channelization approach.

This also means that the output sampling rate is 1/N times the input sampling rate, where N is the number of input data points shifted. If one wants to increase the output sampling rate, the hardware must be modified; this is not as flexible as the software approach discussed in the previous section. A FIR filter design is used in this filter bank, as filter number 0 with $y(0)$ being the output. When the input signal reaches the steady state, the output of the filter contains eight terms. The first and second consecutive filter outputs after the steady state are also listed. The lower line represents the first output, which matches the $y(0)$ output. The upper line represents the second time output where the input data are shifted by 32 points. This result matches the $y(0)$ output. The rest of the filter outputs can be obtained in a similar manner. Now, we will discuss the detailed operation of a poly phase filter. First, the speed of operation is considered. The input data are sampled at 2000 MHz, which is the input data coming out of the ADC at about 0.5 ns per sample. If the input data to the poly phase filter are decimated by 32, the input rate will be 62.5 MHz (2000/32) and the filter operates at this speed. In order to process all the data, there are 32 parallel channels. It is relatively easy to operate at this lower rate. Because this system has two operation rates, it is often referred to as a multirate system. The 32-point FFT following the filters also operates at this low rate; the input data to the filter move 32 points each time. The progression of the data processing is illustrated above, and the input data under the main lobe of the window function are fewer than 50 points. This channelization approach shifts the 256-point window only 32 points. There should be enough data overlapping that all the data contribute to the output in a near uniform manner. Another advantage of shifting the window function 32 points is the fine time resolution. In modern signal sorting algorithms, it is desirable to have fine time resolution.

Above is the basic idea and realization of the channelization process. Although the frequency data resolution (or frequency bin width) calculated through a FFT operation determines the frequency precision measured on the input signal, it is often desirable to obtain better frequency precision than the FFT operation can provide [67–69]. When two signals fall into one frequency output channel, the receiver cannot effectively separate them. Finer frequency estimation and the capability of processing two signals close in frequency are the two main goals of processing after channelization; this can be further explored if interested [70–78]. For example, because the FFT filter has limited dynamic range, if two signals are in one filter, the FFT filter may miss the weaker one. With the advances in digital signal processing, it appears that the wideband receiver can be replaced by a regular FFT operation, which is referred to as the second FFT operation (or chip). This FFT chip will take the channel outputs as input without bit truncation. It performs a 32-point FFT at 125 MHz and generates 32 independent outputs because the inputs are complex. With this kind of arrangement, the receiver should have a higher instantaneous dynamic range for signals in the same digital channel. The number of signals per channel is not limited to two, as in the case of the FFT filters. Theoretically, it should be able to process 32 simultaneous signals per channel. Another potential advantage is that because the inputs to the second FFT are not truncated, the amplitude information on the signals is retained. Thresholds can be set at the outputs of the second FFT, and the detection circuit may no longer be needed at the first channel outputs. The frequency data resolution then generated from the second FFT operation is about 3.9 MHz (125/32). Detecting the signals at these outputs should result in higher sensitivity. It is anticipated that this approach may be considered in future wideband digital receiver designs.

3 Packeting

We have described how the input signals are separated in the frequency domain through channelization. After the frequency channelization and other following arithmetic operations, packeting will be included to packet the data products and send through an Ethernet transmission for further science data processing [79].

The basic packeting includes a packet and a transmission network. A packet is a basic unit of communication over a digital network. A packet consists of control information and signal data, which is also known as the payload. Control information provides data for delivering the payload, such as source and destination network addresses, error detection codes, and sequencing information. Typically, control information is found in packet headers and trailers. A packet is also called a datagram, a segment, a block, a cell, or a frame, depending on the protocol used for data transmission. When data are transmitted, they are broken into similar structures of data before transmission, called packets. The packets are reassembled to the original data chunk once they reach the destination.

The structure of a packet usually depends on the type of packet and the protocol. A packet has a header and a payload. The header keeps overhead information about the packet, the service, and other transmission-related data. For example, data transfer over Ethernet requires breaking down the data into IP packets, which is defined in IP (Internet protocol). An IP packet usually includes the source IP address and the IP address of the digitizing and channelization machine sending the data; the destination IP address and the science data processing machine or device to which the data are sent; the sequence number (e.g., which bins) of the packets, a number that puts the packets in order such that they are reassembled in a way to get the original data back exactly as it was prior to transmission; the time flags; the trace information; other types of services; other technical data; etc. This is the bulk of the packet (the above is considered overhead) and is the data being carried. Packets vary in structure and functionality depending on the protocols that implement the packets. On a local high-speed Ethernet network, for example, data is transmitted in Ethernet frames.

In the IP protocol, the IP packets travel over the Ethernet or Internet through nodes, which are devices and router nodes found on the way from the source to the destination [80]. Each packet is routed toward the destination based on its source and destination IP address. At each node, the router decides, based on calculations involving network statistics, which neighboring node is more efficient to send the packet. This is part of packet switching, which flushes the packets on the Ethernet or Internet, and each packet finds its own way to the destination. This mechanism uses the underlying structure of the internet for free. The transmission control protocol (TCP), commonly referred to as TCP/IP, is one of the main protocols of the Internet protocol suite. It originated in the initial network implementation in which it complemented the Internet protocol (IP). TCP provides reliable, ordered, and error-checked delivery of a stream of octets (bytes) between applications running on hosts communicating via an Ethernet. TCP is connection-oriented, and a connection between the client and server is established (passive open) before data can be sent. A three-way handshake (active open), retransmission, and error detection add to the reliability but lengthen the latency. Applications that do not require reliable data stream service may use the user datagram protocol (UDP), which provides a connectionless datagram service that prioritizes time over reliability. TCP employs network congestion avoidance. Though TCP is a complex protocol, its basic operation has not changed significantly since its first specification. TCP is responsible for ensuring that data transfer is reliable. To achieve that, it checks whether the packets arrived in order, whether any packets are missing or have been duplicated, and whether there is any delay in packet transmission. It controls this by setting a time out and signals called acknowledgments.

As for the second part of basic packeting, a transmission network is a formatted unit of data carried by a packet-switched network. A simple definition of a transmission network is the routing and transferring of data by means of addressed packets so that a channel is occupied during the transmission of the packet only, and upon completion of the transmission the channel is made available for the transfer of other traffic. Packet switching allows the delivery of variable bit rate data streams, realized as sequences of packets, over a computer network that allocates transmission resources as needed using statistical multiplexing or dynamic bandwidth allocation techniques. As they traverse networking hardware, such as switches and routers, packets are received, buffered, queued, and retransmitted (stored and forwarded), resulting in variable latency and throughput depending on the link capacity and the traffic load on the network. Packets are normally forwarded by intermediate network nodes asynchronously using FIFO (first-in, first-out) buffering, but may be forwarded according to some scheduling discipline for fair queuing, traffic shaping, or for differentiated or guaranteed quality of service, such as weighted fair queuing or leaky bucket. Packet-based communication may be implemented with or without intermediate forwarding nodes (switches and routers). In case of a shared physical medium (such as 10BASE5), the packets may be delivered according to a multiple access scheme.

A packet switch has four components: input ports, output ports, routing processor, and switching fabric. Packet switching may be classified into connectionless packet switching, also known as datagram switching, and connection-oriented packet switching, also known as virtual circuit switching. Examples of connectionless systems are Ethernet, IP, and UDP. Connection-oriented systems include X.25, frame relay, multiprotocol label switching (MPLS), and TCP. In connectionless mode, each packet is labeled with a destination address, source address, and port numbers. It may also be labeled with the sequence number of the packet. This information eliminates the need for a preestablished path to help the packet find its way to its destination, but means that more information is needed in the packet header, which is therefore larger. The packets are routed individually, sometimes taking different paths resulting in out-of-order delivery. At the destination, the original message may be reassembled in the correct order, based on the packet sequence numbers. Thus a virtual circuit carrying a byte stream is provided to the application by a transport layer protocol, although the network only provides a connectionless network layer service. Connection-oriented transmission requires a set-up phase to establish the parameters of communication before any packet is transferred. The signaling protocols used for set up allow the application to specify its requirements and discover link parameters. Acceptable values for service parameters may be negotiated. The packets transferred may include a connection identifier rather than address information and the packet header can be smaller, as it only needs to contain this code and any information, such as length, timestamp, or sequence number, which is different for different packets. In this case, address information is only transferred to each node during the connection set-up phase, when the route to the destination is discovered and an entry is added to the switching table in each network node through which the connection passes. When a connection identifier is used, routing a packet requires the node to look up the connection identifier in a table. Connection-oriented transport layer protocols such as TCP provide a connection-oriented service by using an underlying connectionless network. In this case, the end-to-end principle dictates that the end nodes, not the network itself, are responsible for the connection-oriented behavior.

Many FPGA or DSP based hardware platforms are able to sustain Ethernet transmission rates of 10 Gigabits per second (Gbps); however, until recently a single computer was still unable to maintain this data rate. Factors such as the computer's CPU, bus speed, memory allocation, choice of network interface card (NIC), packet size, and how much processing takes place on the received data (for example, data manipulation or disk storage) all affect the received data rate. Although Ethernet is known as a networking and system-to-system protocol, it has been adapted to other applications, including the backplane. Ethernet is a popular protocol choice in FPGAs because of its flexibility, reliability, and performance. Implemented in the latest technology, the FPGA incorporates 10–100G Ethernet MAC and PCS integrated blocks that support CAUI-4, CAUI-10, and switchable CAUI-4/ CAUI-10 modes. As a benefit of integrated high-speed Ethernet MACs, it can reduce design and verification effort. It will free approximately 130 K LUTs, PCS and RS-FEC in the FPGA fabric, reduces power up to 90% over soft implementation and also more easier to implement with the development kit that provides latest libraries, including proven IP and protocol-specific characterization reports.

So, in radio astronomical technology, people prefer to have either transmitted data from FPGA to FPGA or FPGA to switch at 10–40 Gbps, but typically avoided transmitting the full 10 Gbps data rate into a single computer due to computer performance restrictions. We were thus limited to either transmitting at lower rates or resorting to a breakout switch to split the 10 Gbps rate from the FPGA into several computers at 1 Gbps. Modern computers with 10 Gbps PCI Express NICs are able to maintain the full 10 Gbps line rate. The FPGA is loaded with a 10GbE test-suite bitstream that has runtime configurable packet lengths (up to 12 kilobytes + UDP header) and packet rates (0.05 packets/s to 2 million packets/s). The packet payloads contain a continuous 64 bit counter. The data cable is connected through the network under test, either directly or by using a network switch.

Software will also be written to receive the above packets and check for correct payload length and continuous counter sequence. Should there be a break in the counter values, the difference is used to calculate the number of packets dropped. The software by default reports the number of dropped packets per million received. The packet payload length and number of packets to receive are preconfigured for each test case. The software will also report systemwide losses (for example, packets dropped in an upstream network switch will also be reported, of which the kernel would be unaware), record packet losses, and calculate losses explicitly.

Minimal packet data processing takes place in the hardware board with FPGA or DSP chips, which can benchmark the ability of a given platform to manipulate the received data or write it to disk. It is simply a reference against which other systems can be compared for the purposes of absorbing UDP packets at high data rates. Each of these boards will run a UDP server and will respond to requests from remote machines. The interface is very simple in the interests of keeping the network and CPU overhead low.

The idea is that you can implement the server or client in any language you like (Python, Ruby, C, Fortran, etc.) as long as the packet formats remain consistent. As for the UDP packet format, the first byte is the identifier and it indicates the packet type. Thereafter, each packet has its own format such as Type(unsigned char), ID tag(unsigned char), and Data(raw binary).

As for the whole packeting processing, the transmission network will be required as the basic infrastructure that will transfer the data from the dish antenna to the science data processing (SDP) system shown in Fig. 5.5. Observations may be affected if a part of the network in the system is not available, but it is usually not necessary for the system to use redundant routing. The following is a schematic diagram of the main components of the Telescope Network Scheme, which will be further discussed in later chapters.

In summary, digitization, channelization, and packeting are the key procedures in the central signal processing (CSP) systems. The processed signal and data will then be used to from the interfered clearer defect display (visibilities), the pulsar search and pulsar timing data etc., its output then transferred to scientific data

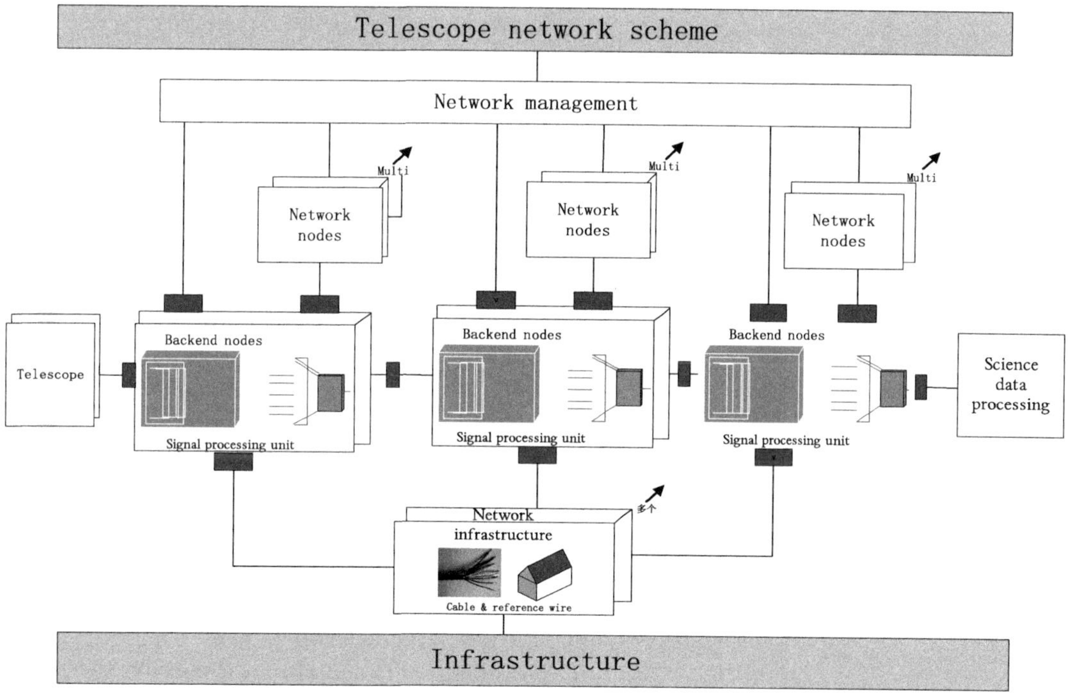

Fig. 5.5 The telescope network scheme.

processing system used for subsequent processing science. Usually the system can process all the input data in real time, which requires the design of special hardware equipment and the overall system architecture.

As of one of the core technologies in radio telescope systems, the CSP element includes the design of the hardware and the associated firmware/software necessary for the generation of visibilities, pulsar survey candidates, and pulsar timing from the telescope arrays. CSP includes the distribution of data within the processor, diagnostic tools, etc., which are necessary for the maintenance and operation of the system. CSP does not include the buildings, cooling, shielding, or power supply to the building. The CSP will utilize the latest generations of high-speed digital processing chips, high-speed/high-capacity memory chips, high-capacity fiber communications, high-speed circuit boards, high-speed modeling software, and electronic test equipment as well as the latest in agile, robust, and intelligent software. The challenge with CSP is that it must process enormous amounts of real-time data, and in so doing produce enormous amounts of output to be consumed by the science data processor. It will be located in remote locations and must be designed to deliver the maximum science possible within a hard cost cap and an aggressive timeline.

The design of the CSP pursues the minimization of total lifecycle costs over the projected design life. Contrary to most other subsystems, the challenge here is not just to satisfy CSP requirements but to do so in the most cost-efficient way, as the CSP can be realized using so many different technological solutions and architectures. For the purpose of this reference design, such efficiency is defined in terms of total lifecycle cost. Also, a low-risk approach is taken considering only already available technologies instead of using predictions about future solutions. Usually, different technological solutions have been assessed independently for the different platforms and machines, and a research collaboration will be established between many parts to develop reference design components. The design reference usually leverages the teams' expertise, as if they are the same team behind some already in use correlator or other CSP system. As a result, the new reference design can usually be based on the same technological solution as the previous design with modifications that have been scaled and adapted to the requirements. It will also require evaluating the different technologies assess for different choices such as GPUs and FPGAs. The first have been coarsely evaluated in several of their different flavors (i.e., general purpose GPUs, system-on-chip (SoC) technologies (NVIDIA's Tegra SoC), and GPU-based server solutions). FPGA implementations have also been studied considering many approaches, such as using on-chip block random access memory (BRAM) or to employ second-generation high bandwidth memory (HBM2). Similar best results are offered by different technologies, with some degree of uncertainty in total cost estimates. This usually prevents the team from making a final choice without further experimental investigation. Consequently, several candidate technologies are proposed for the reference design.

As an example, the CSP is the central processing "brain" of the SKA. It converts digitized astronomical signals detected by SKA receivers (antenna and dipole arrays) into the vital information needed by the science data processor to make detailed images of deep space astronomical phenomena that the SKA is observing. It will also design a "nonimage processor" in order to facilitate the most comprehensive and ambitious survey yet to find new pulsars and precisely time known pulsars. Until now, the SKA phase 1 correlator, beam formers, and pulsar processing systems will be the largest ever built for radio astronomy. Just the raw correlation processing is close to 5TFLOPS arithmetic operations per second, with beamforming and pulsar search each being a similar order of magnitude. The work on these is being undertaken by the central signal processing element of SKA and is broken up into a number of different subelements, each with its own consortium doing the development. All subelements have completed their preliminary design reviews as well as the critical design reviews (CDRs). A greater emphasis is now being put on systems engineering. At CDR, all requirements and interface definitions are in place. These together with a reference design will form the basis for tenders for construction. Work on system availability, logistics, operations, power estimates, cost estimates, etc., will feed into the intergovernment agreements needed to finally build the instrument. The hardware for the SKA CSP system is a combination of GPU and FPGA processing. The GPU systems will rely on continued progress in GPU technology. Two correlators and beam formers are being designed by two different consortia and both will have working ne-t generation FPGA hardware. Work also continues on modeling to verify the algorithms as well as software and firmware to verify the validity of the design.

References

[1] T.S. Bastian, The frequency agile solar radio telescope, Adv. Space Res. 32 (12) (2006) 47–69.

[2] G.I. Langston, et al., Green Bank lunar interferometer for neutrino transients: GLINT, Nucl. Instrum. Methods Phys. Res. 604 (2009) S116–S118.

[3] https://www.skatelescope.org/wbspf/.

[4] P. Bo, J. Chengjin, D. Biao, et al., Continued participation in the world's largest integrated aperture telescope SKA international collaboration, science in China, Phys. Sci. Astron. 42 (2012) 1292–1307.

[5] N. Rendong, 500 m spherical reflector radio telescope FAST, Sci. China Ser. G Phys. Mech. Astron. 35 (5) (2005) 449–466.

[6] B. Klein, S. Hochgürtel, I. Krämer, et al., High-resolution wide-band fast Fourier transform spectrometers, Astron. Astrophys. L3 (2012) 542.

[7] A. Parsons, et al., Digital Instrumentation for the Radio Astronomy Community, astro-ph/0904.1181, April, 2009.

[8] S.M. Kay, Modern Spectral Estimation Theory and Application, Englewood Cliffs, NJ, Prentice Hall, 1988.

[9] A. Siemion, et al., New SETI Sky Surveys for Radio Pulses, astro-ph/0811.3046, November, 2008.

[10] A. Parsons, et al., A New Approach to Radio Astronomy Signal Processing, General Assembly of the International Union of Radio Science, 2005 October.

[11] D. Werthimer, D. Ng, S. Bowyer, C. Donnelly, The Berkeley SETI program: SERENDIP III and IV instrumentation, in: Progress in the Search for ExtrterTrestrial Life, vol. 74, 1995, p. 293.

[12] http://en.wikipedia.org/wiki/Allen_Telescope_Array.

[13] Nan, Pulsar astronomy in China, ChJAAS 6 (1) (2006) http://astronomy.swin.edu.au/pulsar/observing/cpsr2/W.

[14] L. Liyong, A. Isamuddin, et al., Establishment of pulsar coherent de-dispersion observation system at Urumqi astronomical station, Astron. Res. Technol. 4 (1) (2007).

[15] http://www.dlh.pmo.cas.cn/hzjl/gjjldt/201012/t20101224_3049390.html.

[16] J.D. Owens, D. Luebke, N. Govindaraju, M. Harris, J. Kruger, A.E. Lefohn, T. Purcell, A survey of general-purpose computation on graphics hardware, Comput. Graphics Forum 26 (1) (Mar. 2007) 80–113.

[17] https://casper.berkeley.edu/wiki/ROACH2.

[18] https://casper.berkeley.edu/wiki/SNAP2.

[19] http://www.bao.ac.cn/xwzx/zhxw/201104/t20110427_3122692.html.

[20] B. Le, T.W. Rondeau, Analog-to-digital converters, IEEE Signal Process. Mag. 22 (2005) 69–78.

[21] J. Mitola, Software radio architecture evolution: foundations, technology tradeoffs, and architecture implications, IEICE Trans. Commun. E83-B (2000) 1165–1173.

[22] K.R. Stafford, R.A. Blanchard, P.R. Gray, A complete monolithic sample/hold amplifier, IEEE J. Solid State Circuits 9 (6) (1974) 381–387.

[23] O. Mukhanov, D. Gupta, A. Kadin, J. Rosa, V. Semenov, T. Filippov, Superconductor digital-RF transceiver components, in: Proc. of the SDR Technical Conference, San Diego, vol. I, 2002, pp. 227–232.

[24] O.A. Mukhanov, D. Gupta, A.M. Kadin, V.K. Semenov, Superconductor analog-to-digital converters, Proc. IEEE 92 (2004) 1564–1584.

[25] A. Kirichenko, S. Sarwana, D. Gupta, D. Yohannes, Superconductor digital receiver components, IEEE Trans. Appl. Supercond. 15 (2005) 249–254.

[26] E. Hogenauer, An economical class of digital filters for decimation and interpolation, IEEE Trans. Acoust. Speech Signal Process. ASSP-29 (1981) 155–162.

[27] T.V. Filippov, S.V. Pflyuk, V.K. Semenov, E.B. Wikborg, Encoders and decimation filters for superconductor oversampling ADCs, IEEE Trans. Appl. Supercond. 11 (2001) 545–549.

[28] O.A. Mukhanov, S.R. Rylov, D.V. Gaidarenko, N.B. Dubash, V.V. Borzenets, Josephson output interfaces for RSFQ circuits, IEEE Trans. Appl. Supercond. 7 (1997) 2826–2831.

[29] S.V. Rylov, R.P. Robertazzi, Superconductive high-resolution A/D converter with phase modulation and multi-channel timing arbitration, IEEE Trans. Appl. Supercond. 5 (1995) 2260–2263.
[30] I.V. Vernik, D.E. Kirichenko, T.V. Filippov, A. Talalaevskii, A. Sahu, A. Inamdar, A.F. Kirichenko, D. Gupta, O.A. Mukhanov, Superconducting high-resolution low-pass analog-to-digital converters, IEEE Trans. Appl. Supercond. 17 (2007) 442–445.
[31] J. Tsi, C.-H. Cheng, Digital Techniques for Wideband Receivers, third ed., Sci-Tech Publishing, 2016.
[32] M. Win, R. Scholtz, On the robustness of ultra-wide bandwidth signals in dense multipath environments, IEEE Commun. Lett. 2 (1998) 51–53.
[33] T. Mitchell, Broad is the way, IEE Rev. 47 (1) (2001) 35–39.
[34] R. Scholtz, Multiple access with time-hopping impulse modulation, in: Proceedings of IEEE MILCOM '93, Boston, MA, 11–14 October, 1993, pp. 447–450.
[35] M. Win, R. Scholtz, Impulse radio: how it works, IEEE Commun. Lett. 2 (1998) 36–38.
[36] R. Fleming, C. Kushner, Low-power, miniature, distributed position location and communication devices using ultra-wideband nonsinusoidal communication technology, in: ARPA/FBI, Semi-Annual Tech. Rep., July, 1995.
[37] M. Choi, A.A. Abidi, A 6b 1.3 GSample/s A/D converter in 0.35m CMOS, in: Proceedings of IEEE Int. Solid-State Circuits Conference, vol. 438, 2001, pp. 126–127.
[38] S. Velazquez, T. Nguyen, S. Broadstone, J. Roberge, A hybrid filter bank approach to analog-to-digital conversion, in: Proceedings of IEEE-SP International Symposium Time-Frequency Time-Scale Anal., October, 1994, pp. 116–119.
[39] J. Naylor, Testing digital/analog and analog/digital converters, IEEE Trans. Circuits Syst. 25 (7) (1978) 526–538.
[40] Hewlett Packard, Dynamic Performance Testing of A to D Converters. Product note 5180A-2, Hewlett Packard Co., Palo Alto, CA, 1982.
[41] P. Carrier, A microprocessor based method for testing transition noise in analog to digital converters, in: Proceedings of 1983 IEEE International Test Conference, IEEE, New York, 1983, pp. 610–620.
[42] J. Doernberg, H.-S. Lee, D. Hodges, Full-speed testing of A/D converters, IEEE J. Solid State Circuits 19 (6) (1984) 820–827.
[43] IEEE, Trial-Use Standard for Digitizing Waveform Recorders. IEEE Standard 1057, IEEE, New York, 1989.
[44] B. Sklar, Digital Communications: Fundamentals and Applications, NJ, Prentice Hall, 1988.
[45] P.W. Wong, Quantization noise, fixed-point multiplicative round off noise, and dithering, IEEE Trans. Acoust. Speech Signal Process. 38 (2) (1990) 286–300.
[46] W.B. Davenport, W.L. Root, An Introduction to the Theory of Random Signals and Noise, McGraw-Hill, New York, 1958. Reprinted by IEEE Press, 1987.
[47] L.E. Larson, High-speed analog-to-digital conversion with GaAs technology: prospects, trends and obstacles, in: Proceedings of IEEE International Symposium on Circuits and Systems, vol. 3, IEEE, New York, 1988, pp. 2871–2878.
[48] K. Walter, Test video A/D converters under dynamic condition, in: EDN, 1982, pp. 103–112. August.
[49] A. Petraglia, S. Mitra, Analysis of mismatch effects among A/D converters in a time-interleaved waveform digitizer, IEEE Trans. Instrum. Meas. 40 (1991) 831–835.
[50] R. Khoini-Poorard, Mismatch effects in time-interleaved oversampling converters, in: Proceedings of IEEE Int. Symp. Circuits Systems, London, UK, May, 1994, pp. 429–432.

[51] A. Papoulis, Generalized sampling expansion, IEEE Trans. Circuits Syst. 24 (Nov. 1977) 652–654.

[52] J. Cioffi, n.d. Data Signals, Sets, and Detection. ser. EE379a class notes Stanford, CA: Stanford Univ. Press.

[53] R.C. Jaeger, Tutorial: analog data acquisition technology part I—digital-to-analog conversion, IEEE Micro 2 (2) (1982) 20–37. Tutorial: analog data acquisition technology part II—analog-todigital conversion. IEEE Micro 2 (3) (1982) 46–57. Tutorial: analog data acquisition technology part III—sample-and-holds, instrumentation amplifiers, and analog multiplexers. IEEE Micro 2 (4) (1982) 20–35. Tutorial: analog data acquisition technology part IV—system design, analysis, and performance. IEEE Micro 3 (1) (1983) 52–61.

[54] O. Oliaei, Asymptotically perfect reconstruction in hybrid filter banks, in: Proceedings of IEEE International Conference on Acoustics, Speech, Signal Processing, vol. 3, 1998, pp. 1829–1832.

[55] P. Loweborg, H. Johansson, L. Wanhammar, Aclass of two-channel approximately perfect reconstruction hybrid analog/digital filter banks, in: Proceedings of IEEE International Symposium on Circuits and Systems, vol. 1, 2000.

[56] Waveform Measurement and Analysis Committee of the IEEE Instrumentation and Measurement Society, A Guide to Waveform Recorder Testing, IEEE, New York, April 1990. This guide contains four papers: (1) Linnenbrink TE. 'Introduction to waveform recorder testing'; (2) Green PJ. 'Effective waveform recorder evaluation procedures'; (3) Souders TM, Flach DR. 'Step and frequency response testing of waveform recorders'; (4) Grosby PS. 'Waveform recorder sine wave testing-selecting a generator'.

[57] S. Haykin, Adaptive Filter Theory, third ed., Prentice-Hall, Englewood Cliffs, NJ, 1996.

[58] R. Scholtz, R. Weaver, E. Homier, J. Lee, P. Hilmes, A. Taha, R. Wilson, Ultra-wideband radio deployment challenges, in: Proceedings of Personal Indoor Mobile Radio Conference, London, UK, 18–21 September, 2000.

[59] B. Razavi, RF Microelectronics, Prentice-Hall, Englewood Cliffs, NJ, 1997.

[60] P.P. Vaidyanathan, Multirate Systems and Filter Banks, Prentice-Hall, Englewood Cliffs, NJ, 1993.

[61] R. Cramer, An Evaluation of Ultra-Wideband Propagation Channels, Ph.D. dissertationUniversity of Southern California, 2000. December.

[62] N. Jayant, P. Noll, Digital Coding of Waveforms, Prentice-Hall, Inc., Englewood Cliffs, 1984.

[63] F.J. Harris, Time domain signal processing with the DFT, in: D.F. Elliott (Ed.), Handbook of Digital Signal Processing: Engineering Applications, Academic Press, San Diego, CA, 1987.

[64] J.B. Allen, Short term spectral analysis, synthesis, and modification by discrete Fourier transform, IEEE Trans. Acoust. Speech Signal Process. 25 (3) (1977) 235–238.

[65] J.B. Allen, L. Rabiner, A unified approach to short-time Fourier analysis and synthesis, Proc. IEEE 65 (11) (1977) 1558–1564.

[66] F.J. Harris, On the use of windows for harmonic analysis with the discrete Fourier transform, Proc. IEEE 66 (1) (1978) 51–83.

[67] Tran-Thong, Practical consideration for a continuous time digital spectrum analyser, in: Proceedings of IEEE International Symposium on Circuits and Systems, vol. 2, IEEE, New York, 1989, pp. 1047–1050.

[68] R.E. Crochiere, L.R. Rabiner, Multirate Digital Signal Processing, Prentice Hall, Englewood Cliffs, NJ, 1983.

[69] P.P. Vaidyanathan, Multirate Systems and Filter Banks, Prentice Hall, Englewood Cliffs, NJ, 1992.
[70] P.P. Vaidyanathan, Multirate digital filters, filter banks, polyphase networks, and applications: a tutorial, Proc. IEEE 78 (1) (1990) 56–93.
[71] R. Ansari, B. Liu, Multirate signal processing, in: S.K. Mitra, J.F. Kaiser (Eds.), Handbook for Digital Signal Processing, John Wiley & Sons, New York, 1993.
[72] A.V. Oppenheim, R.W. Schafer, Digital Signal Processing, Prentice Hall, Englewood Cliffs, NJ, 1975.
[73] L.R. Rabiner, Theory and Application of Digital Signal Processing, Prentice Hall, Englewood Cliffs, NJ, 1975.
[74] T.W. Fields, D.L. Sharpin, J.B. Tsui, Digital channelized IFM receiver, in: Proceedings of IEEE MTT-S International Microwave Symposium Digest, vol. 3, IEEE, New York, 1994, pp. 1667–1670.
[75] McCormick W. n.d. Professor of electrical engineering Wright State University, Dayton, OH. Private communication.
[76] R.E. Crochiere, L.R. Rabiner, Multirate Digital Signal Processing, Prentice Hall, Englewood Cliffs, NJ, 1983, p. 311.
[77] D.R. Zahirniak, D.L. Sharpin, T.W. Fields, A hardware-efficient, multirate, digital channelized receiver architecture, IEEE Trans. Aerosp. Electron. Syst. 34 (1) (1998) 137–152.
[78] P. Vary, U. Heute, A short-time spectrum analyzer with polyphase-network and DFT, Signal Process. 2 (1) (1980) 55–65.
[79] W.A. Kester (Ed.), Data Conversion Handbook, Newnes, Oxford, 2005.
[80] https://en.wikipedia.org/wiki/Internet_protocol_suite.

6

Processing data of correlation on GPU

Yongxin Zhu[a,b,c], Junjie Hou[b], Yuefeng Song[b], Yu Zheng[b], Tian Huang[d,e], Huaiguang Wu[f]

[a]*Shanghai Advanced Research Institute, Chinese Academy of Sciences, Shanghai, China.* [b]*School of Microelectronics, Shanghai Jiao Tong University, Shanghai, China.* [c]*University of Chinese Academy of Sciences, Beijing, China.* [d]*Astrophysics Group, Cavendish Lab, Cambridge University, Cambridge, United Kingdom.* [e]*Institute of High Performance Computing, Agency for Science, Technology and Research (A*STAR), Singapore, Singapore.* [f]*Zhengzhou University of Light Industry, Zhengzhou, China*

1 Introduction

In the past few decades, radio telescopes have greatly improved in sensitivity and image resolution, and telescopes collect data at very high rates. The Square Kilometre Array (SKA) is the largest comprehensive aperture radio telescope to be built internationally. Compared with the existing radio telescope, the sensitivity of the SKA will be increased by 10–100 times and the measurement speed will be increased by 10,000 times.

As the world's largest comprehensive aperture radio telescope, SKA plans to collect more than 12 Tb of data per second, so massive scientific data processing requires Exa-scale supercomputers. Due to the generation of massive data and the strict requirements of astronomical imaging for real-time processing, it will face a huge challenge in terms of computing power. For now, the computing power and processing speed requirements are constantly improving, and the traditional way to improve the processing power by increasing the CPU frequency is limited by the integrated circuit. In this case, the multicore processor gradually became the mainstream of computing. At present, the new cluster uses multicore to solve computational problems, so it is a trend to use multicore CPUs and multicore accelerators such as Graphic Processing Unit (GPU) to process large amounts of time-sensitive data.

Big Data in Astronomy. https://doi.org/10.1016/B978-0-12-819084-5.00003-1

After the antennas receive analog signals, filter and Fourier transformer engines (F-engines) usually convert analog signals into digital ones and transform them into frequency domain. In the frequency domain, cross-correlator engines (X-engines) carry out correlator calculation to produce visibility data, which will be handled by imaging or nonimaging pipeline later. To implement the X-engines, a few approaches based on both software and hardware have been explored. In this chapter, we focus on software-based approaches, especially GPU-based ones.

ARL (Algorithm Reference Library) is a reference library SKA candidate algorithm library, where gridding algorithm, degridding algorithm, and deconvolution algorithms are ARL in three major modules. ARL plays a vital role in radio astronomy imaging. With the implementation of the SKA project, massive astronomical data have rapidly increased, and how to effectively and quickly process these data has become the focus of research in the computer field. The calculation and optimization of SKA's key algorithms on the GPU architecture can speed up the processing of massive texture data. Gridding algorithms, degridding algorithm, and deconvolution algorithms are three major modules in the ARL as they consume most of execution time. Therefore, in order to speed up the SKA data processing, many astronomers use the GPU architecture to calculate and optimize these three modules in the ARL.

Brouw et al. [1] first proposed the gridding method for the discrete Fourier transform of polar grid sampling. For polar grid sampling, O'Sullivan et al. [2] proposed a fast sinc function grid algorithm for computer tomography (CT) image reconstruction. The ideal convolution method is to convolve with an infinite sinc function, but the amount of computation is too large. The alternative is to use a finite range of convolution functions to get a good approximation of the sinc function interpolation. Meyer et al. [3] applied gridding algorithm to implement coil sampling in MRI (magnetic resonance imaging) application. At the expense of additional calculations, Varbanescu et al. and Humphreys et al. [4, 5] improved spatial locality for better memory performance by sorting and searching data.

However, these methods do not reach 14% of the peak performance. Edgar et al. [6]. implemented grid algorithms on CUDA (computer unified device architecture) [7, 8]. Due to hardware constraints, they associate each grid point with a thread, but in this case the search and sorting operations increase the computational overhead. This method still has an important impact on the design and use of a new generation of radio telescopes. In the astronomical imaging process, gridding and degridding are the two most time-consuming operations. If the processed visibility data are in Exa-byte level or above, the time consumed cannot be mitigated by adjusting the performance of the computer. The traditional gridding is generally run on the CPU, so that the bandwidth of the accelerator on each floating point becomes very small, and the timeliness of the application is relatively reduced, and the algorithm parallelization is an important way to improve the calculation speed as indicated in recent radio astronomy data processing [9].

The current limited number of radio telescope antennas has resulted in incomplete spatial frequency coverage [10], which has an impact on the final image construction as well as another important algorithm, i.e., deconvolution. Since the deconvolution algorithm is able to eliminate this effect, many scholars have done research on the deconvolution algorithm. In 1974, Hogbom et al. [11] first proposed the CLEAN algorithm, a nonlinear deconvolution iterative method to eliminate sidelobe interference. Although the algorithm can eliminate the interference of side lobes, it takes a lot of time and the operation efficiency is quite low. In 2004, Bhatnagar et al. [12] proposed a scale-sensitive deconvolution algorithm (Asp-clean) algorithm for radio interference images, which modeled the image as a set of adaptive scale pixels in a more accurate way, to reconstruct the sky image of an asymmetric structure, at the cost of increasing complexity of the algorithm and computational cost. The calculation time of Asp-clean is more than three times that of the CLEAN algorithm. In 2008, Tim J. Cornwell [13] proposed a multiscale CLEAN algorithm that can better handle extended sources and improve image quality, but the algorithm also requires longer runtime than the Hogbom's CLEAN algorithm. In 2011, Rau, et al. [14] managed to reconstruct sky images by combining multiscale CLEAN algorithm and multifrequency algorithm with higher sensitivity and sampling frequency. In 2016, Zhang et al. [15] proposed an adaptive loop adaptive gain and adaptive scale CLEAN (Algas-Clean) algorithm, which can provide a more accurate model as well as a faster convergence than Asp-clean algorithm, i.e., the Algas-clean algorithm is 50% faster than Asp-clean. To solve the problems of the traditional CLEAN algorithm, Cheng et al. [16]. proposed a wavelet CLEAN algorithm in 2017, optimizing the parameters of the wavelet filter to improve the image quality. Unfortunately, the running time in Ref. [16] was not shortened while the efficiency of the algorithm was not improved. Most scholars are researching on image quality issues, but little research has been done on computational efficiency. Nowadays, with the generation of massive astronomical data, the computational efficiency of the deconvolution algorithm has become an urgent issue.

2 GPU-based cross-correlator engines

2.1 General processing steps of software implementation

There are several software correlators in use in radio astronomy. Here we briefly discuss some of these in relation to the GPU-based solution. In this chapter, we focus on an architecture named FX-correlators that combine a filter- and Fourier transform (F) stage with a cross-correlation (X) stage. This architecture of correlators is used by the Murchison Widefield Array (MWA) [17], the large aperture experiment to detect the dark ages (LEDA) [18], and PAPER [19], i.e., a hybrid FPGA-GPU approach in which the F-stage is implemented in FPGA chips, the X-stage is implemented in GPUs using the xGPU library [20]. A high bandwidth switch connects the F- and X-stages. The open xGPU package was first developed by Clark, La Plante & Greenhill [20].

In the real-time software backend of Giant Metrewave Radio Telescope (GMRT) [21], a structure similar to the MWA correlator is adopted, with nodes dedicated to three specific tasks. The difference is that only CPUs are used in the software backend, with heavy use of off-the-shelf performance-optimized libraries.

Major steps of correlator engines on GPU are illustrated in Fig. 6.1 where conventional correlator implementation on GPU

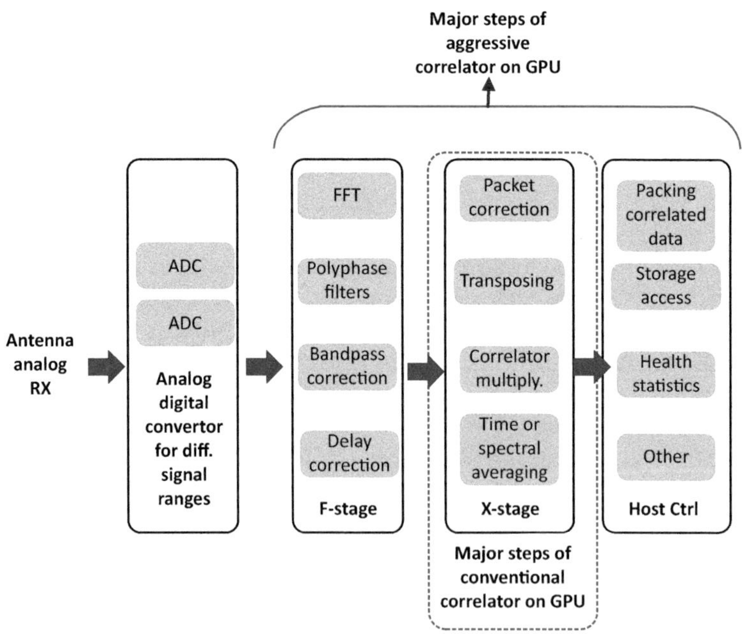

Fig. 6.1 Major steps of correlator engines on GPU.

has much fewer functions than aggressive correlator implementation on GPU.

Before the X stage of a conventional implementation architecture, FPGA executes the synchronous sampling and Fast Fourier Transform of incoming time-series data from ADC (analog digit convertor) in the F stage in a hybrid FPGA/GPU FX correlator [20], data packets are sent to the CPUs, which are the host controllers of GPUs for reordering as the entry of X stage. This practice is similar to real-time processors of LOFAR [22], which identified missing or disordered data packets and tried to correct them if possible. Some data packets had to be treated as lossy data if they missed a short deadline. Once the data packets are set to proper order for each frequency channel, cross-multiplication and integration of products are executed by GPUs as the X stage.

In the X stage, as the common open source library of X phase operations on GPU used in LOFAR [20, 22], the X stage library [20] passed the 8-bit incoming data from F phase through the GPU texture memory, where these 8-bit data were converted from 8 to 32 bits in hardware. Then these data were cross-multiplied by the X stage library executed on GPUs in the form of cross-correlation matrix. After multiplication, the X stage library averaged this matrix on the GPUs before transferring back to host memory controlled by CPUs.

After the X stage, the host CPUs further averaged the matrix in the host memory before writing it back to disk arrays or distributed storage. This step is also known as beamformer generating data of visibilities for further imaging or nonimaging pipeline processing.

Recently, aggressive correlator implementation on GPU was observed in the Low-Frequency ARray (LOFAR) [23] radio telescope, which is often described as one of the first of a new generation of software telescopes. In this implementation, hardware functions in the F stage on conventional FPGA were moved into software on GPU. In fact, LOFAR has pioneered the use of a combined software correlator and beamformer in an operational radio telescope since 2004 [24–26]. In 2004, an IBM Blue Gene/L was installed at the LOFAR central processor. At that time, the LOFAR real-time processor was the fastest supercomputer in the Netherlands, and the second fastest in Europe. Although compute performance of the Blue Gene/L was sufficient, significant research and development was required to achieve the required I/O performance [24].

2.2 Software architecture of GPU-based implementation

In general, software architecture of GPU-based implementation consists of input module, output module, and GPU kernel module. The input module for GPU was usually designed to transfer data from host CPU to GPU devices. Then correlation calculation was carried out by GPU kernel module. The correlation matrix was transferred back to host CPU by the output module.

As the system size grows up, extra data preparation phase is required in the input module to handle missing data packets or out-of-order incoming data packets. For example, as the third generation of LOFAR software correlator and beamformer, Cobalt (COrrelator and Beamformer Application for the LOFAR Telescope) was announced in 2018. The input module handles inputs from antenna fields by fixing data packets and transposing received data.

Similarly, there is a need to adopt additional control in the output module if the correlator system scales up. There were multiple host CPUs running in a distributed cluster instead of a host CPU attached to a centralized disk array. Take Cobalt as the example again, the output module completed the output process on all assigned nodes in the storage cluster consisting of 100 nodes and 2-PB disks. The output module was implemented with concurrent application software coordinating GPU cluster nodes and host CPUs via MPI messages on 10-Gb Ethernet.

In GPU kernel module of conventional architectures [20], only correlation kernel was executed to correlate signals that passed through F stage on FPGAs. To balance the workloads of multiple GPUs in different cluster nodes, proper distribution rules were required to distribute signals for different frequency channels in correspondence to various feeders of antennas.

In GPU kernel module of more aggressive architectures [22], functions in the F stage on conventional FPGA were also carried out on GPU, e.g., frequency transform and polyphaser filter were implemented with FIR filter and FFT kernels on GPU. Additional fine delay compensation and bandpass correction were also implemented along with the essential correlator calculations. The essential correlator calculation is usually implemented with single precision floating point numbers, while mix precision is adopted for delay compensation. To better utilize data locality on GPUs, an early stage of functions of output module, i.e., part of beamformer functionality, was also aggressively implemented in GPU kernel module.

To maintain the operation conditions of the antenna array, health statistics of the array are often collected along with observatory data flow. These statistics can be extracted by either execution module or output module and sent to array controllers.

3 Applying and implementing gridding algorithm after cross-correlator

3.1 Gridding algorithm application in SKA

SKA is an antenna array composed of several antennas. The antennas with similar positions will form a set of baselines, which will sample the electromagnetic radiation in the universe at different frequencies and correlate the sampling data of multiple antennas in a set of baselines. Then a set of data containing information such as location, frequency, etc. is called visibility [27]. The position of each baseline will be scanned along the rotation of the earth along an elliptical curve. The longer the baseline, i.e., farther distance between the antennas, or the higher the sampling frequency, the larger the scan radius will be. By continuously sampling for a certain period of time, a scan of the intensity distribution of the universe can be achieved to obtain the astronomical image to observe.

The observation data after initial integration will appear in three-dimensional coordinates (u, v, w), which is determined by the baseline position, the position of the observed signal source, the frequency of the observed signal, and the observation time. In order to improve the data quality, a series of calibration and elimination of interference have to be carried out. But the generated visibility does not intuitively reflect the characteristics of the observed object, since it is the sampling result in the frequency domain. Therefore, visibility has to undergo iFFT (inverse fast Fourier transform) to get the image of the universe. Since visibility is not two-dimensional (2D) data, it needs to be placed in a 2D grid (UV-grid) to facilitate the inverse Fourier transform. The process of mapping 3D Visibility data into a 2D mesh is called gridding. Fig. 6.2 shows an overall process of converting visibility data into astronomical images.

Therefore, we can understand that gridding algorithm plays a mapping role in SKA science data processing to facilitate subsequent iFFT operations, which is a very important link in astronomical data processing.

There are many ways to implement gridding operations. The early processing method is to simply add data to the mesh in its vicinity [28] or to average the data into the corresponding adjacent mesh [29]. Later, a method of using the weighted sum of the distance between the sampling point (u, v) and the grid point was widely applied to the gridding, and for the first time formed an official gridding method [1]. On top of this method, another

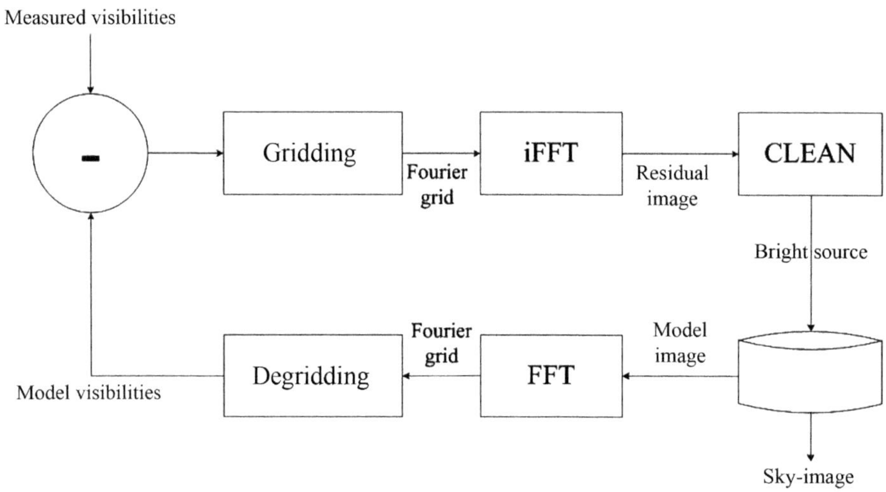

Fig. 6.2 The major steps of imaging pipeline.

gridding method intended to optimize gridding operations with sinc function convolution [2]. However, since the ideal sinc function requires an infinite range, a fixed-size convolution kernel (convolution coefficient matrix) is used instead of the sinc function to complete convolution operations.

Specifically, each consolidated visibility and its corresponding convolution coefficient matrix are convoluted, and the multiplication results of each data and coefficient matrix are accumulated to the corresponding grid. As such, each visibility data set will affect the values of a large grid partition. The size of the convolution coefficient matrix involved and the specific sampling data are determined by factors such as the grid size, the length of the baseline, and the sampling frequency. In practical applications, whenever the observation frequency and the observation target are set, the appropriate baseline length needs to be adjusted. Over time, the grid position of the visibility map will slide along the elliptic curve, i.e., scanning along the elliptic curve. This scanning speed is determined by the baseline length and the grid size. If the speed is too fast, the image quality will be degraded. If the speed is too slow, the time cost will increase, and the area of the convolution coefficient matrix will grow to increase the memory overhead.

In addition, when the observed area is enlarged, we cannot treat the Earth and the observed object as planes. Therefore, the coordinates of the w-axis cannot be regarded as 0, and the mapping cannot be completed using a single convolution coefficient matrix. For the case of large observation area, the most widely used gridding method is called W-projection algorithm [2 2]. Intuitively, a 5D convolution coefficient matrix is generated from data such as sampling frequency, grid size, and baseline length, and then the corresponding convolution kernel is selected according to the w value of the sampling point.

Currently relying on several pathfinder projects of SKA, SKAO member countries have prototyped the implementation of key algorithms in the imaging pipeline. For example, in 2010, Australia SKA pathfinder (ASKAP) project team released standardized test procedures. They also conducted gridding and degridding the test on both CPU and GPU.

In view of gridding algorithm containing relatively high level of parallelism and strong capacity of GPU, quite a few accelerators are implemented on GPUs in the following typical ways:

(1) Romein gridder [30]: This method decomposes the grid according to the width of the convolution kernel, and then assigns the decomposed mesh to the corresponding execution unit for parallel execution.
(2) Merry gridder [31]: On top of Romein gridder method, this method consolidates threads by merging multiple threads into one thread, following the idea of loop unrolling to improve the parallelism of instructions and reduce the number of memory accesses, and thereby further improve the acceleration performance.
(3) NAG gridder: This method decomposes grid into tiles according to the size of the cache, then presorts the sampling data so that data within the same tile can be processed more efficiently.

These methods are designed from high-performance computing (HPC) perspectives to accelerate gridding algorithms. In addition, there are other methods designed from the perspective of image rendering. The idea is sampled data are continuously superimposed to the frame buffer in order to carry out the gridding calculation with tools such as Open GL. This approach also achieves reasonable acceleration results.

Since the data size and grid size of the test set used are different, and the GPU models used are different, it is difficult to compare the acceleration effects of these methods, but they can achieve more than 10 times the acceleration effect compared with the CPU.

3.2 Gridding algorithm analysis

In radio astronomy and medical imaging, the collected signals are usually affected by various factors, and the sampled data often appear in irregularity. In order to reconstruct the image by better exploiting the fast calculation of fast Fourier transform (FFT), the gridding algorithm is adopted to map these sample data. The basic steps of the gridding algorithm can be summarized as:

Step 1. Multiply the sampled data by the density compensation function to compensate for the unevenness of the sampled data.

Step 2. Convolve with the weight data using the selected convolution function.

Step 3. Resample the data to drop the data onto the grid points.

Step 4. Apply FFT.

Step 5. To eliminate the effects of the convolution function, divide the result by the Fourier transform of the convolution function.

In this algorithm, proper choice of sampling density compensation function and the interpolation function (convolution function) impose a major impact on the reconstructed image quality. This is a major theoretical issue of gridding algorithm.

In this chapter, we focus on one kernel of gridding algorithm, i.-e., convolution as we explore GPU acceleration. The core of the gridding algorithm is 2D convolution. The formula for a single sampling point is as follows:

$$\text{grid}(u, v) = \sum_{x=-\text{support}}^{\text{support}} \sum_{y=-\text{support}}^{\text{support}} \text{Samples}(u-x, v-y) \times C(-x, -y) \tag{1}$$

where 2*support+1 is the width of the convolution kernel, C is the coefficient matrix of convolution. If we formatall sampled data and coefficient matrices in single-precision floating-point complex numbers, one core operation of the gridding algorithm requires 2 memory reads and 1 memory write, and then a floating-point complex multiply-accumulate operations. As can be seen from the formula, the calculation of the gridding algorithm is huge, for *Nsamples* sampling points, a total of $Nsamples \times (2 \times \text{support}+1) \times (2 \times \text{support}+1)$ core operations are required.

Referring to the reference code of the standardized test procedure of the gridding algorithm published by the Australian SKA Pathfinder (ASKAP) team in 2010 [30], a rough execution flow can be obtained: the first step of preprocessing includes the generation of the convolution coefficient matrix and the address map lookup. The second step multiplies the sampled data by the corresponding convolution kernel according to the lookup table and then adds them to the correspondinggrid data.

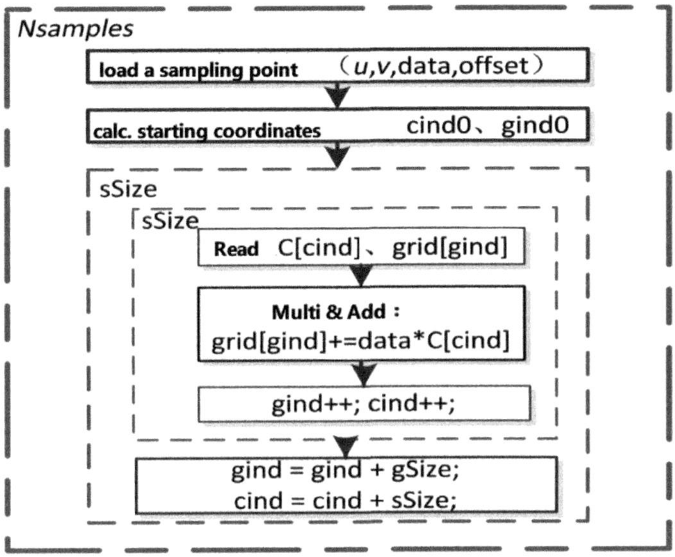

Fig. 6.3 Data flow of gridding algorithm.

The flow chart of the core part (gridding kernel function) is as follows (*cind, gind* are the coordinates of convolution coefficient matrix and the grid) in the following figure (Fig. 6.3).

3.3 Parallel implementation of gridding/degridding algorithms and analysis of experimental results after cross-correlator

Each visibility has coordinates (*uu, vv, ww*) in Fourier space, complex amplitudes, and weights. The (*uu, vv*) coordinates determine that visibility falls on the 2D grid, and the (*ww*) coordinates determine, which kernel in the stack is used to convolve the grid. The grid function simply iterates through the input visibility list and then updates each convolution visibility grid with the appropriate convolution kernel. Once all visibility is convolved onto the grid, a 2D FFT can be performed on the grid. Efficient parallel computing of convolutional grids is the focus of research. The ARL algorithm library is based on CPU operation. However, SKA data are large in scale and need parallel processing. In the whole process of SKA image acquisition and final imaging, gridding algorithm is one of the most computationally intensive and time-consuming algorithms. Parallelization acceleration with PyCUDA can effectively shorten the running time of the entire

ARL. Experimental environment include the Tesla K80 GPU board, the CPU is an Intel Xeon E5-2620 V3, memory size is 816 GB DDR4, the operating system is CentOS 7.0, the python version is 3.5.

3.3.1 Time overhead analysis

The optimization scheme needs to filter the functions in the gridding algorithm for parallelization. We selected several functions in the gridding algorithm that have higher calling frequencies and the longest running time. The relationship between the main functions that can be optimized in the gridding algorithm and the number of times it is called is shown in Fig. 6.7.

As can be seen from Fig. 6.4, the *grdsf* function has the highest number of calls, up to 370 times, *anti_aliasing_calculate* is called 185 times, *convolutional_degrid* is called 47 times, *convolutional_grid* is called 44 times. When the visibility data vis = 10,000, the execution time of these four functions is tested separately. The result is shown in Fig. 6.5.

It can be seen from Fig. 6.5 that the *convolutional_grid* and *convolutional_degrid* functions take the longest execution time,

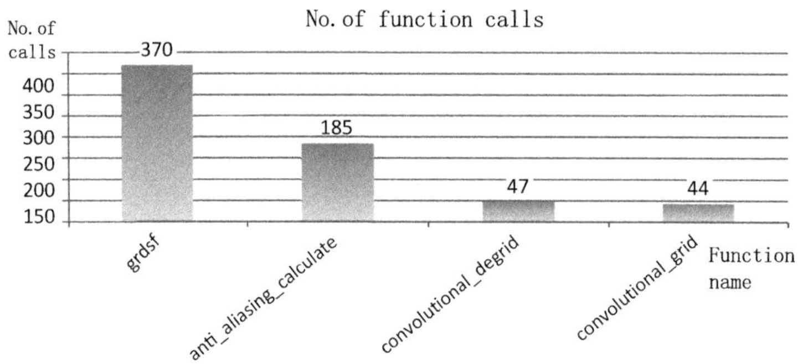

Fig. 6.4 Number of main function calls, which can be optimized in gridding algorithm.

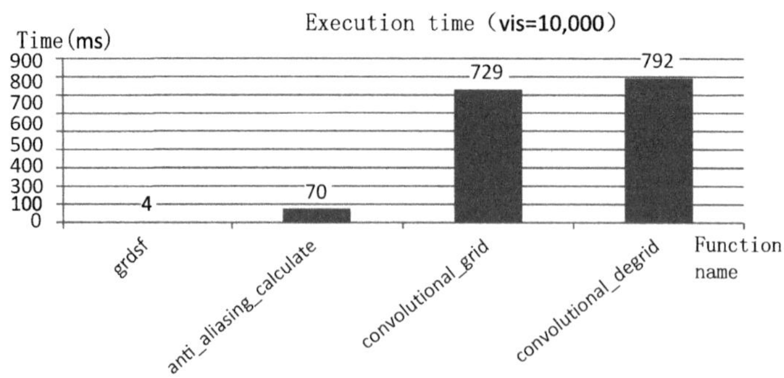

Fig. 6.5 Execution time of main functions that can be optimized in the Gridding algorithm.

while the two functions, *grdsf andanti_aliasing_calculate*, take less time. In other words, the *convolutional_grid* and *convolutional_degrid* functions need to be optimized.

3.3.2 Parallelization implementation of gridding/degridding algorithm

Typically, GPU parallel acceleration is achieved in two ways, one is by reducing cycle on the basis of the serial code to accelerate, other is to redesign the overall algorithm, by transforming the overall module into a plurality of independent submodules that can be distributed to threads for execution in parallel. In this chapter, the first optimization approach was taken.

This chapter performs the corresponding parallelization processing mainly according to the optimization scheme described here. The gridding algorithm convolves the data of visibility in the grid. In dealing with large number of iterations of data, a lot of execution time is taken, hence taking advantage of multithreading to reduce cycle will significantly reduce the execution time. To enable parallel execution, data dependency has to be handled properly. Data dependence is a relationship that represents the mutual constraints between data. It is mainly divided into function dependencies, multivalue dependencies, and connection dependencies. The main use is the function dependency. Before parallelizing the data processing, the data dependencies must be minimized. Then a generic flow is illustrated.

Coordination of CPU and GPU during execution

Require: Convolution kernel must have even number of pixels

1. Copy data from the host to the device
2. Execute convolutional_grid kernel
3. convolutional_grid_kernel<<<block_num, thread_num>>> (params)
4. Copy data from the device to the host

Output uvgrid, sumwt

In an experiment for this chapter, global registers were used to support parallel computation on each thread within a GPU warp. The size of on-board memory is 11,440 MB. All the threads within a warp do not generate an access violation. So we can take advantage of GPU CUDA tools with a function cudaMemcpy () to load data into the GPU on-board memory to be shared by all the threads. Accumulation calculation is done by acquiring visibility data via thread index. After the loop is completed, all threads synchronize their calculation and load the result into memory.

The parallelization of the algorithm is implemented in accordance with the description. Given a large number of iterations, taking advantage of multithreading will reduce significantly the execution time. When performing GPU parallel processing, the pseudocode of the kernel function processed in parallel is as follows:

```
Convolutional_grid_kernel function
Input: kenel-list,uvgrid,vis,visweights,vuvwmap,vfrequencymap,
vpolarisationmap
for pol∈[0, npol] do
obtain thread index i, i∈[0,N]
if i<N then
obtain visibility data, coordinates
for j∈[yy, yy+gh] do
for k∈[xx,xx+gw] do
obtain grid and convolution core index
uvgrid_real[w] ← (kernel0_real[q]*v_real-kernel0_imag[q]*v_imag)+
uvgrid_real[w];
uvgrid_imag[w] ← (kernel0_real[q]*v_imag+
kernel0_imag[q]*v_real)+uvgrid_imag[w];
end for
1sumwt[chan*npol+pol] ← vwt+sumwt[chan*npol+pol]
end for
end if
end for
Output: uvgrid, sumwt
```

The mentioned method to minimize data dependency using global memory is also used to implement the corresponding multithread parallelization of the convolutional_degrid function. The pseudocode of the kernel function processed in parallel is as follows:

```
Convolutional_degrid_kernel function
Input: kernel_list, vshape, uvgrid, vuvwmap, vfrequencymap,
vpolarisationmap
for pol∈ [0, vnpol] do
obtain the thread index i, i ∈ [0,N]
if i<N then
for j∈ [yy, yy+gh] do
for k∈[xx, xx+gw] do
calculate vis, vis=uvgrid* ckernel0
end for
end for
end if
end for
Output: vis
```

3.3.3 Performance of gridding/degridding on GPU after cross-correlator

In this the experiment, the GPU model is Tesla K80, which is dual-core GK210 architecture, containing a total of 4992 CUDA Cores. Each thread block can be allocated with a maximum of 1024 threads according to different amounts of data. For this experiment, the relationship between the thread and the thread block is in the form of (32, 32, 1). The way to use the grid blocks for different data blocks is: (1,1,1), (10,1,1), (10,10,1), (100, 10), (100,100,1), (1000,100,1).

The running time of the convolutional_grid function in ARL and the running time after GPU acceleration and its speedup ratio are shown in Table 6.1.

According to Table 6.2, the acceleration of the convolutional_-grid function after parallelization is realized under different data quantities. As the amount of data increases, the acceleration ratio increases continuously, but the speed of the acceleration ratio decreases. In order to more intuitively express the relationship between the acceleration ratio and the amount of data, a histogram is used to represent the result, as shown in Fig. 6.6.

The foregoing content is the parallel implementation of the convolutional_grid function and its experimental results analysis, followed by the convolutional_degrid function. The results of the convolutional_degrid function according to the experiment are shown in Table 6.2.

From Table 6.2，it can be found the acceleration status of subsequent parallel implementation of convolutional_degrid function given different amount of data. When the amount of data increases speedup also continues to increase, the speedup tends to freeze after the data grow too big. Compared with the

Table 6.1 Convolutional_grid execution time and speedup.

Data volume	convolutional_grid (CPU)	convolutional_grid (GPU)	Speedup
1000	0.040 s	0.180 s	0.22
10,000	0.290 s	0.179 s	1.62
100,000	3.110 s	0.396 s	7.85
1,000,000	27.714 s	2.197 s	12.61
10,000,000	277.703 s	19.539 s	14.21
100,000,000	2886.311 s	182.794 s	15.79

Table 6.2 Execution time and speedup ratio of Convolutional_degrid for different data volumes.

Data volume	convolutional_degrid (CPU)	convolutional_degrid (GPU)	Speedup
1000	0.058 s	0.152 s	0.38
10,000	0.462 s	0.181 s	2.55
100,000	3.292 s	0.351 s	9.38
1,000,000	30.525 s	2.134 s	14.30
10,000,000	272.345 s	17.935 s	15.19
100,000,000	3057.845 s	179.233 s	17.06

convolutional_grid function, the speedup ratio is basically the same, but the speedup ratio is slightly faster. In order to more intuitively express the relationship between the acceleration ratio and the amount of data, a histogram is used to represent the result. The result is shown in Fig. 6.7.

From Figs. 6.6 and 6.7, a small amount of data (vis = 103) can be seen, and the speedup of two functions is 0.22 and 0.38, respectively. In other words, GPU actually slows down due to the overheads. When the data volume is relatively small, the boot time of the GPUdevice, the time of data interaction between the main memory and on-board graphic memory is more time consuming than serial execution on the CPU. Relatively small amount of data

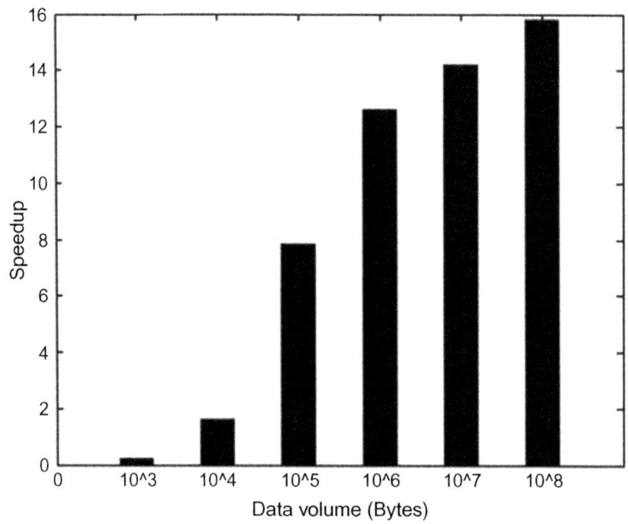

Fig. 6.6 Speedup ratio of convolutional_grid function under different data volumes.

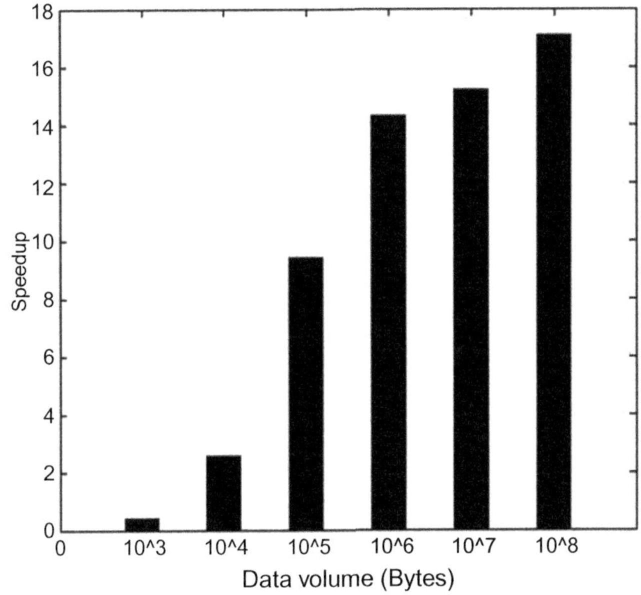

Fig. 6.7 Speedup of convolutional_degrid function under different data volumes.

can be seen, and GPU parallel acceleration is not effective since its internal scheduling is relatively more time consuming. As the amount of data increases, GPU acceleration gradually turns effective, e.g., when the data size grows to 104, speedup starts to be apparent. Therefore, GPU parallel acceleration is suitable for the operation of a large amount of data. The proposed optimization scheme can improve the overall operational efficiency of ARL as the real-time requirements [18] increase computation complexity.

4 Applying and implementing deconvolution algorithm and parallel implementation after cross-correlator

To handle the real-time requirements of massive astronomical data processing in SKA, this chapter proposes a deconvolution algorithm based on GPU parallelization. It is proposed to analyze the time-consuming part of the algorithm and use multithread to design the CUDA kernel function of the algorithm. On GPU, multiple threads simultaneously perform computation, in order to solve the computational efficiency problem of the deconvolution algorithm given massive astronomical data, and efficiently implement the deconvolution algorithm on GPU in a parallel way.

4.1 CLEAN algorithm

The deconvolution algorithm in ARL uses the CLEAN algorithm, which is a nonlinear deconvolution method designed to process incomplete radio data and eliminate side lobes in the image (the antenna pattern usually has two or more lobes, in which the lobe with the highest radiation intensity is called the main lobe, the rest lobes are called the side lobe).

If the sky brightness distribution is represented by $I(x, y)$, the corresponding complex visibility function is represented by $V(u, v)$, where u and v are the spatial frequencies. In the uv plane (the plane where the uv plane represents the baseline projection), u points east and v points north.

Since the number of antennas in the SKA is fixed, the number of uv sampling points is limited, and the distribution function of the uv sampling points is called a sampling function. The Fourier inverse transform is performed on the sampling function to obtain the dirty beam, and the preprocessed data of the radio telescope is inversely transformed by Fourier to obtain the dirty map. The maximum brightness point and its position in the dirty map are found through an iterative process. The result is convoluted with a fitted beam (Gaussian beam) to obtain a restored image. A dirty beam with side lobes is called a point spread function (PSF). The main steps of the CLEAN algorithm are as follows:

Step 1: Perform an inverse Fourier transform on the sampling function to obtain a dirty beam;

Step 2: Perform a Fourier inverse transform on the preprocessed data of the radio telescope to obtain a dirty map, find the maximum point in the dirty map, and record the coordinates;

Step 3: Move the center point of the dirty beam to the position of the maximum point, and multiply the dirty beam by the error term;

Step 4: Subtract the dirty beam from the dirty map;

Step 5: Repeat Steps 2, 3, 4 until the maximum value of the remaining images is less than the given noise level;

Step 6: Convolving a fitted beam and δ function, which is the sum of the maximum points that are subtracted and their corresponding positions;

Step 7: Add the remaining image to the result of Step 6 to get the final clear image.

4.2 Parallel implementation of CLEAN algorithm

The deconvolution algorithm has high requirements on computer memory and CPU. Iterative calculation takes a long time in the data processing process, and parallel deconvolution algorithm on the GPU (graphic processing unit) can greatly improve the operation speed and save time. NVIDIA Corporation proposed a parallel computing architecture the CUDA (computer unified device architecture), the framework combined CPU and GPU advantages by CUDA C programming GPU implemented on parallel computing, computing speed than the CPU by several times or even a hundred times.

Through the analysis of the deconvolution module in ARL, it is known that in the implementation process of CLEAN algorithm, multiple iteration calculations are needed, which takes a long time. Since the result of the last iteration is needed for each iteration, the whole process of the CLEAN algorithm cannot be written as a kernel function, but some of the operations can be written as kernel functions, so that it can be parallelized on the GPU. The parallel computing operation is performed serially on the CPU to improve computational efficiency. According to the characteristics of GPU parallelism, this chapter designs and implements the CLEAN algorithm based on GPU parallelization by using multithreading. The main steps of the parallelized CLEAN algorithm are as follows:

Step 1: Initialize the clean map and pass the dirty beams and dirty maps as formal parameters;

Step 2: Call the dirtyFabsMax kernel function written by CUDA on the GPU, find the maximum point in the dirty map, and record the position coordinates of the point;

Step 3: Move the center point of the dirty beam to the position of the maximum point, and multiply the dirty beam by the factor;

Step 4: Call the subPsf kernel function written by CUDA on the GPU, and subtract the dirty beam from the dirty map;

Step 5: Determine whether the maximum value of the remaining image is less than a given noise level. If the condition is not met, continue the iterative calculation; if the condition is met, the iteration is exited and the execution continues downward.

Step 6: Convolving a fitted beam and δ function, which is the sum of the maximum points that are subtracted and their corresponding positions;

Step 7: Add the result obtained by Step 6 to the remaining images to obtain the final clear image;

Step 8: Pass the processed data on the GPU to the CPU and draw the image.

Since the CLEAN algorithm in ARL is implemented in Python language, this chapter uses PyCUDA to implement parallelization of CLEANalgorithm. Kernel function is a specific implementation using CUDA C programming, the rest still use Python programming, by PyCUDA so Python can invoke parallel computation via Nvidia CUDA API, to achieve parallel execution on GPU. The *Blocksize* and *Gridsize* are passed as parameters to specify the size of the grid and thread blocks. By counting the product of Blocksize and Gridsize, we can determine the number of threads in the grid, and then specify the number of threads created on the GPU that handle multithreaded operations in parallel. We also need *syncthreads* function to ensure thread synchronization; otherwise, due to the disorder and asynchronous nature of thread execution, if there is no synchronization function syncthreads may cause element coverage problems, resulting in incorrect results. We compose a kernel function *dirtyFabsMax* to achieve the above functions. The pseudocode of the kernel function is shown in Table 6.3.

The step 4 of CLEAN algorithm is to subtract dirty beams from the dirty map. This operation may be performed in parallel, because each point in the beam or the dirty map is independent. Simultaneous subtraction operation can improve the computational efficiency of the CLEAN algorithm. On GPU, each thread can be assigned for each element, so subtraction operation of each element can be arranged in parallel with the function call configuration as <<<1024, 1024>>>, which ensures that GPU allocates one thread per data element. These functions are implemented by as a kernel function *subPsf*. The pseudocode of the kernel function is shown in Table 6.4.

Table 6.3 dirtyFabsMax Kernel Function.

dirtyFabsMax kernel

Input dirty
Initialize row←threadIdx.x+blockIdx.x * blockDim.x
col.←threadIdx.y+blockIdx.y * blockDim.y
t←row+col. * blockDim.x * gridDim.x
For stride←blockDim.x * blockDim.y * gridDim.x * gridDim.y, stride>1, stride←stride/2 do
__syncthreads()
If t<stride **Then**
If psf[t]<psf[t+stride]
Then temp←psf[t]
psf[t]←psf[t+stride]
psf[t+stride]←temp
End If
End If
End For

Table 6.4 subPsf Kernal Function.

subPsf kernel

```
Input res, psf, a1o, a2o, mval
Initialize row←threadIdx.x+blockIdx.x * blockDim.x
col←threadIdx.y+blockIdx.y * blockDim.y
If row >=a2o[0] AND row<a2o[1] AND col. >=a2o[2] AND col.<a2o[3]
Then t1←(row+a2o[0]) * blockDim.x * gridDim.x+col. – a2o[2]
t2←row * blockDim.x * gridDim.x+col.
res[t1]←res[t1] – psf[t2] * mval
End If
```

4.3 Performance of CLEAN algorithm on GPU

The data tested were derived from the astronomical test data provided in the ARL. First, the data in the ARL are preprocessed to generate a dirty map and a dirty beam, and then the parallelized CLEAN algorithm is called to process the dirty map to generate a clean map. Clean map and dirty map are shown in Fig. 6.8. Comparing the dirty map and the clean map, it can be seen that a large number of blurred points in the dirty map are cleared, and the clean map clearly reflects the distribution of the brightness of the sky.

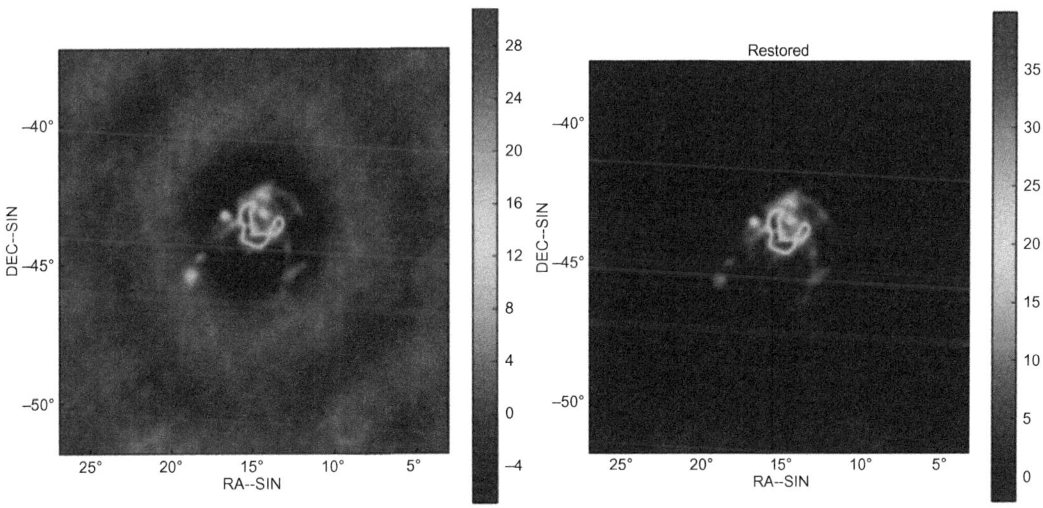

(A) (B)

Fig. 6.8 (A) Dirty map and (B) clean map.

Table 6.5 CLEAN algorithm execution time.

Image size	# of Iterations	Python	PyCUDA	Speedup
512 × 512	10,000	8.875	3.666	2.42
1024 × 1024	10,000	27.289	5.263	5.19
2048 × 2048	10,000	95.240	12.369	7.69
4096 × 4096	10,000	457.186	40.712	11.23
8192 × 8192	10,000	1803.011	170.817	10.56

The CLEAN algorithm in ARL is implemented in PyCUDA language on GPU to evaluate calculation time, then we also implemented CLEAN algorithm with Python language on CPU for comparison. The test is conducted by varying image size under the condition that the number of iterations is 10,000. The test results are shown in Table 6.5.

As shown in above table, the CLEAN algorithm implemented by PyCUDA under the GPU processes the data faster, and the time is much smaller than the time of processing the data by the Python version of the CLEAN algorithm. When the image size is 512 × 512, the amount of data is small at this time, only about 260,000 B, the speedup is not very obvious, only about twice the acceleration effect. However, when the image size increases, especially when it is larger than 4096 × 4096, the amount of data increases to 10 million bytes at this time, the speedup increases sharply, and the acceleration effect is significantly improved. When the image size is 8192 × 8192, the algorithm on the original CPU takes half an hour to run the result, but the algorithm implemented in this chapter only takes more than two minutes to get the result, which greatly saves time and improves the operating efficiency of the algorithm.

The speedup ratio is the ratio of the running time of the CLEAN algorithm under the CPU to the running time under the GPU. The acceleration of the CLEAN algorithm implemented by parallelization under different image sizes is shown in Fig. 6.9.

It can be clearly seen that when the image size is increased, the speedup is larger, and the acceleration effect running on the GPU is more remarkable. Especially when the image size is 4096 × 4096, the amount of data at this time increases to 10 million, and it is obvious that the speedup ratio grows larger.

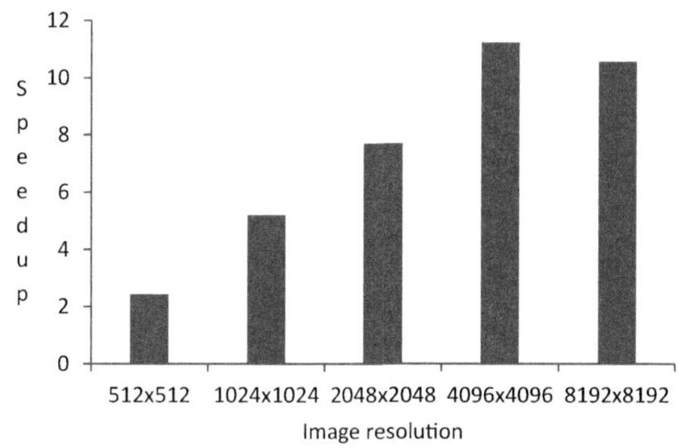

Fig. 6.9 CLEAN algorithm acceleration under different resolution.

5 Summary

To process data of correlation of radio astronomy on GPU, this chapter discusses software pipeline, software architecture and GPU implementation of cross-correlator. The implementation of GPU-based correlator in LOFAR telescope is discussed in detail as a case study. After explaining cross-correlator calculation, the chapter also discusses the optimization of major algorithms in algorithm reference library of SKA project, i.e., gridding, degridding, and deconvolution (CLEAN). To take advantage of CUDA architecture on GPU, these algorithms were implemented using PyCUDA to illustrate the acceleration of most time-consuming functions in these algorithms, e.g., convolutional_grid, convolutional_degrid. When the amount of testing data reaches 10^6B, GPU speedup ratio over the CPU grows more than 12 ×. The acceleration ratio reaches the maximum when the amount of testing data increases to 10^8B. As to the CLEAN algorithm, we also found that the most time-consuming parts lie in calculating the maximum point, and subtracting dirty beams from the dirty maps. These computation-intensive parts were implemented with multithreading functions such as dirtyFabsMax and subPsf kernel functions along with syncthreads function to ensure thread synchronization for speedup over 10 ×.

References

[1] W.N. Brouw, Aperture synthesis, Methods Comput. Phys. 14 (1975) 131–175.
[2] D.J. O'Sullivan, A fast sinc function gridding algorithm for Fourier inversion in computer tomography, IEEE Trans. Med. Imaging MI-4 (4) (1985) 200–207.

[3] C.H. Meyer, B.S. Hu, D.G. Nishimura, A. Macovski, Fast spiral coronary artery imaging, Magn. Reson. Med. 28 (1992) 202–213.
[4] A.L. Varbanescu, On the Effective Parallel Programming of Multi-Core Processors, Universitatea Politehnica Bucuresti Romania, 2010.
[5] B. Humphreys, T. Cornwell, B. Humphreys, Analysis of Convolutional Resampling Algorithm Performance, online at www.skatelescope.org/pages/page_memos.htm, 2011.
[6] J. Cheng, CUDA by Example: An Introduction to General-Purpose GPU Programming, Addison-Wesley Professional, 2010.
[7] Nvidia Corporation, CUDA Programming Guide v0.8, 2006-11-21, Nvidia, 2006, pp. 36–37.
[8] P.B. Broekema, The square kilometre array science data processor preliminary compute platform design, J. Instrum. 10 (2015) 14–25.
[9] E. Van Heerden, A. Karastergiou, S.J. Roberts, O. Smirnov, New approaches for the real-time detection of binary pulsars with the square kilometre array (SKA), in: General Assembly and Scientific Symposium. IEEE, 2014, pp. 1–4.
[10] A. Dabbech, C. Ferrari, D. Mary, E. Slezak, O. Smirnov, J.S. Kenyon, MORE-SANE: MOdel REconstruction by Synthesis-ANalysis Estimators, a sparse deconvolution algorithm for radio interferometric imaging, Astron. Astrophys. 576 (2014) 16.
[11] J.A. Hogbom, Aperture synthesis with nonregular distribution of interferometer baselines, Astron. Astrophys. Suppl. 15 (1974) 417–426.
[12] S. Bhatnagar, T.J. Cornwell, Scale sensitive deconvolution of interferometric images—I. Adaptive scale pixel (Asp) decomposition, Astron. Astrophys. Eur. J. 426 (2) (2004) 747–754.
[13] T.J. Cornwell, Multiscale CLEAN deconvolution of radio synthesis images, IEEE J. Sel. Top. Sign. Proces. 2 (5) (2008) 793–801.
[14] U. Rau, T.J. Cornwell, A multi-scale multi-frequency deconvolution algorithm for synthesis imaging in radio interferometry, Astron. Astrophys. 532 (2011).
[15] L. Zhang, M. Zhang, X. Liu, The adaptive-loop-gain adaptive-scale CLEAN deconvolution of radio interferometric images, Astrophys. Space Sci. 361 (5) (2016) 1–6.
[16] J. Cheng, L. Xu, Z. Lu, D. Zhao, L. Chen, Y. Yan, Application of wavelet clean for Mingantu Spectral Radioheliograph imaging, in: Proceedings of International Symposium on Intelligent Signal Processing and Communication Systems (ISPACS), 2017, pp. 81–86.
[17] S.M. Ord, B. Crosse, D. Emrich, et al., The Murchison Widefield Array correlator, Publ. Astron. Soc. Aust. 32 (2015) e006.
[18] J. Kocz, L.J. Greenhill, B.R. Barsdell, et al., Digital signal processing using stream high performance computing: a 512-input broadband correlator for radio astronomy, J. Astron. Instrum. 4 (2015).
[19] A.R. Parsons, D.C. Backer, G.S. Foster, et al., The precision array for probing the epoch of reionization: eight station results, Astron. J. 139 (April 2010) 1468–1480.
[20] M.A. Clark, P.C. La Plante, L.J. Greenhill, Accelerating radio astronomy cross-correlation with graphics processing units, Int. J. High Perform. Comput. Appl. 27 (2012) 178–192.
[21] J. Roy, Y. Gupta, U.-L. Pen, et al., A realtime software backend for the GMRT, Exp. Astron. 28 (1) (2010) 25–60.
[22] P. Chris Broekema, J.J.D. Mol, R. Nijboer, A.S. van Amesfoort, M.A. Brentjens, G.M. Loose, W.F.A. Klijn, J.W. Romein, Cobalt: a GPU-based correlator and beamformer for LOFAR, Astron. Comput. 23 (2018) 180–192. arXiv:1801.04834v3 [astro-ph.IM].

[23] M.P. van Haarlem, M.W. Wise, et al., A. Gunst, LOFAR: the LOw-Frequency ARray, Astron. Astrophys. 556 (2013).
[24] W. John, P.C.B. Romein, E. van Meijeren, K. van der Schaaf, W.H. Zwart, Astronomical real-time streaming signal processing on a blue gene/L supercomputer, in: ACM Symposium on Parallel Algorithms and Architectures (SPAA'06), Cambridge, MA, 2006, pp. 59–66.
[25] W. John, P.C.B. Romein, J.D. Mol, R.V. van Nieuwpoort, The LOFAR correlator: implementation and performance analysis, in: ACM Symposium on Principles and Practice of Parallel Programming (PPoPP'10), Bangalore, India, 2010, pp. 169–178.
[26] J.D. Mol, J.W. Romein, The LOFAR beam former: implementation and performance analysis, in: EuroPar'11, volume LNCS 6853, Part II, Bordeaux, France, 2011, pp. 328–339.
[27] J.W. Romein, An efficient work-distribution strategy for gridding radiotelescope data on GPUs, in: Proceedings of the 26th ACM International Conference on Supercomputing, ACM, 2012, pp. 321–330.
[28] N.C. Mathur, A pseudodynamic programming technique for the design of correlator supersynthesis arrays, Radio Sci. 4 (3) (2016) 235–244.
[29] D.E. Hogg, G.H. Macdonald, R.G. Conway, C.M. Wade, Synthesis of brightness distribution in radio sources, Astron. J. 74 (10) (1969) 1206–1213.
[30] B. Nikolic, SDP Consortium, Square Kilometre Array Science Data Processing, Exascale Radio Astronomy, 2014.
[31] B. Merry, Faster GPU-based convolutional gridding via thread coarsening, Astron. Comput. 16 (2016) 140–145.

7

Flux calibration for single-dish radio telescopes

Bin Liu, Shenghua Yu
National Astronomical Observatories, Chinese Academy of Sciences, Beijing, China

Astronomers use radio telescopes to collect natural radio emissions from the universe. To reveal the strength of the astronomical source emission, the radio data first need to be calibrated. This chapter describes some basic concepts and methods for calibrating data obtained by single-dish radio telescopes. After introducing the essential concepts of antenna temperature, system temperature, telescope gain, and efficiency, we mainly summarize the flux calibration techniques and give a calibration example with spectral line data.

1 Basic concepts

1.1 Antenna temperature

The most popular units of power or intensity in radio astronomy are temperature and flux. A radio telescope is a device that converts electromagnetic radiation into voltage, which represents the power at a certain frequency. To calibrate the received power intensity in radio astronomy, warm resistors or noise diodes are often used as standard calibrators for antenna systems.

The noise generated by a resistor is indistinguishable from the noise received from an antenna surrounded by black-body radiation of the same temperature. The noise power per unit bandwidth received by a radio telescope is often described in terms of the Rayleigh-Jeans antenna temperature. The antenna temperature is defined as the temperature of an ideal resistor that would generate the same power per unit bandwidth as appears at the

Big Data in Astronomy. https://doi.org/10.1016/B978-0-12-819084-5.00015-8

antenna output. Note that the antenna temperature is not a physical temperature.

Using a telescope to observe a point source with flux density S_ν at frequency ν, we have the power output

$$P = S_\nu A_{eff}/2, \tag{7.1}$$

where the A_{eff} is the effective collecting area of the telescope. Divide by a factor of two because of polarization.

Now suppose that the telescope feed is replaced by a resistor whose temperature, T, is adjusted until the noise power received from the resistor equals that previously received from the point source:

$$P = kT\Delta\nu = S_\nu A_{eff}/2, \tag{7.2}$$

where the Boltzmann constant $k = 1.38 \times 10^{-23}\,\mathrm{J\,K^{-1}}$ and $\Delta\nu$ is the receiver bandwidth. If the temperature of the resistor is T_A, this is known as the antenna temperature due to the source.

1.2 System noise temperature

System noise temperature, T_{sys}, is the total system noise expressed as an antenna temperature. It is the temperature equivalent to the total noise power from all sources referenced to the input of a radio telescope.

T_{sys} is the sum of many contributors to the antenna temperature, including ground temperature T_{ground}, atmosphere temperature T_{atm}, and sky temperature T_{sky} plus the receiver noise temperature T_{rx}.

- T_{ground} is the ground radiation picked up by the feed due to spillover.
- T_{atm} is the contribution from the atmosphere, which will generally increase as frequency increases.
- T_{sky} is the contribution of the radio background such as cosmic microwave background (CMB), the galactic background emission, and other individual sources.

1.3 Telescope gain and effective area

In order to describe the capability of a telescope to collect radio signals, antenna efficiency, effective collecting area, and telescope gain are often used. The antenna efficiency η is defined as the ratio of radiated power to input power. Effective area (A_{eff}) is the equivalent area over which radiation would have to be fully collected to give the same telescope gain as the actual telescope. Telescope gain (G) is the telescope response to a point source. Radio astronomers usually express G as the antenna temperature, T_A, due to a point source of flux density 1 Jy lying at the peak of the telescope beam, and it has units of $\mathrm{K\,Jy^{-1}}$.

To observe a point source of flux density S_ν, the received power of the telescope is given by Eq. (7.2). We have $A_{eff} = \eta\, A_g$, where A_g is the physical collecting area of the telescope.

The telescope gain can be expressed as

$$G = \frac{T_A}{S_\nu} = \frac{A_{eff}}{2k}. \tag{7.3}$$

Another common definition of telescope gain is equivalent to the directivity of a telescope, D. It is the ratio of the radiated intensity at the beam center to the average radiated power. D is related to the effective telescope area, A_{eff}, by

$$D = \frac{4\pi A_{eff}}{\lambda^2} = \frac{8\pi kG}{\lambda^2} \tag{7.4}$$

1.4 System equivalent flux density

System equivalent flux density (SEFD) is the system temperature expressed as an equivalent point-source flux density. It has the units of Jy. It is not an independent quantity, but depends on the system temperature and telescope gain by

$$\mathrm{SEFD} = \frac{T_{sys}}{G}$$

It is to be noted that SEFD can be estimated rather accurately by measurements of a point source of known flux density without recourse to secondary standards such as noise-diode calibrations. The major source of error in the measured SEFD will usually be the uncertainty of the flux density of the point source.

2 Flux calibration

In this section, we describe the calibrating methods to the most common ON/OFF point source observation.

The antenna temperature of the source is given by

$$T_A = \frac{\mathrm{ON} - \mathrm{OFF}}{\mathrm{OFF}} T_{sys}, \tag{7.5}$$

where the ON and OFF stand for the measured values of the relative power level obtained from the receiver back-end. ON is the total signal when pointing the telescope directly toward the source, and OFF is the signal collected from an adjacent off-source position. The purpose of calibration is then to convert the measured values to flux density accurately.

2.1 Calibrating with SEFD

As shortly mentioned above, the SEFD can be obtained by observing the standard calibrator of known flux density, S_ν, then

$$\mathrm{SEFD} = S_\nu \times \frac{\mathrm{OFF}}{\mathrm{ON} - \mathrm{OFF}}, \tag{7.6}$$

Conversely, with the ON/OFF observation to a targeted source, we can derive its flux density from the measurement with the known SEFD of the observing system:

$$S_\nu = \mathrm{SEFD}\frac{\mathrm{ON} - \mathrm{OFF}}{\mathrm{OFF}}, \tag{7.7}$$

2.2 Calibrating with noise diode

Using a noise diode to inject a calibrating signal of known temperature, T_{cal}, during the ON/OFF observation, we can measure the CAL_ON and CAL_OFF and then derive the antenna temperature

$$T_A = T_{cal}\frac{\mathrm{CAL\_OFF}}{\mathrm{CAL\_ON} - \mathrm{CAL\_OFF}}, \tag{7.8}$$

Therefore,

$$S_\nu = \frac{T_{cal}}{G}\frac{\mathrm{CAL\_ON} - \mathrm{CAL\_OFF}}{\mathrm{CAL\_OFF}}\frac{\mathrm{OFF}}{\mathrm{ON} - \mathrm{OFF}}, \tag{7.9}$$

2.3 Gain curve

As for most radio telescopes, the gain is usually a monotonic function of the zenith angle (ZA) due to the gravity deformation of the primary reflector. T_{sys} also varies with ZA because the spillover contribution from the ground emission changes. In order to reach a more accurate flux density calibration, the system parameters such as G, T_{sys}, and SEFD should be modified with the curve of responses. In practice, gain curve can be obtained by observing a set of flux calibrators across the telescope's visible sky.

3 Processing spectral line data

3.1 Example of SIGGMA data

A survey of ionized gas in the galaxy, made with the Arecibo telescope (SIGGMA), uses the Arecibo L-band Feed Array (ALFA) to fully sample the galactic plane observable with the telescope in radio recombination lines (RRLs) [1,2].

RRL maps with 3.4′ resolution and a line flux density sensitivity of <1 mJy are obtained, which enable us to identify new HII regions, measure their electron temperatures, study the physics of photo dissociation regions (PDRs) with carbon RRLs, and investigate the origin of the extended low density medium (ELDM). Twelve $H_{n\alpha}$ lines fall within the 300 MHz bandpass of ALFA; they are resampled to a common velocity resolution to improve the signal-to-noise ratio (SN) by a factor of 3 or more and preserve the line width.

As an example, we present the observing and data reduction techniques in detail. SIGGMA uses the ALFA (see http://www.naic.edu/alfa for more information) receiver on the Arecibo telescope. The ALFA receiver has seven independent beams, each recording two orthogonal linear polarizations. The beams are arranged in a hexagonal pattern such that there is one central beam (Beam 0) surrounded by six outer beams (Beams 1–6). Due to the optics of the telescope, the outer beams are projected onto an ellipse in the sky. This ellipse is centered on Beam 0 and has semiaxes of 6.4′ in zenith angle by 5.5′ in azimuth. The orientation of the long axis of the ellipse changes with the parallactic angle. Although the array is derotated during observations, the positions of the outer beams change on the sky due to the elliptical projection. This change of outer beams with respect to Beam 0 is typically less than the FWHM of the beam width.

The average FWHM of the beams varies between 3.3 and 3.4′ across our frequency range (the source of the average FWHM values as well as of other parameters given in this paragraph is Arecibo Observatory's performance and calibration measures, which are made at 1420 MHz. The errors on the beam area introduced by using the average FWHM and a circular beam rather than the true elliptical beam are at around 1 part per thousand, and thus negligible compared to other sources of uncertainty and to the change in beam size over our frequency range).

The footprint of all the beams on the sky, out to the ellipse enclosing the FWHM of the outer beams, is (15.3–18.6′) × (13.6–16.5′), covering an area of 163–241 sq. arcmin. The area falling within the FWHM of the beams is 53–78 sq. arcmin., or around a third of the total footprint.

The gain of Beams 1–6 is ~8.5 K Jy^{-1} while Beam 0 has a gain of ~11 K Jy^{-1}. All seven beams have measured system temperatures of 27–28 K at low zenith angles (<15 degrees.) where most of our observations will occur, giving a system equivalent flux density of ~2.5 Jy for Beam 0 and ~3.2 Jy for Beams 1–6.

The receiver has a bandwidth of 300 MHz, covering 1225–1525 MHz.

Spectra are recorded once per second using the mock spectrometer, a Fourier-transform device that has two groups, each of 14 boards. This enables it to process data from the seven ALFA beams, each of which is divided into two IF subbands. The first group is used in a high time-resolution, low spectral-resolution mode to obtain data for the commensal "P-ALFA" pulsar survey [4] while the second group is used to acquire data simultaneously for the RRL survey. Each IF subband covers 172 MHz, with the first IF centered at 1450 MHz and the second IF centered at 1300 MHz. Together, these cover the entire 300 MHz ALFA bandpass, with the filter roll-off being either in the overlap region between the two subbands or outside the band of the receiver.

For the RRL survey, data are accumulated for 1 s before being Fourier transformed to form spectra of 8192 channels per IF subband. This gives a spectral resolution of 21 kHz, equivalent to 4.2–5.1 $\mathrm{km\,s^{-1}}$ for the $H_{n\alpha}$ recombination lines within the ALFA bandpass. The required integration time is then built up from multiple 1 s spectra.

The survey uses a leapfrog-style observing technique [5] with a grid of points that repeats along lines of constant declination (see Fig. 7.1). Each observation integrates on each position for 270 s (180 s in the outer galaxy). Combined with the slew time, this gives a spacing of nearly 5 min between observations. By choosing pointings from the grid with the same declinations and separated by 5 min in right ascension, we track the same azimuth and zenith angle in consecutive points. This allows consecutive points to be used as ON-OFF pairs in order to form a bandpass-corrected (ON-OFF)/OFF spectrum.

The 5-min separation in time between ON and OFF points provides the best baselines because ripples caused by internal reflections in the antenna, ground pick up, and atmospheric effects mostly cancel out in the corrected spectrum. The survey tiling pattern is made up of clusters of three pointings that are offset from each other by one beam FWHM beamwidth, labeled as the north (N), central (C), and south (S) pointings in Fig. 7.1. This is a modification of the original P-ALFA pulsar survey tiling pattern [3], with the axis of the pattern (and of the ALFA receiver) rotated by 19 degrees to celestial north so that adjacent clusters of pointings fall on lines of constant declination.

By leapfrogging through this tiling pattern, the survey covers approximately 10 sq. deg. of sky in 100 h of telescope time. In the outer galaxy, where SIGGMA is commensal with the ALFA Zone of Avoidance survey (ALFA ZOA), integration times per pointing are shorter but the pattern is observed three times with

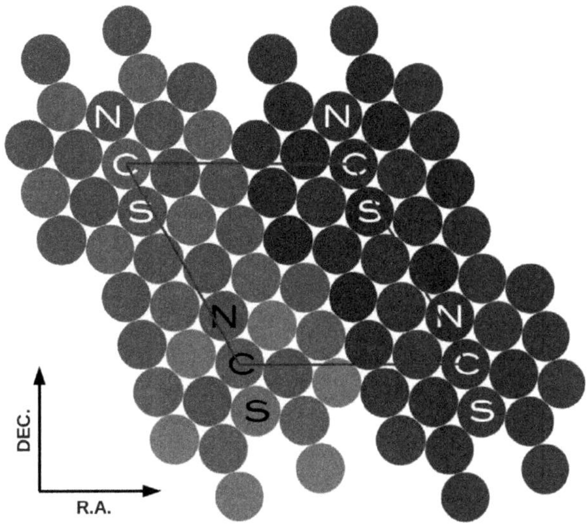

Fig. 7.1 The grid of points used for the RRL survey, with RA horizontal and Dec vertical. *Circles* indicate the size (FWHM) of the ALFA beams. Circles with the same fill color are observed in the same ALFA pointing; circles with similar fill colors are part of the same pointing cluster, with N, C, and S indicating the north, central, and south pointing of the cluster. The *red* (dark gray color in print version) rhomboid indicates the repeating pattern of the pointing clusters, illustrating that they repeat along lines of constant declination. Developed from Fig. 5 of Freire, P.C.C. Optimal Tiling for Non-Drifting ALFA Surveys. ALFA Tech Memo Series. Arecibo Observatory, Arecibo, 2003, http://www.naic.edu/alfa/memos/.

a small offset ($\sim$1.9′) in order to make a Nyquist-sampled map of 10 sq. deg. of sky to a similar depth in 200 h of telescope time.

The entire 300 MHz bandpass of the RRL survey observations is divided into two IF subbands. Examples of the spectra of the two subbands are shown in Figs. 7.2 and 7.3. The two orthogonal polarizations are averaged for each subband pass, thereby achieving a $\sim$40% increase in SN but rendering the data insensitive to polarization.

After matching the ON/OFF pointing pairs, we perform bandpass correction to the raw ON/OFF spectral pairs via the customary position switched method of (ON-OFF)/OFF × T_{sys}. Bandpass-corrected spectra are shown in Figs. 7.4 and 7.5. It is clear that the RFI in the higher frequency IF subband is considerably less serious than in the case of the lower IF band.

We separate each 300 MHz spectrum into 12 segments, each of which covers $\pm$300 km s^{-1} centered between the rest frequencies of the $H_{n\alpha}$ and $He_{n\alpha}$ lines such that the $H_{n\alpha}$ and $He_{n\alpha}$ lines are offset by +61 km s^{-1} and $-$61 km s^{-1}, respectively, from the band center, and the $H_{n\alpha}$ and $C_{n\alpha}$ line is offset $-$88 km s^{-1} from the band center. The velocity range, with respect to the local standard of

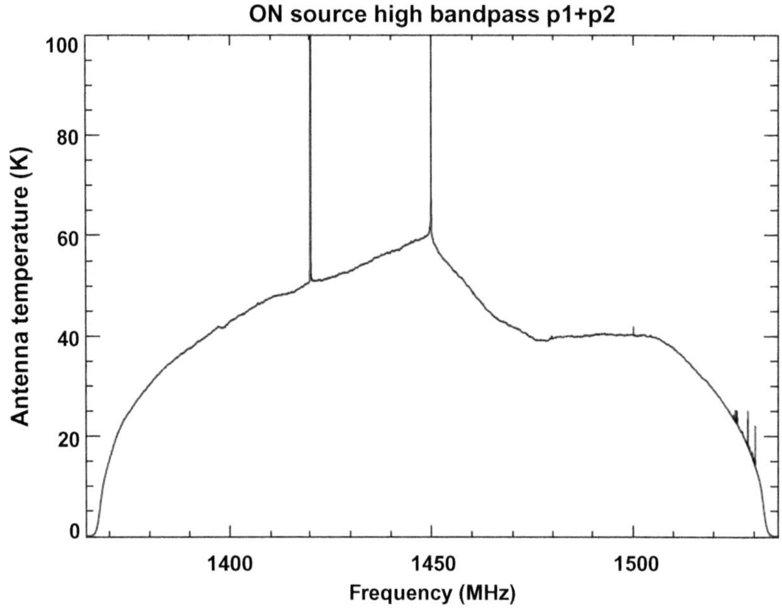

Fig. 7.2 The spectral shape of the raw data for the higher IF subband when the antenna beam is on the source S255. This spectrum is produced by averaging more than 270 s exposure. The orthogonal polarizations have also been averaged. The antenna temperature includes about a 12$\sim$K continuum flux from the source and the spikes are due to RFI.

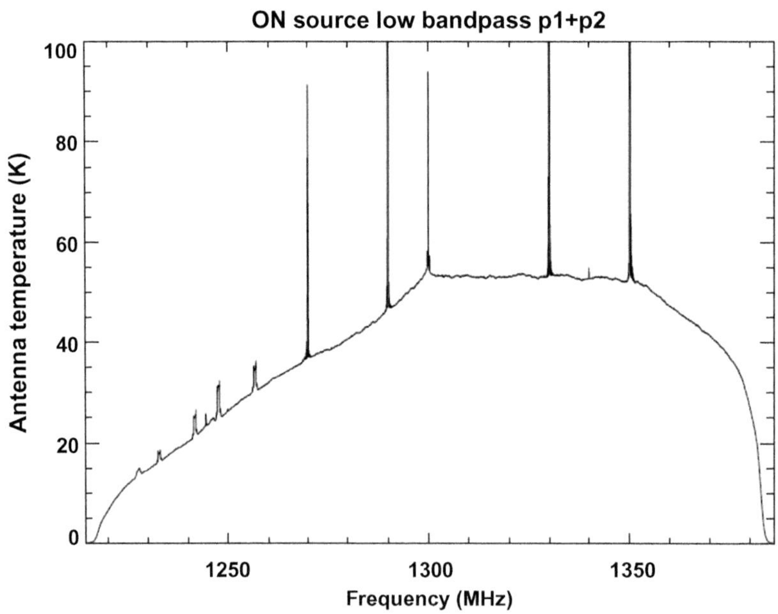

Fig. 7.3 Same as Fig. 7.2 but for the lower IF subband.

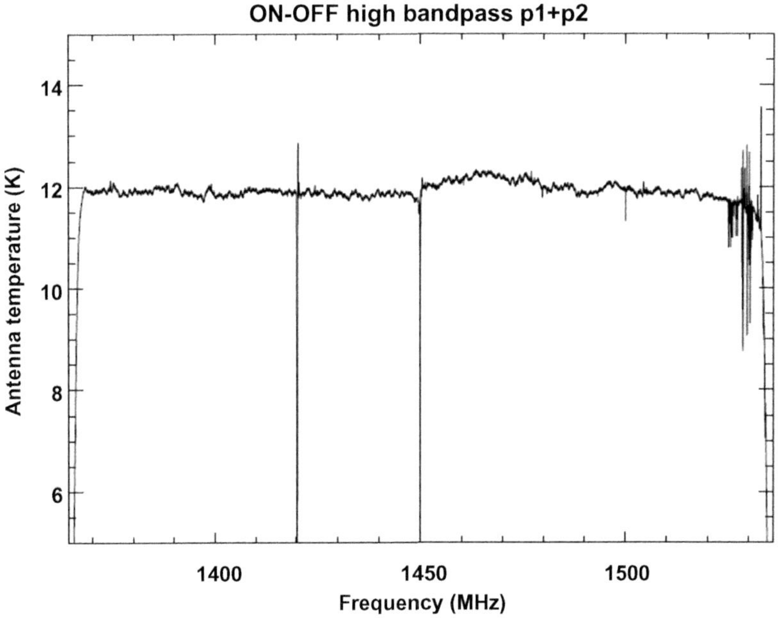

Fig. 7.4 The spectrum after (ON-OFF)/OFF bandpass correction: the higher IF band. The antenna temperature includes about a 12 K continuum flux from the source and the spikes are due to RFI.

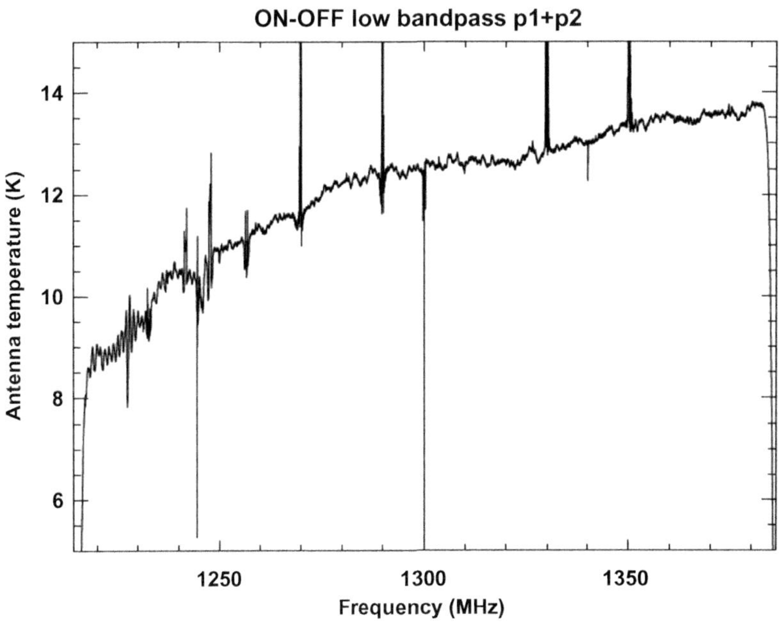

Fig. 7.5 Same as Fig. 7.4 but for the lower IF subband.

rest (LSR), spans the entire velocity of galactic gas toward each pointing.

After performing a third-order polynomial fitting individually, we resample the 12 segments to a common velocity grid of $5.1\,\mathrm{km\,s^{-1}}$ resolution and then average all segments to achieve a higher signal-to-noise ratio. The stacked spectra frequently have irregular baseline ripples, especially when there is a strong continuum emission within the beam. We fit a fifth-order polynomial baseline to all stacked spectra after masking data from velocities between -50 and $+150\,\mathrm{km\,s^{-1}}$. As an example, we show the final $H_{n\alpha}$ spectrum toward the center of the W49A in Fig. 7.6.

There are about 33,000 raw stacked spectra in total. Fig. 7.7 shows the rms noise distribution of the stacked spectra, which peaks at about $0.65\,\mathrm{mJy\,beam^{-1}}$ ($\sim$6.5 mK, obtained by applying an average gain of $10\,\mathrm{K\,Jy^{-1}}$). This agrees well with the expected rms of $\sim 0.5\,\mathrm{mJy\,beam^{-1}}$. The extension toward higher values in the rms distribution is caused by positions with bright continuum emissions.

Fig. 7.8 gives the rms distribution as a function of galactic longitude and latitude. Spectra within $|b| < 0.5$ degree are more likely to have high rms due to the higher density of continuum sources. We see this same pattern in the plot of rms versus galactic longitude, as the higher rms values correspond to locations of large massive star formation complexes.

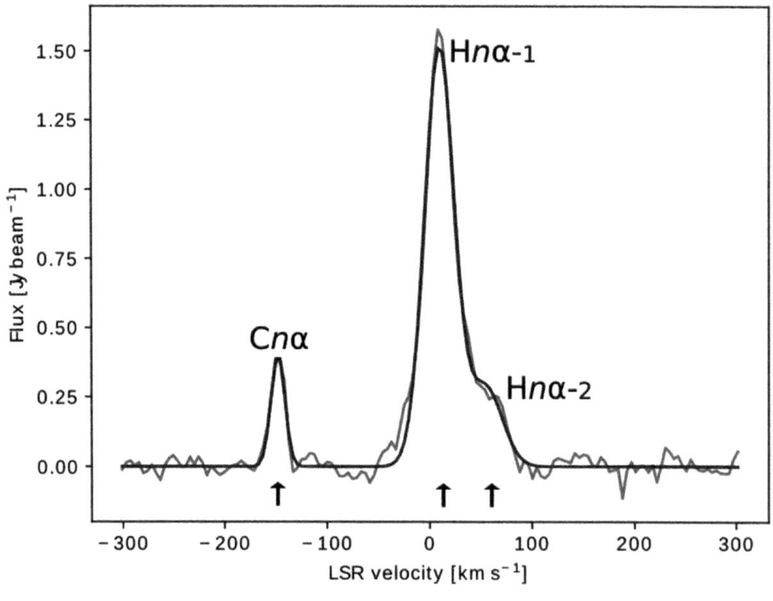

Fig. 7.6 The stacked $H_{n\alpha}$ spectrum from the peak W49A region, with Gaussian fits. We label the two velocity components of $H_{n\alpha}$ by $H_{n\alpha}$-1 and $H_{n\alpha}$-2.

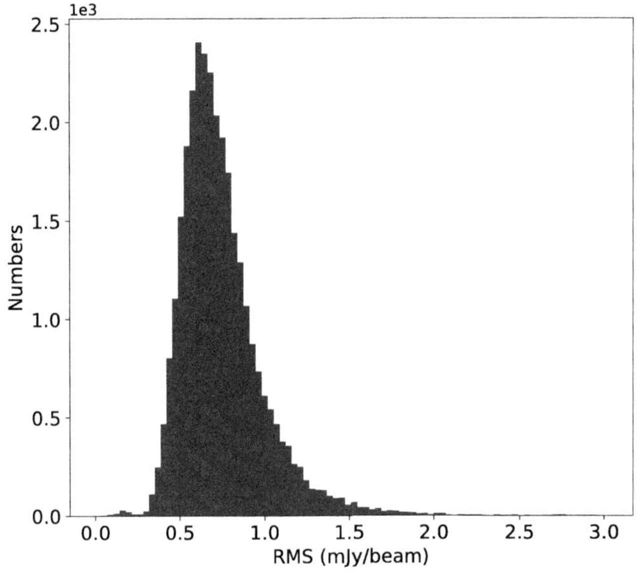

Fig. 7.7 The rms distribution of the final spectra, calculated from the RFI- and line-free channels of spectra (−300 to −200 km s^{-1} and +200 to +300 km s^{-1}).

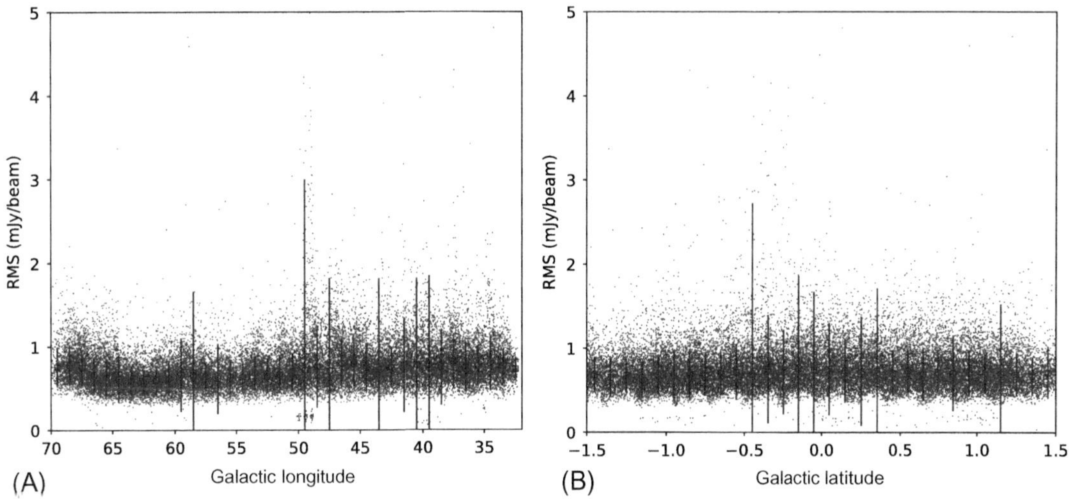

Fig. 7.8 The rms distribution versus galactic longitude (A) and latitude (B). The *open red* box (dark gray in print version) is the median of the 0.1 degree binned rms values and the *red* bars (dark gray in print version) are their standard deviations.

After stacking, we use the software package Gridzilla [6] to grid the stacked spectra into data cubes. Each cube is 5×3 degrees with a pixel size of 1×1 arcmin, and has 5.1 km s^{-1} channel spacing covering −300 to +300 km s^{-1}. The beam weighting is set to be proportional with the beam FWHM of 3.4 arcmin. We apply a

Gaussian smoothing with the kernel FWHM of 6 arcmin and ~2 beam widths to fill in gaps where we lack usable data. The spectrum of each pixel is then generated from the weighted median of the stacked spectra within a cut-off radius of 4 arcmin. These cubes of stacked RRL emissions are the primary data product of SIGGMA.

RFI is prevalent in the SIGGMA data, and RFI removal is the first step of our data reduction process. We first excise strong and broad RFI from the whole bandpass by removing spectral values more than five times the spectral rms about the median value, which is obtained using a median filter of width 300 channels. The spectral rms is calculated from RFI free channels in the velocity range of -300 to $-200\,\mathrm{km\,s^{-1}}$ and $+200$ to $+300\,\mathrm{km\,s^{-1}}$.

During line stacking, we apply a second median filter to the 12 subbands. This is to remove any weak and narrow RFI that remains. The filter parameters are set to 15 channels in width and 10 times of the spectral rms criterion. The RRL signal remains unaffected with this configuration.

Some spectra show strong baseline ripples or abnormal noise not identified by the median filter method. We set a rms threshold to remove such spectra from further analyses. Spectra with strong continuum sources also have high rms noise. So as to not exclude such spectra, which also frequently have true line emission, we examine the difference of the rms from the central portion of the spectra (from -200 to $+200\,\mathrm{km\,s^{-1}}$) and to that of the edges.

3.2 Example of FAST data

Here, we also give a simple example of spectral line calibration with data obtained by the 500-m Aperture Spherical Radio Telescope (FAST). Fig. 7.9 shows the calibrated bandpass with FAST L-band data. The observation is performed in position switch mode with the noise diode fired on and off every second. Thus, we applied the calibrating method given in Section 2.2.

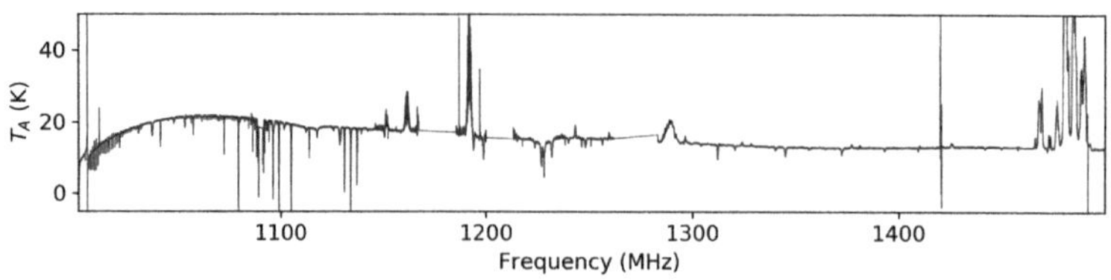

Fig. 7.9 The calibrated bandpass with FAST L-band data. The observation is performed in position switch mode with the noise diode fired on and off every second.

4 Observations of a brown dwarf by Arecibo single dish

In this section, we present radio observations of a brown dwarf (also known as an ultracool dwarf) TVLM 513 in time and frequency. The observations were conducted using the single dish at the Arecibo Observatory, Puerto Rico. Some background information and an outline of previous observations are summarized in Section 4.1. Information on the single dish, introduction to the observation, and data reduction are briefly shown in Section 4.2. The results are provided in Section 4.3. The observation data in this section were taken by G. Hallinan, A. Antonova, and G. Doyle. References in this section can be found in [7].

4.1 Basic information and previous observations of TVLM 513

TVLM 513 (2MASS J15010818+2250020) may be a young, active M8.5V dwarf with a bolometric magnitude of $\log(L_{bol}/L) = -3.65$, effective temperature $T_{eff} = 2200\,K$, and situated at a distance of $d = 10.6$ pc. From the theory of the formation and evolution of ultracool dwarfs and the absence of lithium in TVLM 513, it is reasonable to infer values for the mass and radius of this star of ~0.07 Msun and ~0.1 Rsun respectively. In addition, a high-resolution spectrum observed by the Keck Observatory indicates that TVLM 513 has a relatively high rotational velocity, 60 km s^{-1}. The Hα emission of this dwarf is quite weak, with an equivalent width measurement of 1.7–3.5 Å. With the VLA, a highly right-circularly polarized (~65%) radio event from TVLM 513 was detected with a flux density of ~1100 μJy as well as a persistent variable emission at 8.46 GHz. A multifrequency VLA observation of TVLM 513 was conducted at 8.4, 4.8, and 1.4 GHz in 2006, using a strategy that involved time-sharing a single 10 h observation between the various frequency bands. TVLM 513 was detected at each frequency band with only marginal confirmation of variability and no detection of flares or strong circular polarization. VLA observed persistent and periodic radio emissions from TVLM 513 at 8.44 GHz and 4.88 GHz simultaneously, with a period of ~2 h. Subsequently, extremely regular periodic bursts ($p = 1.96$ h, up to ~4 mJy) of high brightness and highly circularly polarized radio emissions were reported by Hallinan et al. in 2007. Multiple bursts of both left and right 100% circularly polarized emissions were detected. Interestingly, the radio emission can switch states from nearly 100% left polarization to 100% right polarization in each phase. Another radio burst with a flux density up to ~4 mJy was presented by Berger et al. in 2008 from a period of simultaneous radio, x-ray, ultraviolet, and optical spectroscopic observations. Steady quiescent radio emissions superposed with multiple, short-duration, highly polarized bursts were observed, but these authors reported a nonperiodicity in the pulses/flaring activity. Only marginal x-ray emission was detected. In a reanalysis of this data, plus data taken ~40 days later (June 2007), Doyle et al. in 2010 reported the 1.96 h periodicity in both datasets, deriving a more accurate period.

The recent observations performed by the Very Large Array in June 2007 show no trace of radio emission from TVLM 513 at 325 MHz, placing an upper flux limit of 795 μJy at 2.5σ level by Jaeger et al. in 2011.

4.2 Instrument, observation, and data reduction

The single dish with a diameter of 306 m at the Arecibo Observatory was used to carry out the observations for the active M8.5 dwarf TVLM 513–46546. A dynamic spectra of the radio waves were observed by the wide-band Arecibo pulsar processor (WAPP) back-end on the Arecibo dish spanning the frequency range 3050–9250 MHz. More information about the instrument can be found at http://www.naic.edu/.

The noise level of the observations, or the expected root mean square (rms) noiseσ, can be calculated via

$$\sigma = \frac{T_{sys}}{G\sqrt{2\beta\tau}}\left(\frac{\omega}{P-\omega}\right)^{0.5} \text{Jy}, \tag{7.10}$$

where T_{sys} is the system temperature, including the contributions from the celestial background and any continuum emission from a host supernova remnant, etc., if applicable; G is the telescope gain (in unit of K Jy^{-1}); β is the system bandwidth; τ is the total integration time of the observation; P is the pulsar period; and ω is the effective duration of the pulse. Please see the Arecibo single-dish guide for details.

The observations of TVLM 513–46546 were carried out on nine successive nights in June 2009 and three successive nights in May 2008. Dynamic spectra of the periodic pulses on nearly every night were obtained, with a high degree of variability in the brightness of individual pulses over the observation. The WAPP spectrometer was set up with 100 MHz bw/band, three-level sampling, 2048 channels, and full stokes. The data were recorded in 1 s intervals. The RMS noise level per time bin was ~0.7 mJy. For the observations on each day in May 2008, 67 s of calibrations were done after every 600 s of integration time on the source. This observation process was repeated 13 times each day. The observations in June 2009 consisted of 67 s of calibration after every 900 s of integration time on the source, repeated nine times each day.

Some standard subroutines in IDL (a software called Interactive Data Language) were used for the data reduction. First, the data files were read and converted from cross correlations to uncorrected stokes. Then an RFI (radio frequency interference) mask was created for the calibration, which removed the bandpass. The subroutine 0corblauto0 was used to exclude channels that were more than 3σ from the mean. Subsequently, the subroutine "corstokes" was used to scale to temperature (Kelvin) with the calibrations and then phase correct the signal path using the calibration data. The data were then converted from temperature (Kelvins) to flux (Janskys) using the gain curve for the C band (4.7 GHz). The total power was then averaged from the array of 2048 channels over 600 s. The subroutine "corrms" was called to make a mask where there was no RFI. Most of the information in this section is from the observation guidance and instructions made by Phil Perillat at the Arecibo Observatory in May 2008. In all observation results (figures) below, we use upward and downward directions for the right and left circular polarizations, respectively. We neglect all bad data in the results.

4.3 Results

The observations provide us several important physical variables and their combination, such as time, time interval or relative time, frequency, frequency bandwidth, relatively spatial position, and energy flux density. We usually record the data of other physical variables as a function of time. Frequency usually indicates the different oscillation modes of a system in a time period, which is strongly associated with the physical process between one object and another object. So we use frequency as another independent variable to describe the evolution of a system. In this section, we give the observed results of some important physical variables. Instead of being concerned about the quiescent radio emission, we only concentrate on the radio pulses.

4.3.1 May 2008

Fig. 7.10 shows the observed flux density as a function of time and frequency on May 18, 19, and 20, 2008 (UTC). The observation on May 18 is shown in the bottom panel, which consists of two obvious pulses with a left circularly polarized wave followed

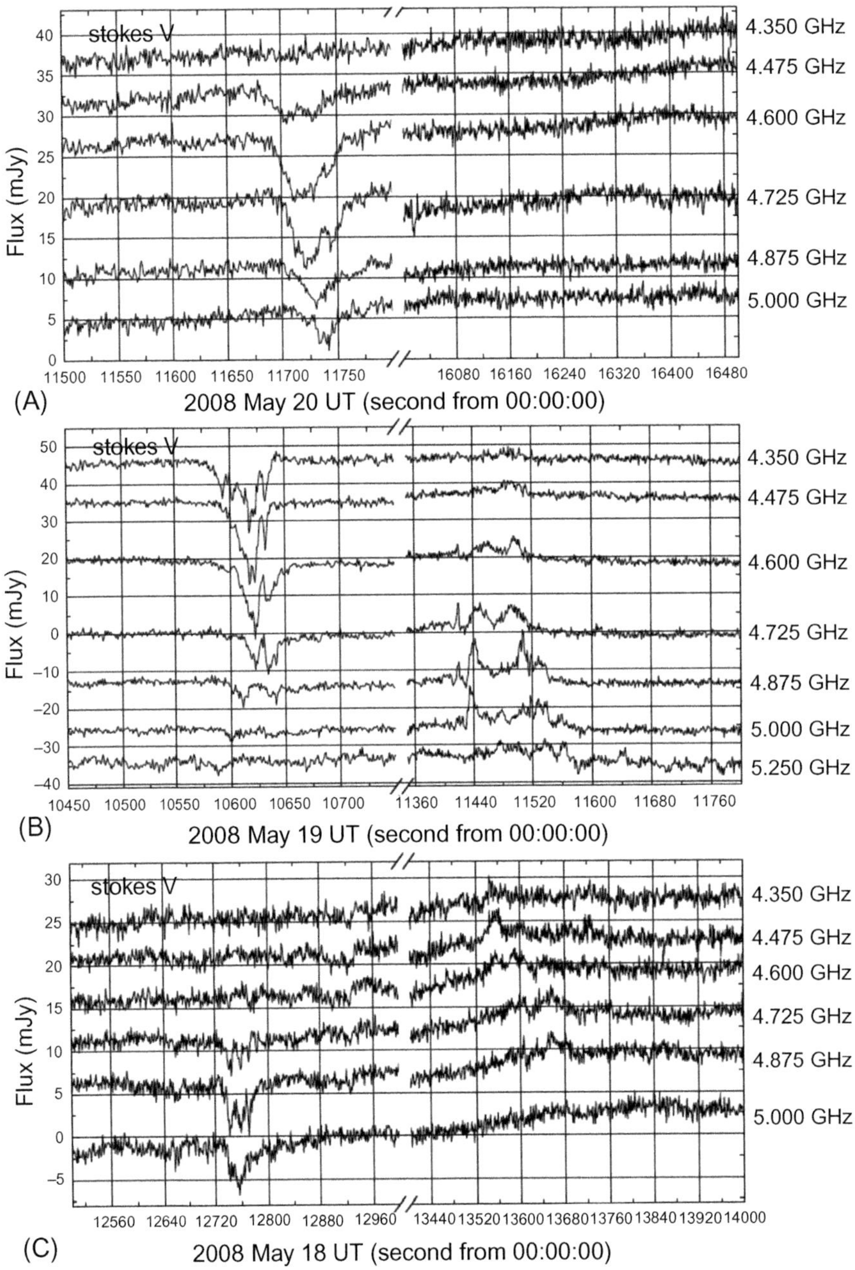

Fig. 7.10 The observation for TVLM 513-46546 from the Arecibo Observatory in May 2008. In each panel, the curves at different frequencies are shifted for some value in the flux direction, and the frequencies for each curve are shown on the right sides of the panels. *Upward* and *downward* directions indicate the right and left polarization. For conveniently comparing the peak value of the observations with the x-y scale, we also plot *blue* grids (dark gray in print version) in this figure.

by a right circularly polarized wave. The observation on May 19 is similar to that on May 18, consisting of two obvious pulses with an opposite circular polarization direction. However, only one left circularly polarized pulse was observed on May 20. Also, we can see the peak values of the pulses from the panels in the figure. A large degree of variability of the flux density is noticed.

4.3.2 June 2009

A significant data corruption occurred on June 8 and 10, 2009, so the observation results in 2009 include June 9 and June 11–16 (UTC). The flux density in all the observations includes a large degree of variability due to poor weather, but some obvious pulses can be seen. There are two pulses on June 9 (Figs. 7.11 and 7.12) with the same polarization direction. As shown in Figs. 7.13–7.15, three pulses appeared on June 11. Again, we see the opposite polarization from the latter two pulses. These calibrated data can provide us many information for the physical properties of plasma in the radio emission region of brown dwarfs.

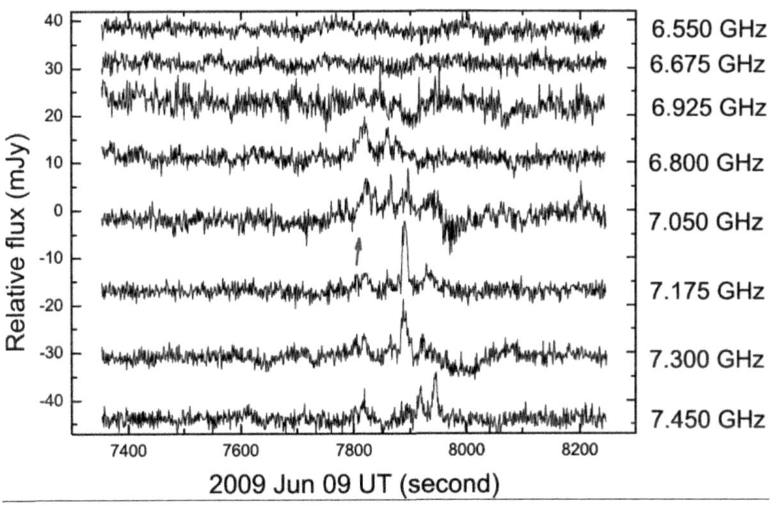

Fig. 7.11 Same as Fig. 7.10 but for the observation on 09 June 2009. *Red arrow* (dark gray in print version) denotes the approximate position of the pulse.

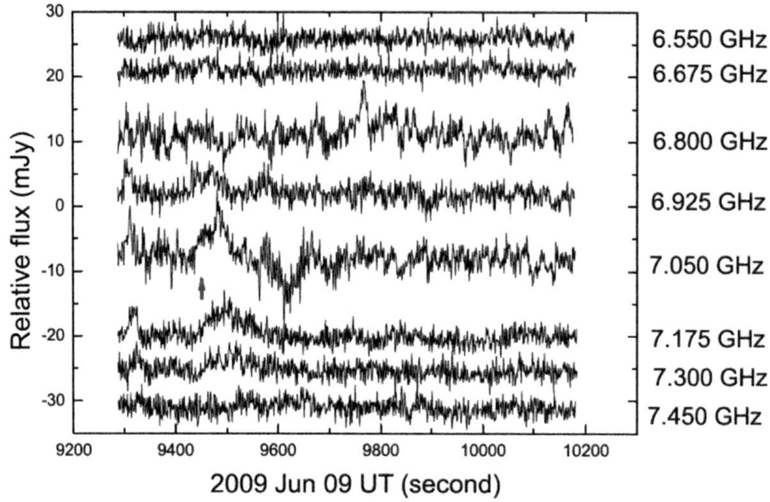

Fig. 7.12 Same as Fig. 7.11 but for different observation time (see x-scale). *Red arrow* (dark gray in print version) denotes the approximate position of the pulse.

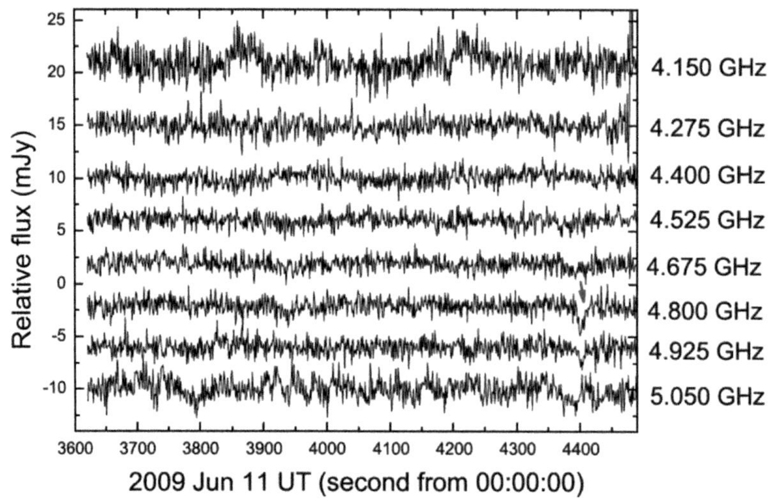

Fig. 7.13 Same as Fig. 7.11 but for different observation time (see x-scale). *Red arrow* (dark gray in print version) denotes the approximate position of the pulse.

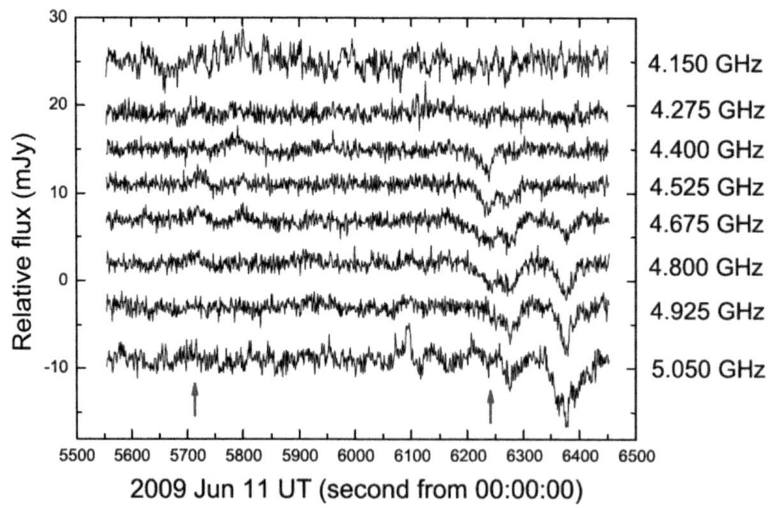

Fig. 7.14 Same as Fig. 7.13 but for different observation time (see x-scale). *Red arrow* (dark gray in print version) denotes the approximate position of the pulse.

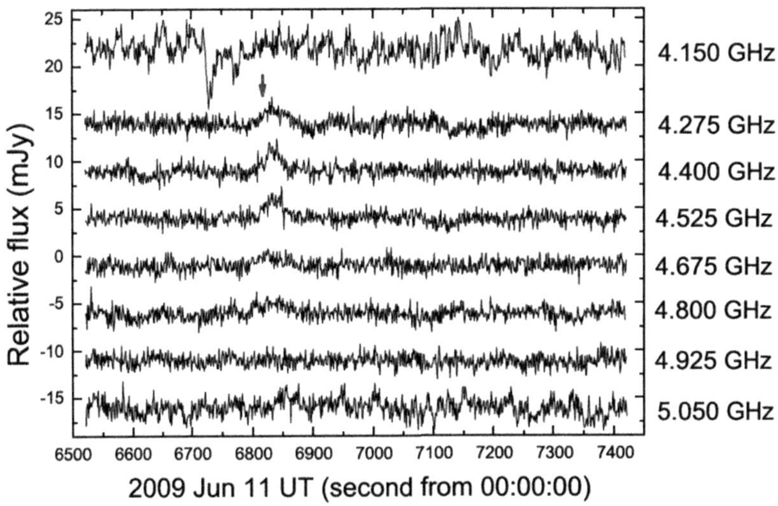

Fig. 7.15 Same as Fig. 7.13 but for different observation time (see x-scale). *Red arrow* (dark gray in print version) denotes the approximate position of the pulse.

References

[1] B. Liu, T. McIntyre, Y. Terzian, R. Minchin, L.D. Anderson, E. Churchwell, M. Lebron, D.A. Roshi, SIGGMA: a survey of ionized gas in the galaxy, made with the Arecibo telescope, Astron. J. 146 (4) (2013) 80, https://doi.org/10.1088/0004-6256/146/4/80.

[2] B. Liu, L.D. Anderson, T. McIntyre, D. Anish Roshi, E. Churchwell, R. Minchin, Y. Terzian, Survey of ionized gas of the galaxy, made with the Arecibo telescope (SIGGMA): inner galaxy data release, Astrophys. J. Suppl. Ser. 240 (1) (2019) 14, https://doi.org/10.3847/1538-4365/aaef8e.

[3] P.C.C. Freire, Optimal Tiling for Non-Drifting ALFA Surveys, ALFA Tech Memo Series, Arecibo Observatory, Arecibo, 2003. http://www.naic.edu/alfa/memos/.

[4] J.M. Cordes, P.C.C. Freire, D.R. Lorimer, F. Camilo, D.J. Champion, D.J. Nice, R. Ramachandran, et al., Astrophys. J. 637 (1) (2006) 446–455.

[5] J.G. Spitzak, S.E. Schneider, Astrophys. J. Suppl. Ser. 119 (1998) 159–180.

[6] D.G. Barnes, L. Staveley-Smith, W.J.G. de Blok, et al., Mon. Not. R. Astron. Soc. 322 (2001) 486.

[7] S. Yu, G. Hallinan, J.G. Doyle, et al., Astron. Astrophys. 525 (2011) 39.

8

Imaging algorithm optimization for scale-out processing

Haoyang Ye[a], Peter Hague[a], Stephen F. Gull[a], Sze Meng Tan[b], Bojan Nikolic[a]

[a]*Astrophysics Group, Cavendish Lab, Cambridge University, Cambridge, United Kingdom.* [b]*Picarro Inc., Santa Clara, CA, United States*

The previous chapter investigated popular data calibration techniques for astronomical big data. In the present chapter, we focus on the imaging process in radio interferometry. A radio interferometer is an array of multiple antennas at separated locations. It can achieve the same resolution as a single dish telescope with a diameter equal to the longest distance between the antennas.

1 Imaging process

1.1 Make a dirty image

In radio interferometry, the fundamental relationship between the observed visibility data V and the sky brightness distribution I, or intensity distribution, can be written in the following form [1]:

$$V(u, v, w) = \iint \frac{\mathrm{d}l\mathrm{d}m}{\sqrt{1-l^2-m^2}} I(l, m) \exp\left[-i2\pi\left(ul + vm + w\left(\sqrt{1-l^2-m^2}-1\right)\right)\right] \tag{8.1}$$

where (l, m) are direction cosines between -1 and 1. Each (u,v,w) coordinate is in units of wavelength and corresponds to a baseline, which connects two antennas. The visibility data are assumed to have been already edited and calibrated.

If the field of observation is very small, and is close to the phase center $(l, m) = (0, 0)$, then the so-called w-term in Eq. (8.1), which involves the coordinate w can be neglected. The visibility function can then be written as

Big Data in Astronomy. https://doi.org/10.1016/B978-0-12-819084-5.00005-5

$$V(u,v) = \iint \mathrm{d}l\mathrm{d}m I(l,m)\exp[-i2\pi(ul+vm)] \tag{8.2}$$

The intensity distribution $I(l, m)$ is then simply the inverse two-dimensional Fourier transform of the visibility function $V(u,v)$

$$I(l,m) = \iint \mathrm{d}u\mathrm{d}v V(u,v)\exp[i2\pi(ul+um)] \tag{8.3}$$

However, this can be evaluated only when the visibility function is known at all baseline spacings. In practice, the number of antennas of an interferometer is very limited, making the reconstruction of $I(l,m)$ using the inverse Fourier transform infeasible. The antennas, often located at a relatively small number of fixed locations on the surface of the earth, form a limited set of baselines, which are unable to fully sample the whole (u,v) plane even allowing for the rotation of the earth during the observation. Fig. 8.1 shows the relative locations of 27 antennas in the very large array (VLA) [2, 3], which is a Y-shaped radio interferometer located in New Mexico, United States. Each *red dot* (*gray* in print version) represents one antenna with its number written next to it [4].

The collection of all sampled baselines on the (u,v) plane is called the "*uv*-coverage." Fig. 8.2 shows an example plot of the *uv*-coverage of the VLA A-array produced by the common astronomy software applications package (CASA [5]). This is for a 5-h

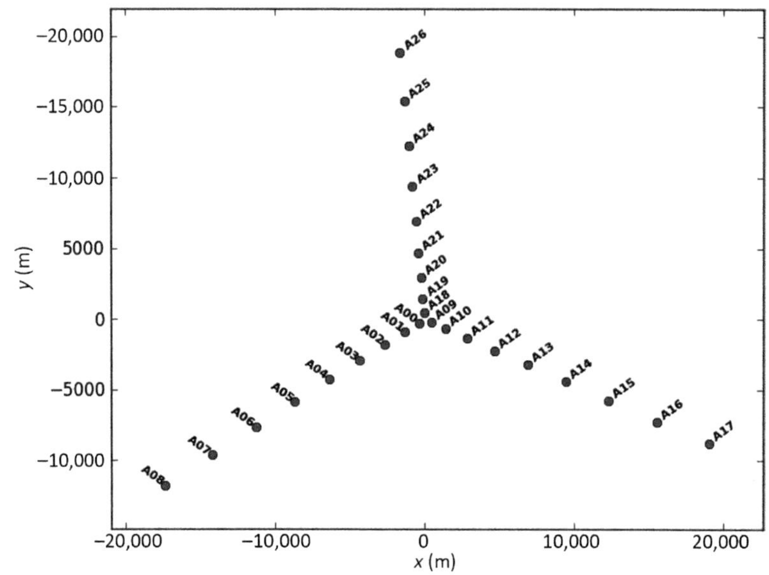

Fig. 8.1 The relative locations of 27 antennas of the VLA. Each *red dot* (*gray* in print version) represent an antenna with its number written next to it. The *x*- and *y*-axes are centered at the intersection of the three arms of the Y-shaped array.

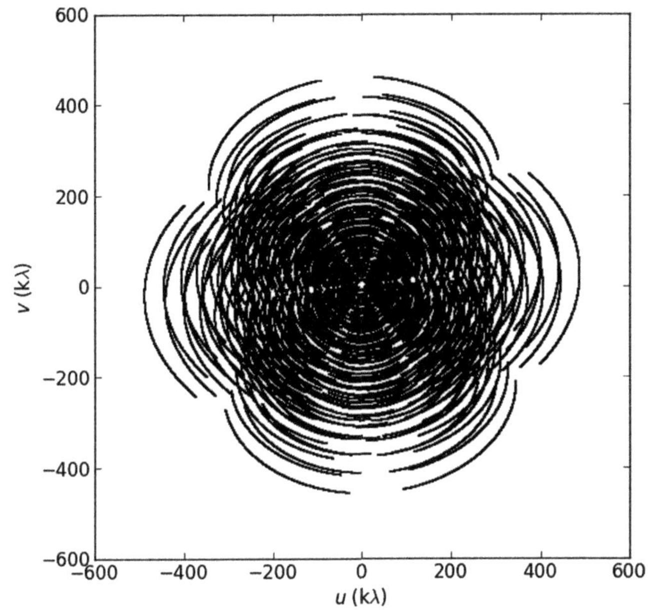

Fig. 8.2 *uv*–coverage of a VLA A-array 5-h observation at a declination of 45 degrees. λ represents the observation wavelength, so *u* and *v* axes are in units of 1000λ, written as kλ.

simulated observation at a declination of 45 degrees [4]. With time, each baseline moves along with the rotation of the earth, tracing out two arcs in the (u,v) plane. However, as shown in Fig. 8.2, there still exist many gaps in the *uv*-coverage, making it impossible to recover the sky intensity distribution directly from the measured visibility function via a two-dimensional inverse Fourier transform [4].

We next consider the effects of taking the two-dimensional inverse Fourier transform of the available visibilities, rather than of the entire set. A sampling function $S(u,v)$ is introduced to represent the sampling of the baselines on the (u,v) plane.

$$S(u,v) = \sum_{k=1}^{M} \delta(u-u_k)\delta(v-v_k) \tag{8.4}$$

The values (u_k, v_k) denote the kth baseline at which the visibility is measured, and M is the number of baselines. The product of $S(u,v)$ and $V(u,v)$ has a Dirac delta function at the location of each measurement and represents the sampled visibility function.

If we apply the two-dimensional inverse Fourier transform to the sampled visibility data $S(u,v)V(u,v)$, we obtain a quantity known as the "dirty map" or "dirty image" defined by

$$I_D(l,m) = \iint \mathrm{d}u\mathrm{d}v\, V(u,v)S(u,v)\exp\left[i2\pi(ul+vm)\right] \tag{8.5}$$

In this chapter, the words "image" and "map" are interchangeable unless specified. Since the Fourier transform is invertible, the information content in the dirty map and the measured visibilities are identical, provided that I_D is calculated for all values of l and m.

1.2 The relationship between the dirty image and the sky intensity distribution

According to the convolution theorem, the inverse two-dimensional Fourier transform of the product of the sampling function and visibility function is the two-dimensional convolution of the inverse Fourier transforms of the individual functions. As stated in Eq. (8.3), the inverse Fourier transform of the full visibility function is the desired sky intensity distribution, whereas the inverse Fourier transform of the sampling function is referred to as the "synthesized beam," or the "dirty beam," denoted by $B(l,m)$. In this way, Eq. (8.5) can then be written as

$$I_D(l,m) = I(l,m) \times B(l,m) \tag{8.6}$$

The dirty image is thus the two-dimensional convolution between the true sky intensity image and the dirty beam. The process of trying to recover the true sky intensity distribution from the dirty image according to Eq. (8.6) is called "deconvolution," and will be briefly introduced here.

The deconvolution problem arises from the fact that the full visibility function is sampled only at a limited set of baselines. In theory, the visibility values at the unsampled baselines can have any arbitrary value, leading to an infinite number of possible solutions for $I(l,m)$. Existing deconvolution algorithms, such as CLEAN and the maximum entropy method (MEM), select from among the infinite set of possibilities by favoring images, which conform to some more or less well-defined criterion for image generation.

The CLEAN algorithm was proposed by Högbom [6], and was followed by many variants, such as Clark CLEAN [7], Cotton-Schwab CLEAN [8], and Multiscale CLEAN [9]. CLEAN is based on the assumption that the sky intensity may be well represented as a sum of a number of point sources, so that even extended sources are approximated by a collection of point sources. It expresses the dirty image as the convolution of the dirty beam and a list of point sources with various intensities at various locations (called the "CLEAN model"), together with random noise. This model is built up one source at a time until the noise term is comparable with the expected measurement noise, thus favoring reconstructions consisting of fewer sources.

The Högbom CLEAN algorithm can be summarized as follows:

(1) At the start of the algorithm, initialize the "residual" map to be equal to the dirty map and set the CLEAN model to be an array of zeros.

(2) Find the "peak pixel" with the highest value in the residual and use this to update the intensity and location of the corresponding pixel in the CLEAN model. The model pixel is incremented by an amount equal to the product of the maximum pixel value in the residual and a factor called the "loop gain," which is usually chosen to be a few tenths.

(3) The residual map is updated by subtracting a scaled version of the dirty beam shifted to the location of the peak pixel, the scaling factor being equal to the amount by which the model pixel was incremented in the previous step.

(4) Go back to step 2 and repeat the procedure, continuing until the maximum value of the residual is below a chosen threshold, or the number of iterations exceeds a chosen maximum. The final residual should then consist mainly of noise.

(5) Convolve the final CLEAN model with a "CLEAN restoring beam" (usually a Gaussian), and then add the final residual to obtain the so-called CLEANed image.

By construction, the dirty map is equal to the convolution of the CLEAN model and the dirty beam to within the noise. This means that the CLEAN model (like the true sky brightness) is a solution of the deconvolution equation (Eq. 8.6). When we take the Fourier transform of the CLEAN model (which consists of a collection of point sources), the result is a sum of complex exponentials extending over the visibility plane. This agrees with the visibilities at the measured baselines (to within the noise), and also fills the gaps in the (u,v) coverage.

When the sky contains extended sources, the true sky intensity will not be well represented by the CLEAN model, which consists of a sum of point sources. The CLEAN restoring beam introduced in step 5 of the algorithm attempts to smooth out the discrete nature of the CLEAN model to better accommodate such sources. However, by doing so, the CLEANed image is no longer a solution of the deconvolution equation, since the effect of the convolution with the restoring beam is to apply a taper in the (u,v) plane, causing the Fourier transform of the CLEANed image not to agree with the measured visibilities.

Apart from the original Högbom algorithm, CLEAN variants divide the process of updating the residuals into two phases, called major and minor cycles, largely for computational efficiency. Repeatedly subtracting shifted and scaled versions of the dirty beam from the residual can become expensive, so the

Clark CLEAN algorithm uses a patch consisting of the central portion of the dirty beam instead to update the residuals during the minor cycles. The scaling factor applied to the patch is the same as that used in the Högbom CLEAN algorithm. After a number of these minor cycles, the cumulative effect of approximating the beam becomes significant and the residuals are no longer accurate. During a major cycle, the visibilities associated with the current CLEAN model are calculated at the baseline locations (using a direct Fourier transform) and subtracted from the measured visibilities. This gives an accurate residual in the visibility plane, which can be made into a new set of image plane residuals via an inverse Fourier transform for use in a subsequent series of minor cycles. The process is repeated until the termination condition is met.

The Cotton-Schwab CLEAN algorithm is one of the most commonly used variants. It differs from the Clark algorithm mainly in the way that the visibilities of the current CLEAN model are computed at the baseline locations during a major cycle. Instead of using a slow direct Fourier transform, a combination of the fast Fourier transform (FFT) and degridding is used for faster computational speed. The original dirty map and the image domain residuals are generated using a combination of gridding and the FFT algorithm. These two processes will be discussed in detail in Section 2.

In the maximum entropy method, which was introduced and implemented by Gull and Daniell in 1978 [10], an "entropy" function is defined, which allows model functions to be ranked. By finding the solution of the deconvolution equation which maximizes the entropy, the algorithm finds a smooth sky image with positive pixel values, which is also consistent with the measured visibility data.

Many variants of the entropy function have been proposed. In Gull and Daniell's work, the entropy function was chosen to be

$$f = -I \log I \tag{8.7}$$

where I is the model image. In the absence of any data other than the total intensity, f is maximized when I is the flat image. On the other hand, if it is desired that the maximum should be attained when I is proportional to a "default" or "prior" image I_0, the entropy function can be generalized to [11]

$$f = -I \log(I/I_0) \tag{8.8}$$

where I_0 is the initial model. If no prior information is provided, I_0 would be a flat image.

At each iteration of the MEM algorithm, the model image is updated and the difference between the model visibilities and the measured visibilities is found. A χ^2 statistic is used to quantify how closely the model fits the measurement. The algorithm seeks to maximize the entropy of the model image subject to some criterion on the proximity between the predictions of the model and the measurements. Although calculation of the visibilities and the residuals associated with the model image conceptually requires the Fourier transform of the model at the locations of the baselines to be evaluated, a variety of techniques can be used to avoid the cost of the slow transform and make this computation more efficient.

In "Historic MaxEnt," Gull and Daniell proposed that iteration should be halted when χ^2 equals the number of visibilities. A Bayesian version of MEM, called "Classic MaxEnt," was given by Gull in 1989 [12]. Both the historic and Classic MaxEnt are implemented in the licensed software package *MemSys5* [13].

In both deconvolution methods, the Fourier transform operation is vitally important, especially in producing the image and visibility models.

2 Gridding and degridding

Eq. (8.5) can be rewritten explicitly in terms of the measured visibilities as

$$I_D(l, m) = \sum_{k=1}^{N_v} V(u_k, v_k) \exp\left[i2\pi(u_k l + v_k m)\right] \tag{8.9}$$

Since the values of u_k and v_k do not lie on a regular grid, evaluating the dirty map on a collection of $N_x \times N_y$ pixels by direct evaluation of Eq. (8.9) requires $O(N_v N_x N_y)$ operations. This is termed the direct Fourier transform (DFT).

It is known that a Fourier transform can be performed more rapidly using the fast Fourier transform (FFT) algorithm if the points in the image and visibility domains lie on regular grids. For an image of size $N_x \times N_y$ pixels, this algorithm requires $O(N_x N_y \log(N_x N_y))$ operations. However, in order to work with the irregularly distributed visibilities, an additional "gridding" step is required to place the visibilities onto the grid. This can be designed to take $O(N_v)$ operations, leading to an overall complexity of $O(N_x N_y \log(N_x N_y)) + O(N_v)$. If $N_v \gg \log(N_x N_y)$, the algorithm based on the fast Fourier transform and gridding can be advantageous over the direct Fourier transform.

2.1 Gridding process

In order to apply the FFT to measured visibility data, which are not located on a uniform Cartesian grid, the visibility data must first be spread among the points in the regular grid via a process known as "gridding." Perhaps the simplest approach is the nearest-neighbor method, in which each measured visibility is assigned to the nearest gridpoint and summed together with any other visibility also assigned to that gridpoint. The FFT algorithm is then used to construct a dirty map. However, this procedure produces significant artifacts and leads to a dirty map, which differs considerably from that found using the DFT [4].

Gridding is most often performed using the convolutional gridding procedure, of which the nearest-neighbor method is a simple example. More generally, a "gridding function" of compact support is chosen. The measured visibilities are regarded as a collection of delta functions located at the baselines on which they are measured. These are convolved with the gridding function and the result is sampled on the regular grid for use as the input of the FFT [14]. The size of the support of the gridding function determines the number of grid points which each measured visibility affects [15]. In the case of the nearest-neighbor method, the gridding function is constant over a rectangle equal in size to the gridpoint separation, so that its support in each dimension is equal to the gridpoint separation in that direction. The gridding function is also known as the "gridding convolution function."

The gridding function $C(u, v, w)$ is usually chosen to satisfy the following conditions [16]

(1) The gridding function is separable: $C(u,v,w) = Cu(u)Cv(v)Cw(w)$, where Cu, Cv, and Cw are usually chosen as the same function.

(2) The gridding function is real and even: $C(-u) = C(u)$. The one-dimensional form $C(u)$ is used for simplicity from now on.

(3) $C(u)$ should have limited width (or support), since the time taken for gridding is proportional to the number of gridpoints which each visibility affects.

By satisfying these conditions, the three-dimensional gridding function $C(u, v, w)$ is completely determined by $C(u)$ for $u \geq 0$, which greatly reduces the cost of computing and storing the gridding function.

Fig. 8.3 illustrates the process of gridding one visibility with a chosen gridding function $C(u)$ in a simplified one-dimension example. The visibility $V(u_0)$ is denoted by the blue cross, at which the gridding function is centered. In the example, the

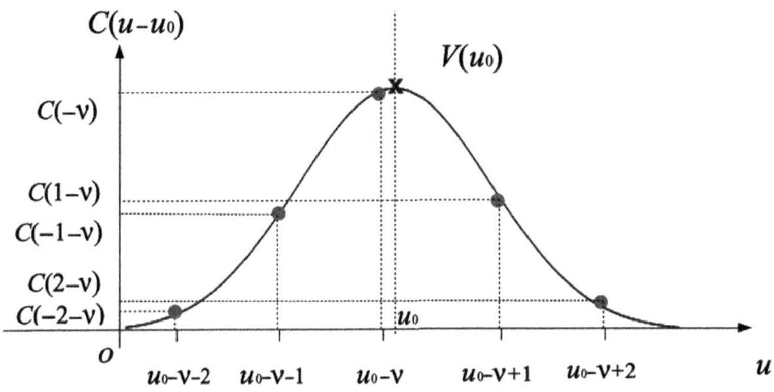

Fig. 8.3 The visibility (denoted by a *blue cross; light gray* in print version) at u_0 ought to be assigned to the nearby 5 grid points (denoted by *red dots; dark gray* in print version). The gridding function $C(u)$ is centered at $u = u_0$. The value assigned to each point equals to $V(u_0)$ multiplied with the grid point's corresponding gridding function value, which is dependent on the distance between the grid point and u_0.

width of the gridding function is taken to be $W=5$. The grid-points, which this visibility affects, are those integer values of u, which satisfy $|u - u_0| \leq W/2$. The visibility is multiplied by $C(u-u_0)$ before being added to the gridpoint. These values are denoted by the *red dots* (*gray* in print version) in the figure.

In the figure, v is the distance between u_0 and $\lfloor u_0 \rfloor$, the nearest gridpoint to the left of u_0. If $v<0.5$ there are fewer gridpoints to the right of u_0 than to the left. The reverse is true if $v>0.5$. This asymmetry occurs when the width W is chosen to be odd. If W is even, there is always an equal number of gridpoints on each side of u_0, which are affected by the value of the visibility measured there.

The following list summarizes the steps needed to make the dirty map using gridding and the fast Fourier transform:

(1) Grid the visibilities with a chosen gridding function on the uv plane.
(2) Apply the inverse FFT to the gridded visibility data to obtain an image.
(3) Crop the outer edge of the image.
(4) Multiply the cropped image by the grid correction function to obtain the desired dirty image.

The first two steps have been discussed earlier. Multiplication by the grid correction function in step (4) compensates for the effect of convolving the measured visibilities by the gridding function. By using a correction function, it is possible to use a gridding function with compact support and still get good agreement with the results of the DFT. To good approximation, the correction function is equal to the reciprocal of the Fourier transform of the gridding function.

Even with a careful choice of gridding convolution and correction functions, the agreement with the DFT gets considerably worse near the edges of the image compared with near the center. It is thus useful to crop the outer edge of the image in step (3), discarding the regions of large misfit. As we shall demonstrate, it is possible to achieve very small misfits in the portion of the image that is retained, at the computational cost of performing a somewhat larger FFT. Cropping the image is also beneficial in reducing the effects of aliasing introduced by the use of the FFT. Aliasing is an unwanted phenomenon in which source intensity noise outside the field of interest is shifted into the field of interest, since the fast Fourier transform is intrinsically periodic. This topic will be further discussed in Section 2.2.

We thus obtain the FFT dirty image, which is an approximation to the DFT dirty image. To visualize the steps, Fig. 8.1 demonstrates these two different ways to make a dirty image, one through DFT, the other via gridding and FFT. The *red crosses* (*gray* in print version) represent visibilities not measured on the Cartesian grid (denoted by *black dots*).

Unlike the traditional convolutional gridding process, van der Tol [17] proposed a different approach to gridding called "image domain gridding." The visibilities are partitioned into small blocks and each block is imaged using the inverse direct Fourier transform to give a (possibly undersampled) low-resolution image extending over the same region of the sky. The collection of low-resolution images from all the blocks are multiplied by a taper function and the results are transformed back onto regular grids on the visibility plane using the fast Fourier transform. If the block sizes are small enough, the cost of calculating the collection of direct inverse Fourier transforms and fast Fourier transforms may be acceptable. The gridded visibilities are then imaged using an inverse FFT followed by gridding correction as before. With this approach, the effect of the gridding is approximately to convolve the measured visibilities with the Fourier transform of the taper function, but without the need for explicit storage and evaluation of the convolution function.

2.2 Aliasing

Use of the FFT not only makes the extra gridding operation necessary, but also causes aliasing in the image plane due to its periodic nature. In this section, we give an example of aliasing caused by the FFT. Fig. 8.5 shows the DFT dirty image (left) and the FFT dirty image (right) from the same simulated VLA A-array data. There is only one simulated point source, which is located just outside of the field of interest. The sidelobes of the beam shape centered at the point source can be clearly seen within the imaging area in both images.

However, in contrast with the DFT dirty image, the point source from outside the field of view has been aliased into the FFT dirty image at the position shown by the *dashed red arrow* (*gray* in print version).

The periodicity of the FFT may be understood in terms of the sampling operation onto the regular (u, v) grid following the convolution operation performed during the gridding step. Sampling involves multiplication by a rectangular array of delta functions (i.e., the two-dimensional analogue of the shah function) in the (u, v) plane, whose inverse Fourier transform in the image domain is another rectangular array of delta functions whose periods are inversely related to those on the (u, v) plane. In the image plane, it is necessary to convolve with this periodic array of delta functions, which leads to periodic replication of the image, or aliasing.

The unwanted alias at the image edge is removed by the image cropping operation. Aliasing, which occurs closer to the center of the image center, can usually be suppressed adequately by a well-chosen gridding function, as will be discussed in Section 3.

Despite the need to perform gridding and the care required to minimize aliasing, the process outlined in Fig. 8.4 is still the most commonly used practice in radio interferometric imaging, rather than a direct Fourier transform. This is largely because the computational cost of the FFT and gridding processes become

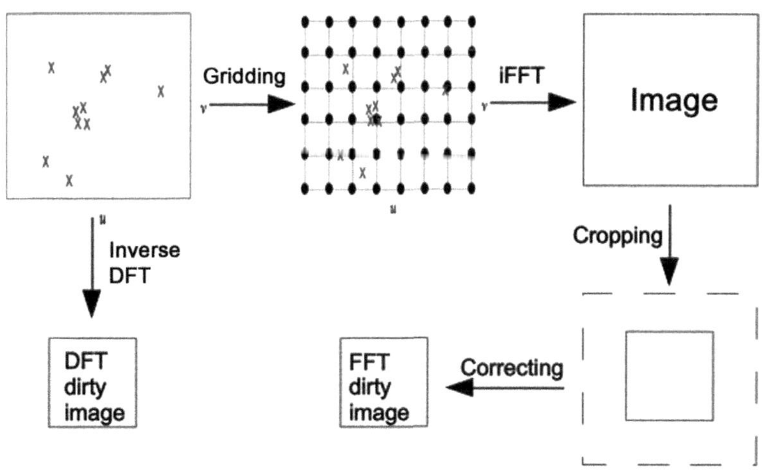

Fig. 8.4 The process of making a dirty image through DFT and FFT.

comparatively cheaper relative to the DFT as the number of visibility measurements increases. Also, aliasing effects can be greatly suppressed by choosing the right gridding function.

2.3 Degridding

Gridding and degridding are mathematically transpose operations. While gridding seeks to place irregularly sampled visibility data onto a regular grid in such a way that a good approximation to the DFT of the visibility data can be calculated from an FFT of the gridded data, the aim of degridding is to calculate the visibilities corresponding to a model image on the irregular set of points at which the actual visibilities are measured. This is achieved using the following steps:

(1) Multiply the model image by the grid correction function.
(2) Zero pad the grid corrected model image to the size required for the fast Fourier transform.
(3) Apply the FFT to the zero-padded, corrected image, giving an array of values on the visibility plane.
(4) For each measured baseline, use the gridding convolution function centered at that position to take a weighted linear combination of the values on the visibility plane from step (3). This is (an approximation to) the desired visibility on that baseline.

A block diagram of this process is shown in Fig. 8.6.

Degridding is essential to deconvolution methods. For example, in the major cycle of Cotton-Schwab CLEAN, it is necessary to generate accurate residuals associated with the model image calculated during the previous minor cycle. This is done by calculating the visibilities for the model image on the measured baselines using FFT and degridding. The difference between the measured and model visibilities (i.e., the visibility-domain residual) is then used to calculate the accurate image-domain residual for the next minor cycle. In MEM, the model image found during each iteration is processed using FFT and degridding to produce the associated visibilities on the measured baselines. The visibility-domain residual is used to determine whether the model is sufficiently close to the data, or if further iterations are required to update the model. Additionally, degridding is also used during the self-calibration procedure. Since each of a relatively small number of antennas contributes to the visibility measurements for a relatively large number of baselines, it is possible to regard the gain and phase shift of the individual antennas as calibration parameters whose values may be adjusted to improve the fit of the measured visibilities to those predicted by a model image.

This model image is usually generated by an image reconstruction process (such as deconvolution) based on the measured visibility data. The FFT and degridding procedures are used to calculate the model visibilities, from which the antenna-based calibration parameters may be extracted. The recalibrated measured visibilities are then used to update the model image in an iterative approach.

3 The choice of the gridding function in the era of big data

The gridding convolution function is required in both gridding and degridding procedures, and its choice consequently influences the quality of the images obtained, and all analyses based on them.

As early as 1961, Elizabeth Waldram in the Radio Astronomy Group at the Cavendish Laboratory, Cambridge, was using gridding functions that included the Gaussian, the sinc function, and Gaussian times sinc function, according to her work notes, although this work was unpublished. Brouw [18] later used a Gaussian gridding function with the east-west synthesis array at Westerbork. Many gridding functions have since been explored.

3.1 Spheroidal function and least-misfit gridding function

Today the most commonly used gridding function is probably the spheroidal function. It was proposed by Schwab [19] in the early 1980s with the aim of minimizing the effects of aliasing. In 1975, Brouw [20] proposed a quantity to measure the extent to which a convolutional gridding procedure suppresses aliasing in terms of integrals in the image domain. By generalizing this quantity to include a weighting factor over the map (parameterized by a constant α) and optimizing this quantity, Schwab identified the spheroidal function as the "optimal" gridding function [21]. The spheroidal function is widely implemented in imaging pipelines such as AIPS (short for Astronomical Image Processing Software [22]) and CASA. Fig. 8.7 displays a collection of spheroidal functions (for $\alpha = 1$) with different gridding width W.

Instead of using functions, which are chosen to be optimal with respect to a criterion based on alias suppression, we propose a parameterized family of "least-misfit gridding functions," [23] which minimize an upper bound on the difference between the DFT and FFT dirty images, which we call the "misfit." The misfit is minimized not over the entire map but only over the central portion defined by a parameter $x_0 \in (0, 0.5]$. In one dimension, a map is taken to extend over $-0.5 \le x \le 0.5$, and the misfit is minimized over $-x_0 \le x \le x_0$, so that, for example, taking $x_0 = 0.5$ indicates that the whole map is considered, whereas taking $x_0 = 0.25$ means that the central half of the map (in each direction) is considered while the rest of the map is cropped away and discarded. For a given choice of x_0 and of the width W of the convolving function, an optimal least-misfit function may be calculated.

It turns out that introducing x_0 is key to achieving good performance, since the misfit near the edge of the map ($x = \pm 0.5$) is always large. By retaining only the central portion of the map, it is possible to get very small misfits and also very good alias suppression within the retained portion (even though this was not explicitly optimized over). Indeed, when compared with the widely used spheroidal function for the same value of W, the least-misfit gridding functions can provide more than 100 times better alias suppression and RMS (root mean square) misfit reduction over the portion of the dirty map which is retained. In other words, by attempting to suppress aliasing out toward the edges of the dirty map, the spheroidal function significantly sacrifices performance near the map center.

By minimizing the upper bound of the misfit, optimal gridding convolution and correcting functions will be obtained. A collection of the least-misfit gridding functions optimized with $x_0 = 0.25$ and different widths W are shown in Fig. 8.8.

For the purposes of illustration, we compare the performance of the spheroidal functions and the least-misfit functions for convolutional gridding using simulated data for a VLA "snapshot." The image misfits for both families of functions relative to the ideal DFT dirty image are computed [4, 23]. Fig. 8.9 shows the RMS value of the image misfit using the least-misfit gridding functions ($x_0 = 0.25$) with differing values of W. The horizontal axis represents normalized image plane coordinate x, with 0 representing the image center and 0.5 the edge. For a given value of x, the RMS value of the image misfit within the central portion of the map $[-x, x]$ in both dimensions is plotted [4]. It can be seen that the RMS misfit within the central half of the image ($x_0 = 0.25$) that is retained remains essentially constant at a small value but that the misfit rises rapidly outside this region as the edge of the image is approached. This is because the difference between the DFT and FFT dirty image is minimized within the retained area, while the unavoidable FFT approximation errors near the edge of the map are confined to the area where the image will be cropped.

The final FFT image therefore differs from the ideal DFT image by only a very small amount across the entire portion of the image, which is retained. For these simulated data, the least-misfit function with $W=7$ already makes the image misfit negligible at the level of single precision floating point arithmetic.

When the value of x_0 is changed, the optimization will lead to different gridding convolution and correction functions. In general, the achievable image misfit within the optimization region increases with x_0 and decreases with W. Since the cost of the computation increases with W but decreases with x_0 (a larger x_0 means that a larger fraction of the map is retained, possibly allowing for the use of a smaller FFT), it is possible to trade off the cost of computation against the achievable misfit. A more detailed analysis of the operation count shows that for a given desired level of misfit, an optimal choice of x_0 and W can be made depending on the number of measured visibilities.

When the spheroidal function is used on the same simulated data, a similar plot can be made and this is shown in Fig. 8.10 [4, 23]. The RMS misfit values increase from the image center to the edge. For both families of gridding functions, larger W produces less image misfit, which shows that for high accuracy, each visibility data point has to be assigned to a large number of gridpoints, increasing the computational cost and time for gridding.

Compared to Fig. 8.9, the image misfit in the range $-0.25 \leq x \leq 0.25$ using the least-misfit gridding function is at least 100 times better than that using the spheroidal function with the same width W. In order to make the image misfit negligible at the level of single precision arithmetic, the least-misfit gridding function needs only a support width of $W=7$ with $x_0=0.25$, whereas the spheroidal function requires a width $W=10$ if only points within $x_0=0.2$ are retained. By way of comparison, if the CASA package is used to make the dirty image, the RMS of the image misfit over the portion of the image retained is around 10^{-4}–10^{-3}.

We next test the performance of the gridding functions by examining how well they suppress aliasing. A simple example is considered where only a single point source exists outside of the field of view, which is otherwise empty, as shown in Fig. 8.5. A series of VLA A-array simulated data is produced with the point source being shifted from very close to the edge to further away. The position of the source relative to the edge of the image in the x direction was varied, while its position in the y direction is kept constant. The normalized brightness of the aliased source in the dirty map calculated using the FFT is recorded [4, 23].

Figs. 8.11 and 8.12 show the normalized brightness of aliases versus the normalized horizontal distance from the source to the edge of the image, for the two families of gridding functions. Since

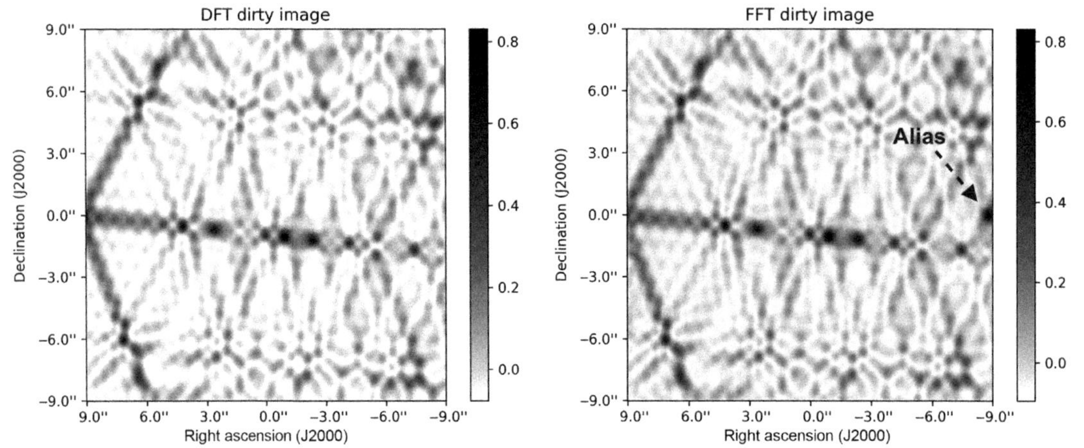

Fig. 8.5 Demonstration of the aliasing effect due to the periodicity of FFT. A DFT dirty image (left) and a FFT dirty image (right) made from the same simulated data are shown. The alias is obvious in the FFT dirty image. Taken from P.R. Hague, H. Ye, B. Nikolic, S.F. Gull, MNRAS 484 (2018) 574.

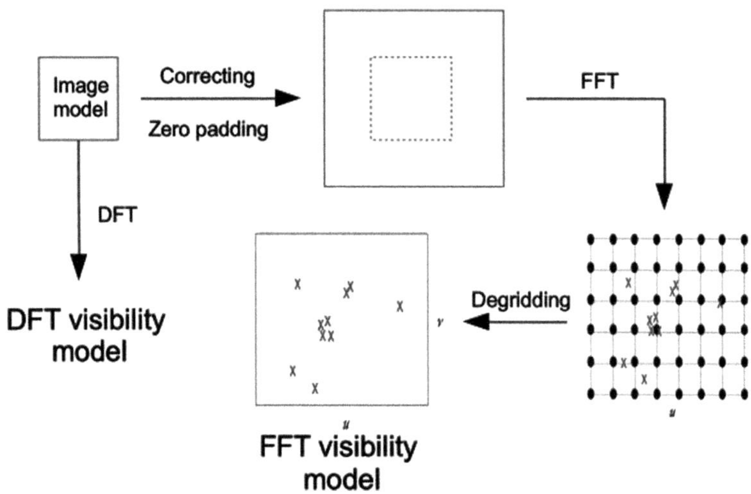

Fig. 8.6 The process of making the visibility model through DFT and FFT.

there is no aliasing in the DFT image, for this simple example, we expect the misfit between the DFT and the FFT maps at the location of the aliased source to be related to the degree of suppression of aliasing. Taking into consideration the position of the aliased source, it is not surprising that these figures show the same behavior as Figs. 8.9 and 8.10. For the least-misfit gridding functions, sources that are aliased into the central half of the image are well suppressed by a factor of at least 10^7 with $W=7$. As expected, if

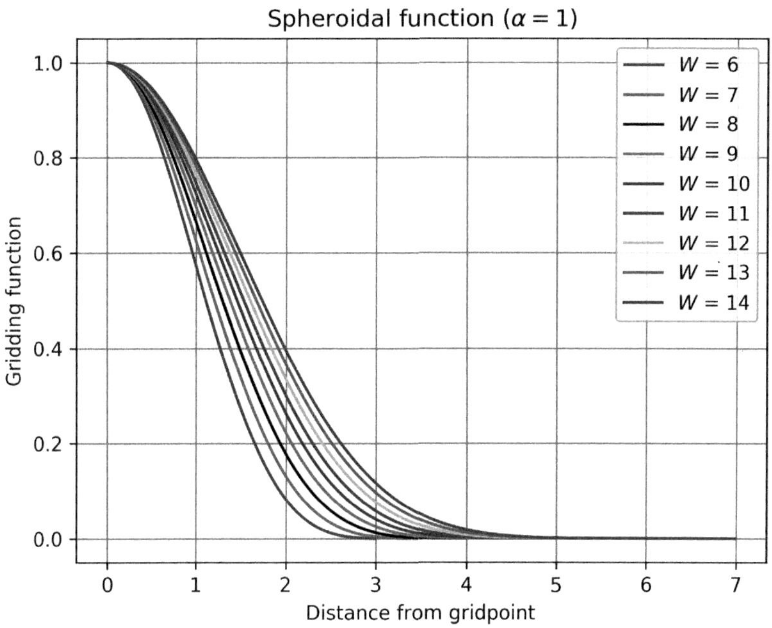

Fig. 8.7 Spheroidal functions ($\alpha = 1$) with a width from 6 to 14.

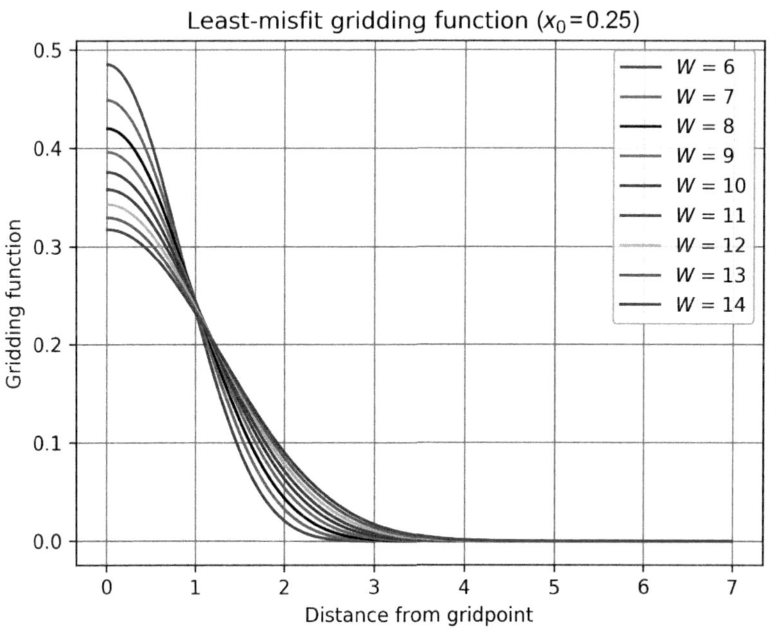

Fig. 8.8 Least-misfit gridding function with $x_0 = 0.25$ and different width W.

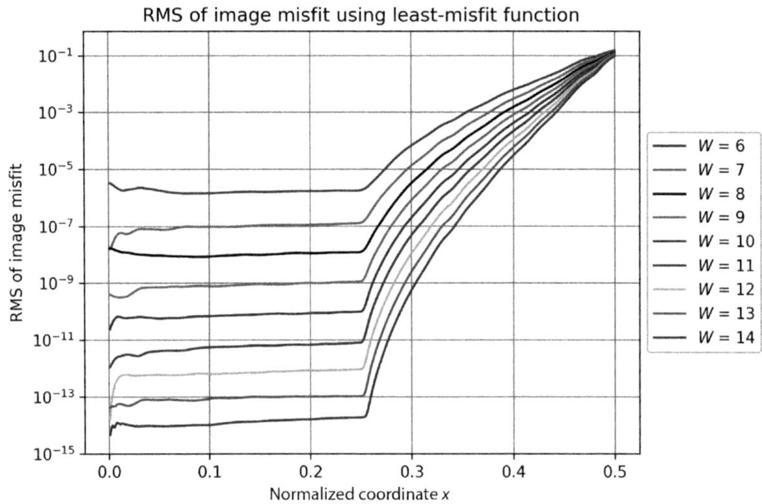

Fig. 8.9 RMS value of the image misfit for VLA simulated snapshot using the least-misfit functions with differing values of *W*.

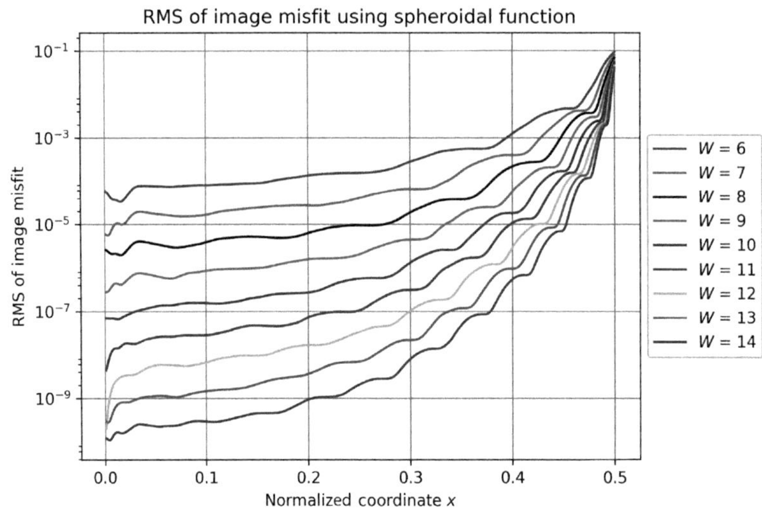

Fig. 8.10 RMS value of the image misfit for VLA simulated snapshot using the spheroidal functions with differing values of *W*.

the aliased source is in the outer half of the image, the suppression is much poorer. The spheroidal functions were optimized for suppression of aliasing over the entire image, but we see that the performance when the aliased source is shifted close to the map center is considerably worse than for the least-misfit function.

In summary, although the spheroidal functions are optimized for alias suppression over the whole map, the performance near the edges of the map is still so poor that, in practice, it is still usual to crop away the outer portions of the map. Given that only a portion of the map is retained, the least-misfit function, which can be

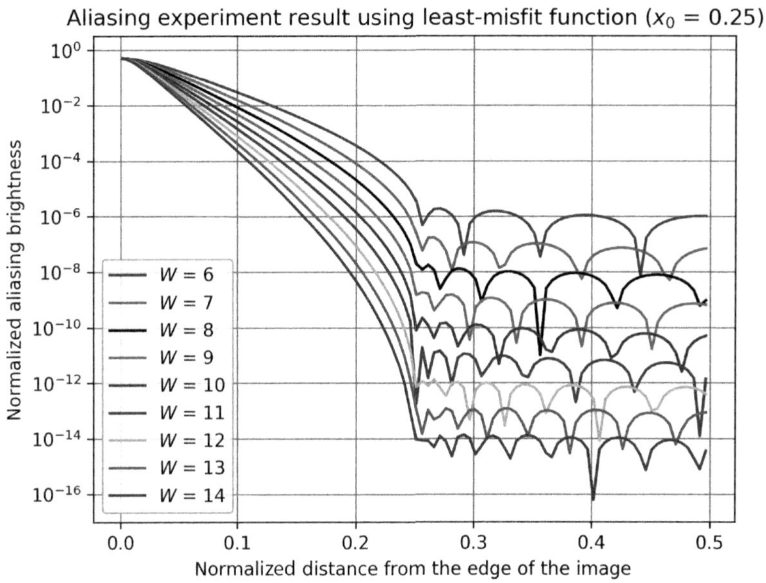

Fig. 8.11 Normalized brightness of alias versus the horizontal distance from the point source to the edge of the field of view. Least-misfit functions with differing values of *W* are used.

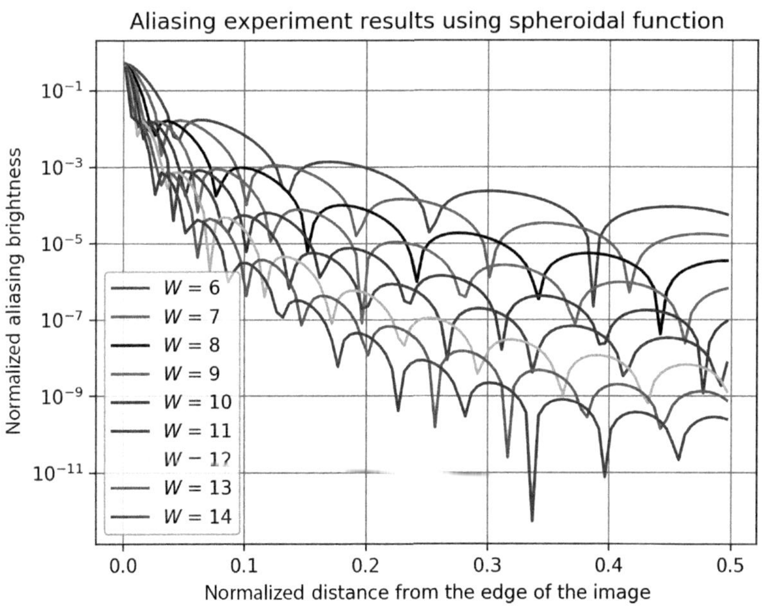

Fig. 8.12 Normalized brightness of alias versus the horizontal distance from the point source to the edge of the field of view. Spheroidal functions with differing values of *W* are used.

optimized for good performance within that region, is able to perform very well, so that the alias suppression can be two orders of magnitude better than for the spheroidal function within the portion retained, with the same value of *W*.

In wide-field imaging algorithms, such as the w-projection method [24] and the w-stacking method [25], the least-misfit gridding function can potentially replace the spheroidal function for better alias suppression. In the w-stacking method, if a three-dimensional gridding is adopted rather than the original two-dimensional gridding, and the gridding function is changed to our least-misfit gridding functions, the difference between the DFT and FFT dirty images for wide-field observations can become negligible to single precision by using $W=7$ and to double precision by using $W=14$ [4].

It is also possible to evaluate the performance of the spheroidal and least-misfit functions for use in degridding. Given a model image, the visibilities on an arbitrary collection of baselines can be found by explicit calculation of the Fourier transform or alternatively by using a combination of the FFT and convolutional degridding. A visibility plane misfit may be defined as the difference between the visibilities calculated by these two methods.

We find that when the least-misfit function with $W=7$ and $x_0=0.25$ is used for degridding, the visibility plane misfit is negligible at the level of single precision arithmetic (provided, of course, that the model image lies within the region with normalized coordinates between $-x_0$ and x_0 in both directions). In this way, if these functions are used during major cycles in the Cotton-Schwab CLEAN algorithm, no extra error is introduced by the degridding process. Under the same conditions, we have found that the spheroidal function with the same support leads to an RMS visibility plane misfit which is at least two orders of magnitude larger [4, 23].

In summary, it is recommended that the least-misfit gridding function with $W=7$ and $x_0=0.25$ be used for both gridding and degridding when single precision arithmetic is adequate.

3.2 Tabulating the gridding function for image processing

Once the gridding function is chosen, a look-up table is usually precalculated and used when the function needs to be evaluated. The reciprocal of the separation between the values at which the continuous one-dimensional gridding function are tabulated is called the "sampling rate." During the gridding process with a convolutional gridding function of width W, each baseline requires a lookup of W convolving weights from the stored table in each dimension. Since the coordinate of the baseline can have any value, a very large sampling rate may be needed in order to recover accurate values of the convolving weights. It is possible to interpolate between the tabulated values of the gridding function, but often the nearest-neighbor rule is used to retrieve the nearest available function value.

When the nearest-neighbor rule is used to retrieve gridding values from the look-up table, for the least-misfit function with $W=7$ and $x_0=0.25$, it is necessary to use a sampling rate of 10^6 in order to make the effects of using a lookup table indistinguishable from performing an exact calculation at the level of single precision arithmetic [4, 23].

Fig. 8.13 and Table 8.1 show the effects of using nearest-neighbor table lookup to retrieve values of the least-misfit gridding function for $x_0=0.25$ and $W=7$. The various lines show the effects of different numbers of sample points within the lookup table. The image misfit in Fig. 8.13 should be compared with the graph for $W=7$ in Fig. 8.9 where we see that the misfit when using the exact convolution function is approximately 10^{-7} over the range $|x| \leq 0.25$.

Further investigation shows that the accuracy achieved when using a look-up table with $M_s=100$ and $W=3$ gives results comparable to that achieved by CASA [4, 23]. This suggests that the algorithm in CASA would benefit from the use of the least-misfit convolution functions.

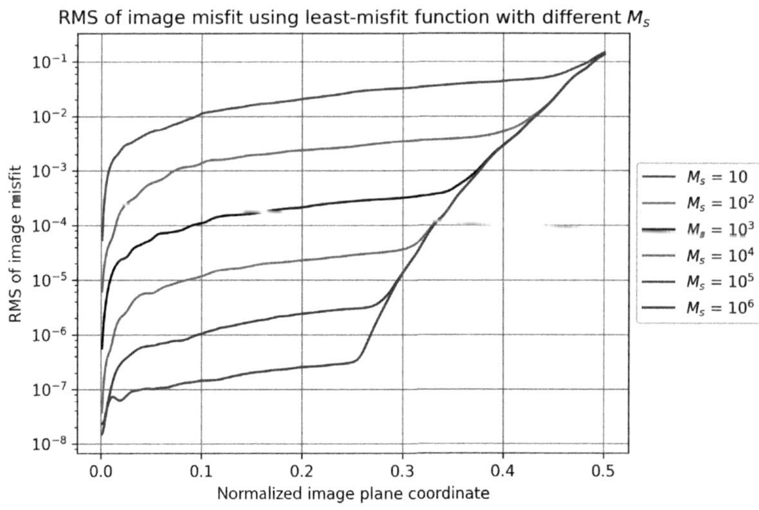

Fig. 8.13 RMS value of the image misfit achieved using look-up tables of the least-misfit function ($W=7$ and $x_0=0.25$) with differing sampling rates.

Table 8.1 RMS value of the visibility misfit using the least-misfit gridding ($W = 7$ and $x_0 = 0.25$) look-up table with differing sampling rates M_s.

M_s	RMS of the visibility misfit
10^2	6.23×10^{-3}
10^3	4.48×10^{-4}
10^4	4.40×10^{-5}
10^5	4.62×10^{-6}
10^6	4.85×10^{-7}
No sampling	2.98×10^{-7}

3.3 Gridding computational cost in the big data era

Although it is possible to reduce the errors introduced in the gridding or degridding process to the level of machine precision, accuracy comparable to that achieved by CASA requires only a look-up table with 300 entries and a support width of 3, allowing a greatly reduced computation cost for a given performance.

As is widely known, gridding is one of the most computationally expensive processes in radio interferometric imaging. The number of visibilities is denoted as Nv, and each visibility comprises both real and imaginary parts. Firstly, for each visibility, there will be two multiplications with the corresponding real weight. When this visibility is to be gridded onto both u and v *grids*, 2 W^{-2} multiplications with the gridding values and 2 W^{-2} additions on the corresponding grid points will be performed, a total of (4 W^{-2+2}) Nv operations. If the one-dimensional gridding function is precomputed, an extra W^{-2} multiplications will be added to find the coefficients, increasing the number of operations to (5 W^{-2+2}) Nv [23].

In the era of SKA, the number of visibilities will surge to an amount that even the best supercomputer in the world today would face difficulty in processing with speed. By using the least-misfit gridding function, the gridding width can be reduced from the current $W = 6$ or 7 to $W = 3$, eliminating approximately 75% of gridding operations. In addition, the storage and cache needed during the gridding process will also be greatly reduced due to a much smaller look-up table. As a result, the least-misfit gridding function can reduce the computational cost of wide-field imaging methods such as the w-stacking method [25].

4 Bayesian source discrimination

Image analysis in interferometry is a compromise between the accuracy of working with visibilities and the speed of working with the processed maps. As the data scale increases, it becomes more and more infeasible to work with the visibilities at all—they likely cannot even be stored, and must be gridded in one pass. It is therefore important to minimize the information loss during this process. The CLEAN algorithm is an example of a processing technique that does not do this; it removes information in order to produce maps that are easily understandable to the human eye, and uses methods such as Gaussian source extraction that are designed principally for optical maps. The output of CLEAN is made up of the residual of the initial extraction process, where the dirty beam is repeatedly subtracted from the brightest point of the map in proportion to some gain value, with a Gaussian restoring beam then inserted at every point where the dirty beam was extracted. The only possible information that can be found in such data products is the CLEAN model. Regardless of what algorithm is used to fit Gaussians to the CLEANed map, the actual business of finding sources in the map has already been done by CLEAN. Any method that is intended for use on interferometric maps should treat them as what they actually are, instead of trying to manipulate them to resemble another data product that is considered easier to deal with. Algorithms that make direct inferences from dirty maps, on the other hand, are subject only to information loss due to gridding. We have developed one such algorithm in the form of BaSC. In comparison, source extraction packages such as SExtractor [26] would only work on CLEANed images when dealing with interferometric observations. SExtractor is a typical tool for extracting sources from images; it fits 2D Gaussians to areas on an image, until the image minus all such Gaussians are noise. This makes sense for a particular application; it works well for direct imaging instruments (i.e., ones that measure in the spatial domain rather than the frequency domain, such as CCDs) and is widely used for this reason. There are other tools that work in a similar way, and even examples of tools that use Gaussian fitting in interferometric images—such as Aegean [27]. Aegean is designed to solve other problems in radio astronomy, but it is still subject to the limitations of CLEAN. In this section, the performance of BaSC and SExtractor will be compared.

4.1 An application: Bayesian source discrimination

A situation in which relevant information is lost by using cleaned maps is shown here. In Fig. 8.14 we see that parts of the dirty beam have been left in the map, creating a spurious "bridge" of flux between the two points, which causes SExtractor to fit the Gaussians with closer centers [28]. The two point-source data are simulated based on the VLA A-array.

If there are two nearby sources of flux, at a distance comparable to the width of the central peak of the dirty beam, then a Bayesian method, which deals with the dirty map, is able to distinguish between two points at a smaller distance than by using Gaussian source fitting on a cleaned map. Also in this case, the CLEAN model does not provide any useful data.

In terms of accurately locating sources, this process is close to the theoretical Bayesian resolving limit. This can be calculated by assuming a map $D(x)$ consisting of a single source of flux F, and having a dirty beam $B(x)$.

$$D(\mathbf{x}) \simeq FB(\mathbf{x}) = F\sigma^2 \sum_n \frac{1}{\sigma_n^2} \cos\left(2\pi \mathbf{u}_n \cdot \mathbf{x}\right) \tag{8.10}$$

The uncertainty of the position of $\mathbf{x}$ can be determined by taking the Taylor expansion of the expression for the beam, with the second-order term yielding a Gaussian approximation near $\mathbf{x}$

$$P(\mathbf{x}) \propto \exp\left[-2\pi^2 F^2 \sum_{nml} \frac{(u_n)_m (u_n)_l}{\sigma_n^2} x_m x_l\right] \tag{8.11}$$

From this expression, we can see the covariance is

$$< \delta x_m \delta x_l > = \frac{\sigma^2}{F^2} \Delta x_{ml}^2, \tag{8.12}$$

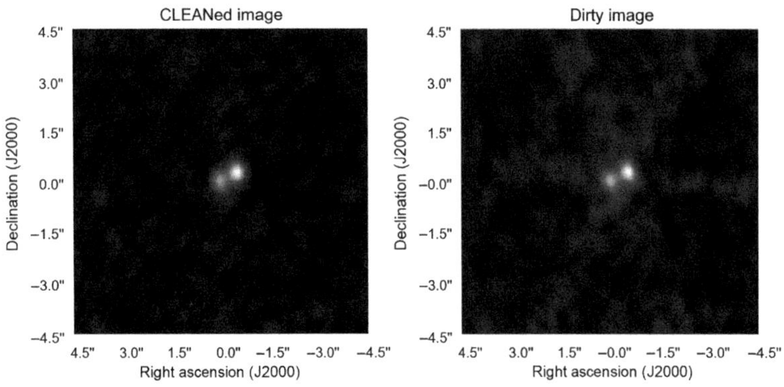

Fig. 8.14 Example of a CLEANed image (left) and a dirty image (right) from the same VLA simulation.

where

$$\left(\Delta x_{ml}^2\right)^{-1} = -\partial_{ml}^2 B(\mathbf{x})\Big|_0 = 4\pi^2\sigma^2 \sum_n \frac{(u_n)_m (u_n)_l}{\sigma_n^2} \tag{8.13}$$

This is the characteristic width of the central peak of the dirty beam. The resolving ellipse is therefore just $a/S, b/S$, where S is the signal-to-noise ratio F/σ.

Practical resolving power diverges as the predicted limit approaches the cell size, however; clearly at this level the position uncertainty must tend toward $\frac{1}{\sqrt{12}}$ as subgrid placement of proposed sources is meaningless. The implication is that an oversampled grid is necessary in order to make the best possible inferences from the same data. This is not a claim that oversampling can add information; it simply enables certain techniques, which are better at extracting information from the underlying data.

Fig. 8.15 shows the improvement in performance at this task compared with CLEAN/SExtractor. The test maps were simulated ALMA images, with a central point and a second point, of the same flux, at a random distance and direction. Some cases had zero distance and thus were a single point with twice the flux.

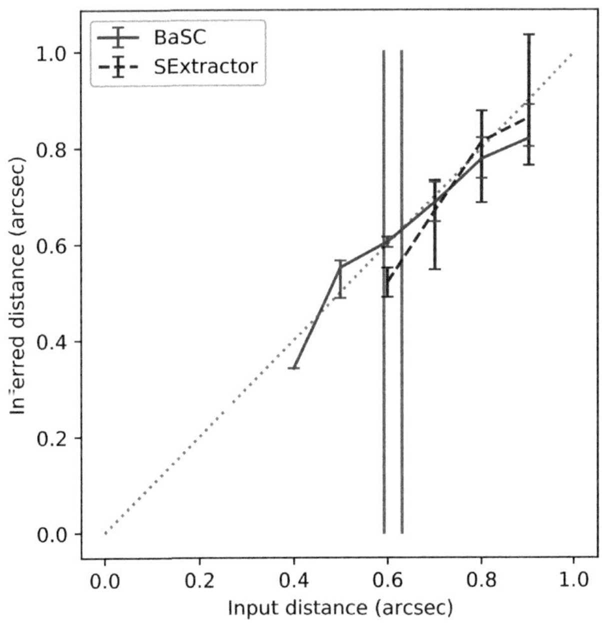

Fig. 8.15 Plots of the separation of sources in the input file versus the separation inferred using each different source finders. Taken from P.R. Hague, H. Ye, B. Nikolic, S.F. Gull, MNRAS 484 (2018) 574.

4.2 Applications to big data radio interferometry

The value of this method in a big data context is that the inferences work on the earliest data product in the pipeline that can be reasonably stored: the dirty map (which contains information equivalent to the gridded visibilities). Furthermore, instances where a single bright source has to be removed from a map may be well served by this technique, because this cannot be done at the visibility level.

Compared to Gaussian-seeking approaches, the Bayesian method has the perhaps counterintuitive property that, in the case of a dim source near to a bright source, the constraint on the position on the dim source (and also how close it can be to the bright source and still be detected) is improved. This is because constraints on position are smaller for higher signal-to-noise ratios, and misplacement of a proposed point during the MCMC process acts like an additional source of measurement error for other points, according to the contribution of the dirty beam at the relative position of that point. A brighter point is more tightly constrained in its position, and therefore the error it introduces in points the same distance away is smaller.

References

[1] A.R. Thompson, in: G.B. Taylor, C.L. Carilli, R.A. Perley (Eds.), Synthesis Imaging in Radio Astronomy II, Astronomical Society of the Pacific Conference Series, vol. 180, 1999, p. 11.

[2] A.R. Thompson, B.G. Clark, C.M. Wade, P.J. Napier, ApJS 44 (1980) 151.

[3] P.J. Napier, A.R. Thompson, R.D. Ekers, Proc. IEEE 71 (1983) 1295.

[4] H. Ye, Accurate Image Reconstruction in Radio Interferometry (Doctoral thesis), (2019). https://doi.org/10.17863/CAM.39448.

[5] J.P. McMullin, B. Waters, D. Schiebel, W. Young, K. Golap, in: R.A. Shaw, F. Hill, D.J. Bell (Eds.), Astronomical Data Analysis Software and Systems XVI, Astronomical Society of the Pacific Conference Series, vol. 376, 2007, p. 127.

[6] J.A. Högbom, A&AS 15 (1974) 417.

[7] B.G. Clark, A&A 89 (1980) 377.

[8] F.R. Schwab, W.D. Cotton, AJ 88 (1983) 688.

[9] T.J. Cornwell, IEEE J. Sel. Top. Sign. Proces. 2 (2008) 793.

[10] S.F. Gull, G.J. Daniell, Nature 272 (1978) 686.

[11] S.F. Gull, J. Skilling, in: J.A. Roberts (Ed.), Indirect Imaging. Measurement and Processing for Indirect Imaging, Cambridge University Press, 1984, p. 468, ISBN-10: 9780521262828; 13: 978-0521262828.

[12] S.F. Gull, Developments in Maximum Entropy Data Analysis, Springer Netherlands, Dordrecht, 1989, pp. 53–71. https://doi.org/10.1007/978-94-015-7860-8_4.

[13] S. Gull, J. Skilling, Quantified Maximum Entropy MemSys5, User's Manual Maximum Entropy Data Consultants, http://www.mrao.cam.ac.uk/~steve/malta2009/images/MemSys5_manual.pdf, 1991.

[14] J.D. O'Sullivan, IEEE Trans. Med. Imaging 4 (1985) 200.

[15] A.R. Thompson, J.M. Moran, G.W. Swenson Jr., Interferometry and Synthesis in Radio Astronomy, third ed., Wiley, 2017, https://doi.org/10.1007/978-3-319-44431-4.
[16] E. Greisen, VLA Scientific Memorandum No. 131, National Radio Astronomy Observatory, 1979.
[17] S. van der Tol, B. Veenboer, A.R. Offringa, A&A 616 (2018) A27.
[18] W.N. Brouw, Leiden Observatory, Leiden University, Leiden, The Netherlands, 1971 (PhD thesis).
[19] F.R. Schwab, VLA Scientific Memorandum No. 132, National Radio Astronomy Observatory, 1980.
[20] W.N. Brouw, Methods Comput. Phys. 14 (1975) 131.
[21] F.R. Schwab, in: J.A. Roberts (Ed.), Indirect Imaging. Measurement and Processing for Indirect Imaging, Cambridge University Press, 1984, p. 468, ISBN-10: 9780521262828; 13: 978-0521262828.
[22] D.C. Wells, in: V. di Gesu, L. Scarsi, P. Crane, J.H. Friedman, S. Levialdi (Eds.), Data Analysis in Astronomy, Springer, 1985, p. 195.
[23] H. Ye, S.F. Gull, S.M. Tan, B. Nikolic, arXiv (2019). arXiv:1906.07102.
[24] T.J. Cornwell, K. Golap, S. Bhatnagar, IEEE J. Sel. Top. Sign. Proces. 2 (2008) 647.
[25] A.R. Offringa, et al., MNRAS 444 (2014) 606.
[26] E. Bertin, S. Arnouts, A&AS 117 (1996) 393.
[27] P.J. Hancock, C.M. Trott, N. Hurley-Walker, Publ. Astron. Soc. Australia 35 (2018).
[28] P.R. Hague, H. Ye, B. Nikolic, S.F. Gull, MNRAS 484 (2018) 574.

PART

C

Computing technologies

9

Execution framework technology

Ying Mei[a], Rodrigo Tobar[b], Chen Wu[b], Hui Deng[a], Shoulin Wei[c], Feng Wang[a,c]

[a]Center for Astrophysics, Guangzhou University, Guangzhou Higher Education Mega Center, Guangzhou, China. [b]International Center for Radio Astronomy Research (ICRAR), The University of Western Australia, Crawley, Perth, WA, Australia. [c]Kunming University of Science and Technology, Chenggong District, Kunming, China

1 Introduction

In the most recent decade, the information from cosmic perceptions has exponentially expanded. The Sloan Digital Sky Survey (SDSS) telescope, for instance, delivers roughly 200 GB information consistently, adding to a database that was around 50 TB in 2012 [1]. Moreover, the Large Synoptic Survey Telescope has a three-billion-pixel advanced camera and produces 5–10 TB of information every night.

The square kilometer array (SKA) [2, 3] will be the biggest radio telescope on the planet. The main period of the task—SKA1 [4]—will consist of hundreds of dishes and hundreds of thousands of antennas, empowering the surveying of the sky in phenomenal detail and speed, with a second phase expanding these capabilities to at least an order of magnitude.

In view of its gigantic size, only one SKA1 science venture will deliver connected information at a pace of 466 GB/s [5] for the low-frequency telescope (SKA1-Low) and 446 GB/s [6] for the mid-frequency part (SKA1-Mid). This related interferometry data will be injected into the science data processor (SDP) [7], a many-task computing [8, 9] center, responsible for reducing observational data, and producing science-level products continuously. The SKA1 will have constrained power allocations [10] to process observations as they are performed in real time. This poses considerable challenges to manage, process, and store such large datasets.

Big Data in Astronomy. https://doi.org/10.1016/B978-0-12-819084-5.00004-3

The working experience of building the data system [11] for the SKA-low precursor telescope [12] shows that the overhead related with data migration, reading/writing, release, and format conversion are progressively ruling the overall pipeline execution costs.

In addition, the complexity of data processing in radio astronomy derives not only from the simple algorithmic components (e.g., FFT, gridding, deconvolution, and so on) but also from the diverse combinations of ways in which these components access their input, output, metadata, and intermediate information. It makes applying a "one-size-fits-all" strategy (e.g., data reorganization, I/O overlapping, intelligent caching, etc.) very challenging for a global optimum across multiple pipeline stages on a distributed computing environment. Because the current state-of-the-art astronomy data processing systems are designed to handle data approximately two to three orders of magnitude smaller than the SKA1 [5, 6], a new data execution framework is much needed.

At present, defining workflow components statically in scripts is the most regular way to reduce radio astronomy data. These scripts (i.e., the source codes) can then either be run sequentially on a local computer or wrapped into job scripts submitted to job scheduling systems such as PBS or SLURM in a high-performance computing platform. These application-driven workflow models have several drawbacks. First of all, most astronomical projects involve many institutes across the world. It means that astronomers often have to refine or even redesign their workflow codes in order to make them work—compiling, deploying, running, monitoring, and so on—on a single machine or computer clusters with heterogeneous hardware or software architectures. This happens whenever there is an upgrade on the workflow, hardware, or telescope configuration, leading to considerable cost. Furthermore, there is no real-time monitoring and control for the execution of the process. For example, in many situations, users cannot easily determine the pipeline execution status (e.g., success, failure, etc.) until the entire process is completed or a significant amount of computational and storage resources has been consumed. For SKA-scale data processing (with tens of millions of concurrent tasks), this is not only infeasible, but it is extremely expensive to delay fault detection and subsequent recovery actions (e.g., reexecution). Last, users still need to restart the entire job for reexecution even though failures or exceptions can be informed at an earlier stage. Because a workflow driven by "processing" rather than "data" cannot adjust task execution dynamically based on whatever intermediate datasets (e.g., from previous runs) and resources happen to be available. More "intelligent" workflow management often requires substantial ad hoc human interactions, which become impractical for SKA-scale data reducing workflows.

2 OpenCluster

The OpenCluster is a Python-written lightweight execution framework. We introduced OpenCluster here because it is easy to understand the concept and principle of OpenCluster. In the meantime, OpenCluster has been deployed in a modern radio telescope's real computing environment [12a].

2.1 Fundamental model

Scientific data processing is similar to manufacturing industrial products. In common, before it is put on the market, a product faces multiple stages such as manufacturing, packaging, and shipping. Fig. 9.1 shows a schematic of production. The product procedure is controlled by many managers who supervise a large group of employees. When a manufacturing process is complete, the product is transferred to the next procedure. In production, all

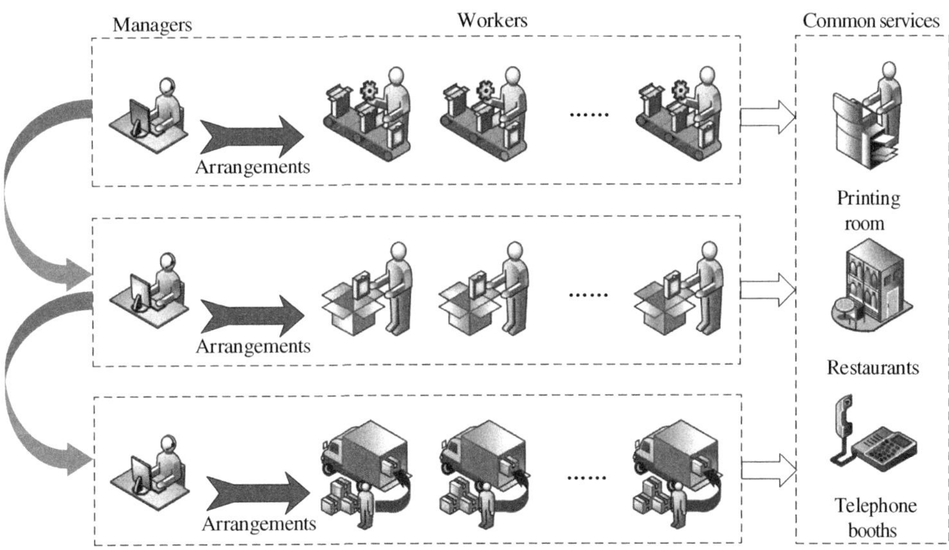

Fig. 9.1 Abstract process of traditional production. Reprinted with permission from S. Wei, F. Wang, H. Deng, C. Liu, W. Dai, B. Liang, Y. Mei, C. Shi, Y. Liu, J. Wu,OpenCluster: a flexible distributed computing framework for astronomical data processing, Publ. Astron. Soc. Pac. 129 (972) (2016) 024001. Copyright (2016) IOP Publishing.

employees should be provided with several public services such as printing, telephone use, and supply of meals.

The modern industry model can be referenced to build a distributed computing framework for massive data processing. The processing of data is, in essence, very close to industrial production. First, raw observational data are provided by the astronomical telescope. In a sequence of steps, these data should be processed.

OpenCluster is developed to provide a cohesive framework within a computer cluster setting that can easily support massive distributed computing. An OpenCluster-based system can be built in a simple and efficient way. OpenCluster's overall objective is to enable a set of heterogeneous machines on a network to be viewed as a concurrent computing tool for general purposes.

With minimal software requirements, OpenCluster was built in-house. In addition, it does not depend on a specific middleware or analysis framework, thus promoting greater installation flexibility. For this function, OpenCluster also provides an integrated monitoring system with no reliance on external resources. Such properties make OpenCluster an effective yet very lightweight tool and expand its range beyond astronomical applications.

To devise an abstract model for OpenCluster, we further adapt components from the modern industrial assembly line. The OpenCluster has five major elements in its abstract model: factory, workshop, manager, worker, and service (see Fig. 9.2).

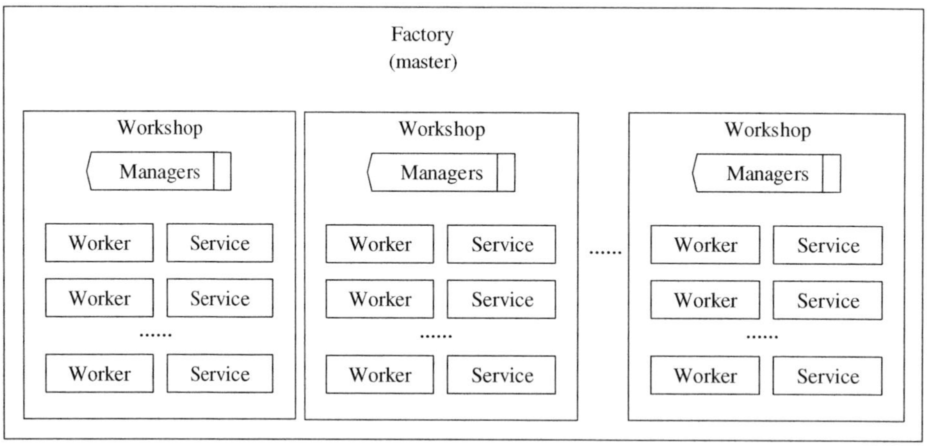

Fig. 9.2 Relational diagram of OpenCluster components. Reprinted with permission from S. Wei, F. Wang, H. Deng, C. Liu, W. Dai, B. Liang, Y. Mei, C. Shi, Y. Liu, J. Wu,OpenCluster: a flexible distributed computing framework for astronomical data processing, Publ. Astron. Soc. Pac. 129 (972) (2016) 024001. Copyright (2016) IOP Publishing.

- Factory—A logical registry that contains all information about workshops, managers, workers, and services. It adopts the master/slave model to avoid a single-point failure.
- Workshop—A computational node that refers to a real computer in a common area. A daemon process runs in the workshop to communicate with the factory instance and to report information and status updates.
- Manager—A particular task or job manager that requests services and assigns tasks to workers.
- Worker—A process that runs in a workshop (computer). The workers wait for the call from managers.
- Service—A process similar to that of a worker. The service can be invoked by managers, workers, or even external applications.

2.2 OpenCluster implementation

The OpenCluster was ultimately developed by using Python. As a script language, Python is low performance in developing data processing software. However, Python is widely used in current astronomical data processing because of its simplicity and elegance. Moreover, as open-source software, Python has a user and developer community, and many mature data processing packages are available for it. Thus, many emerging and mature astronomical data processing systems have been developed for scientific computations written by astronomical scientists. These programs can be seamlessly and easily integrated into OpenCluster. Aiming at this goal, we implemented five main base classes for the functions of factory, workshop, manager, worker, and service. The other classes are inherited from these five base classes.

2.2.1 Factory

Factory is the core component of the OpenCluster execution framework, which administers one or more workshops, and all workshops, managers, workers, and services should first register with the factory as a registration center. We can generally launch at least two instances at the factory: One master and some slaves, which can run on a specified node or various physical nodes. The master will synchronize the updated information to all slaves if any connection information changes. Thus, theoretically, all slaves have the same information with the master. A slave worker will be selected as a new master when a master factory failure occurs. Thus, no factory single-point failure exists in the system.

2.2.2 Workshop

A real cluster computing server is described as a workshop that includes several employees or instances of operation. Node Daemon, which is responsible for managing staff, tracking their resource uses (CPU, memory, and network), and reporting factory updates, runs in the background on each cluster node, respectively. Node Daemon's system interfaces can be used to launch or stop a new instance of the worker/service. Node Daemon registers with the factory on startup and sends node resource data. To monitor the connection between Node Daemon and the factory, a heartbeat mechanism is used. If the factory does not receive the heartbeat, it can determine the workshop's problems. The factory must disable the timeout Node Daemon registration after a specific time period. If the heartbeat is not received by Node Daemon, it will determine that there are problems with the current factory leader and a new leader will be required.

2.2.3 Worker

The worker is assigned to perform a task specified by the manager and returns the results to the manager. After the worker instance is successfully started, it attempts to connect to the leader factory. It then submits the information, including connection information, types of workers, and so on. After the information communication, the worker enters a listening state and awaits a call from the manager.

OpenCluster provides only a base class of workers (worker.py), which includes a function ("perform") for executing tasks and setting attributes, such as the host, port, and work type (defined in the constructor). Workers defined by the user must inherit from the base class and override the function. The function perform has an argument with the *WorkPiece* type and returns a *WorkPiece* value.

These functions are included in the set of methods described by this base class, but are not limited to this. There are also additional features such as workplace contact, breathing, and correspondence with supervisors, all of which are hidden within the base worker class's higher hierarchy. The client is not concerned about the function component. Every instance of the worker is run as an independent operation as well as a socket server. It allows multiple processes of the worker to run in a single node. The number of processes generated in one node, however, will work in harmony with the core number of the CPU. If we set a larger number of processes than the core number, the processing speed will only be reduced.

2.2.4 Manager

It is the manager's responsibility to create a job to run, query the factory, and request the factory to provide its workers or services with metadata that matches the provider's endpoint address and port. It also finds a variety of appropriate workers who can run the job and dispatch the work to these workers. The manager looks up the list of registered workers in the factory and dispatches the work to the available workers after it starts. The workers then accept this request, perform the related jobs, and return the results to the manager. The entire task can deal with multiple managers. As with the worker, the scheduling logic of the manager is determined by software developers. Developers can freely define and adjust the scheduling process and define different strategies. The manager acts as an execution entry of the computing tasks or an initial stage of a chain process.

The worker instances can be invoked by the manager in either an asynchronous or a synchronous way. In OpenCluster, a lightweight thread pool has been implemented so that users can easily implement multithreading. Additionally, OpenCluster provides a base class of manager (manager.py), which includes scheduling task methods (a function called "schedule" (task)). User-defined managers are required to inherit the base class and override the schedule function. The schedule function contains only the *WorkPiece* type argument. The work is done when it returns, and the manager must inform the factory of the relevant statistics of the project.

2.2.5 WorkPeice

DataItem is added to OpenCluster to simplify programming, which is inherited from the Python dictionary [13] and provides a set of convenient functions. DataItem can be treated as an unordered set of key/value pairs indexed by keys, which can be any immutable types such as strings and numbers. The value can store arbitrary data formats. Thus, any data structures such as text and binary data can be sent freely. This means that DataItem can accept any number or type of input parameters. It can also return any number or type of output results. Considering that OpenCluster is only a framework, it should not assume the type and quantity of input and output parameters that are user-defined.

2.3 Deployment and operation

It is easy to deploy an OpenCluster system. All we need to do is copy files into a directory and modify the files with related permissions. A web application was developed using the web.py [14] framework to control the starting and stopping of OpenCluster while displaying the status of all operational nodes, managers, services, and workers. During execution, the status updates and statistics are reported by the clients to the factory. They provide useful information for monitoring the processing progress and for detecting errors. The updates include status changes and information about the execution host as well as manager statistics. The web server can run as a standalone daemon or as a CGI script within a more robust web server. The authenticated users can use the web interface to control worker and service instances. Other features of the interface include graphs displaying resource usage and the number of managers in various states. Fig. 9.3 shows a screenshot of the OpenCluster interactive interface.

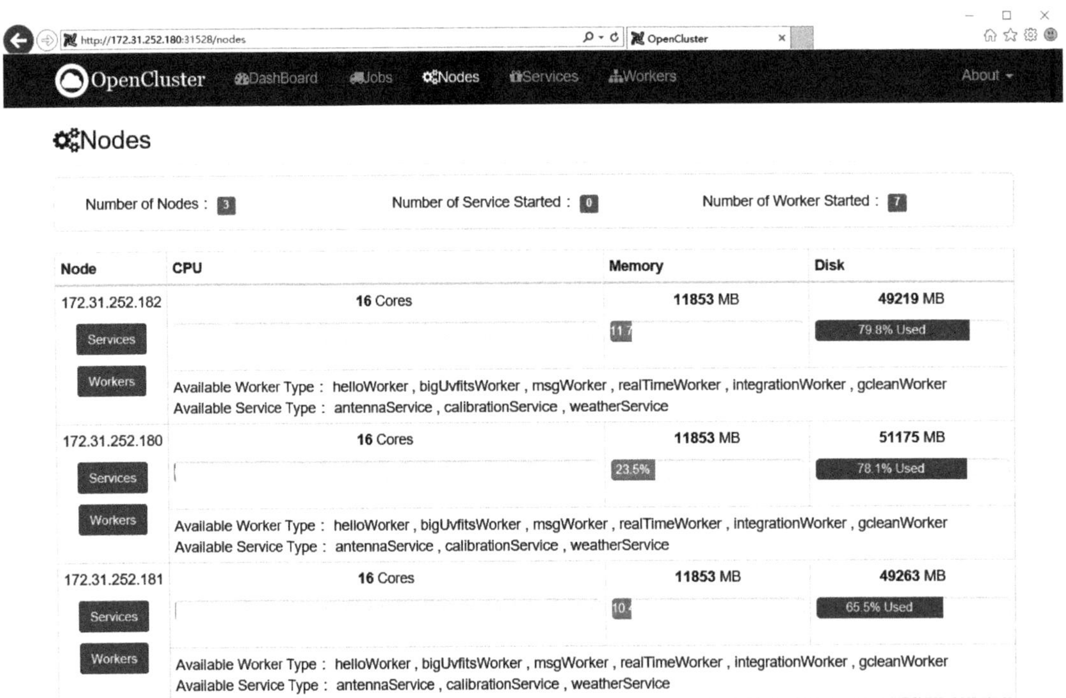

Fig. 9.3 OpenCluster monitoring interface. Reprinted with permission from S. Wei, F. Wang, H. Deng, C. Liu, W. Dai, B. Liang, Y. Mei, C. Shi, Y. Liu, J. Wu,OpenCluster: a flexible distributed computing framework for astronomical data processing, Publ. Astron. Soc. Pac. 129 (972) (2016) 024001. Copyright (2016) IOP Publishing.

In addition, OpenCluster has a number of configurations to fine-tune plant, manager, and worker behaviors. Every configuration has a default value specified in the "config.ini" file, where users can override them. Users must first code the creation of a factory object and execute the "start" function of the factory to run a cluster.

2.4 Stability

Several mechanisms have been considered to enhance the stability of OpenCluster. At least one node will serve as a leader responsible for cluster management and scheduling due to distributed system constraints. Furthermore, when the cluster operates at different stages, it is necessary to ensure that in the event of malfunctions there is a leader to provide services. When a failure occurs, a new leader must be selected by the system.

OpenCluster's leader selection approach is distinct from the one suggested by Paxos. Paxos is a system protocol to reach consensus on a series of proposals. Paxos adopts a significant majority approval process, similar to a parliamentary vote, in terms of maintaining the integrity of leader selection or variable modifications. For example, the selection of the leader is known as a bill requiring more than half the claims of the vote. After bearing the bill, every node has a record number. When the node receives another leader candidate's request again, it refuses the request as it already has the record number.

OpenCluster designs a lightweight courtesy method in leader selection of factories. When a factory instance is started, it asks other factory instances if they would like to be a leader. If there is no leader, the factory performs as the leader. If a factory was already the leader, the factory will perform as a slave. The benefit of this courtesy approach with no competition possibly avoids conflict and maintains a consistency between factories. Fig. 9.4A shows the process of leader selection of a factory. The leader automatically duplicates updates to other factory instances via a background thread. A factory is responsible for providing a registration and lookup for connections of workshops, workers, services, and managers (categorized as clients), just like a directory service, whereas it does not dispatch the computing tasks or data. Thus, much less stress occurs on the CPU or memory in a factory instance. Once the factory leader is determined, the clients deal with the leader. Therefore, the clients must know who the leader is. Fig. 9.4B illustrates the client operational processes before detailed works.

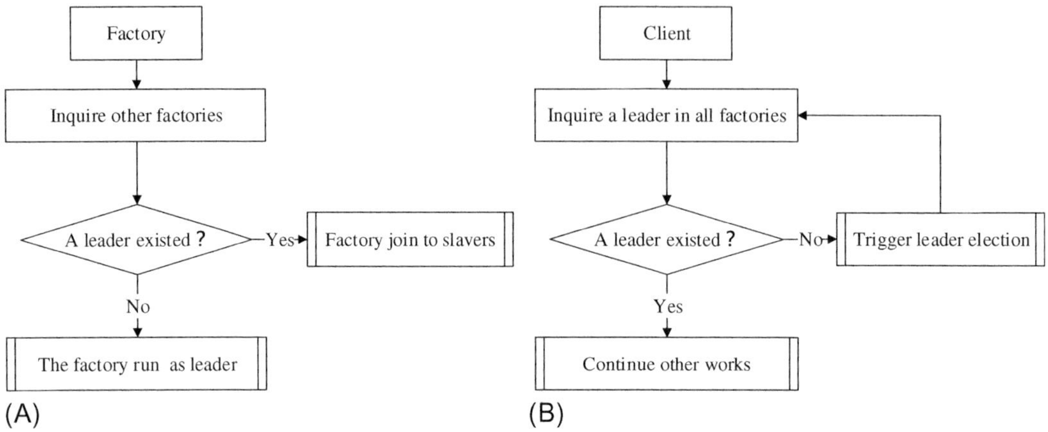

Fig. 9.4 (A) Process of leader selection of a factory; (B) client operational processes before other works. Reprinted with permission from S. Wei, F. Wang, H. Deng, C. Liu, W. Dai, B. Liang, Y. Mei, C. Shi, Y. Liu, J. Wu,OpenCluster: a flexible distributed computing framework for astronomical data processing, Publ. Astron. Soc. Pac. 129 (972) (2016) 024001. Copyright (2016) IOP Publishing.

2.5 Hybrid resource scheduling

On a computing platform that has CPU-only and CPU/GPU hybrid machines, OpenCluster can use hybrid hardware resources. CPU-oriented tasks can be performed on general-purpose nodes (all computers) with the OpenCluster schedule, whereas time-consuming image processing tasks can be performed on GPU-enabled devices (e.g., CPUs/GPUs). This feature is very useful when designing high-performance pipelines, particularly locally executed pipelines in an observatory. Implementing high-performance imaging tasks without supporting GPU is very difficult. Nonetheless, only parts of computers are fitted with qualified GPU cards due to the power consumption limit. OpenCluster supports scheduling mechanisms for hybrid resources owing to its worker registration approach. The worker must register the type of task that he can perform at the factory during start-up. Managers can therefore distinguish worker varieties and assign various tasks to specific workers operating on the correct hardware.

3 DALiuGE

The Data-Activated Liu (Flow) Graph Execution Engine (DALiuGE) is an execution framework designed for the SKA1 SDP consortium. To handle the intensive, continuous, and time and power bounded SKA science data, the DALiuGE execution framework aims to provide a distributed and scalable data processing platform [14a]. Compared to existing distributed execution frameworks, the main advantages of DALiuGE are as follows:

1. It decouples the logical graph of a data processing pipeline/task explicitly from the run-time realization, which separates concerns of various stakeholders such as telescope operators, astronomers, and pipeline developers. What is more, this enables multilayer data processing to be collectively implemented in a homogeneous manner, thus allowing the system to optimize implementation plans using resources and profiling knowledge.
2. It is based on an extended dataflow programming paradigm. Compared to conventional data flow models describing information as tokens moving between nodes around directed edges, we are instead modeling data as graph nodes, elevating it to manageable entities that are implemented in memory as active objects and can be tracked and maintained if necessary. Streaming-oriented applications are integrated to the model to manage continuous data flow, which is the main concept of the paradigm. The generalized nodes in the execution framework are termed **Drops**, specifically, **Application Drops** (tasks) and **Data Drops** (data) are defined during pipeline design.
3. The Drop enables both data and applications to receive and trigger events that pass through the edges of the graph, and thus cascaded parallel processing tasks are allowed to be performed. This fulfills the idea of "data-activated" execution in which Data Drop determines whether and when to trigger the following tasks. This completely decentralizes the orchestration of execution, making the framework scalable to at least tens of millions of nodes.
4. It integrates a data lifecycle management component within the execution engine, which keeps track of Drops, automatically migrating or deleting them when necessary.

3.1 DALiuGE works mode

Typically, there are six stages to develop a data processing pipeline using DALiuGE. Fig. 9.5 gives an explanation.

1. Logical Graph design,
2. Logical Graph template composition,
3. Logical Graph Selection,
4. Translate the Logical Graph to Physical Graph Template,
5. Deployment of Physical Graph, and
6. Physical Graph Execution.

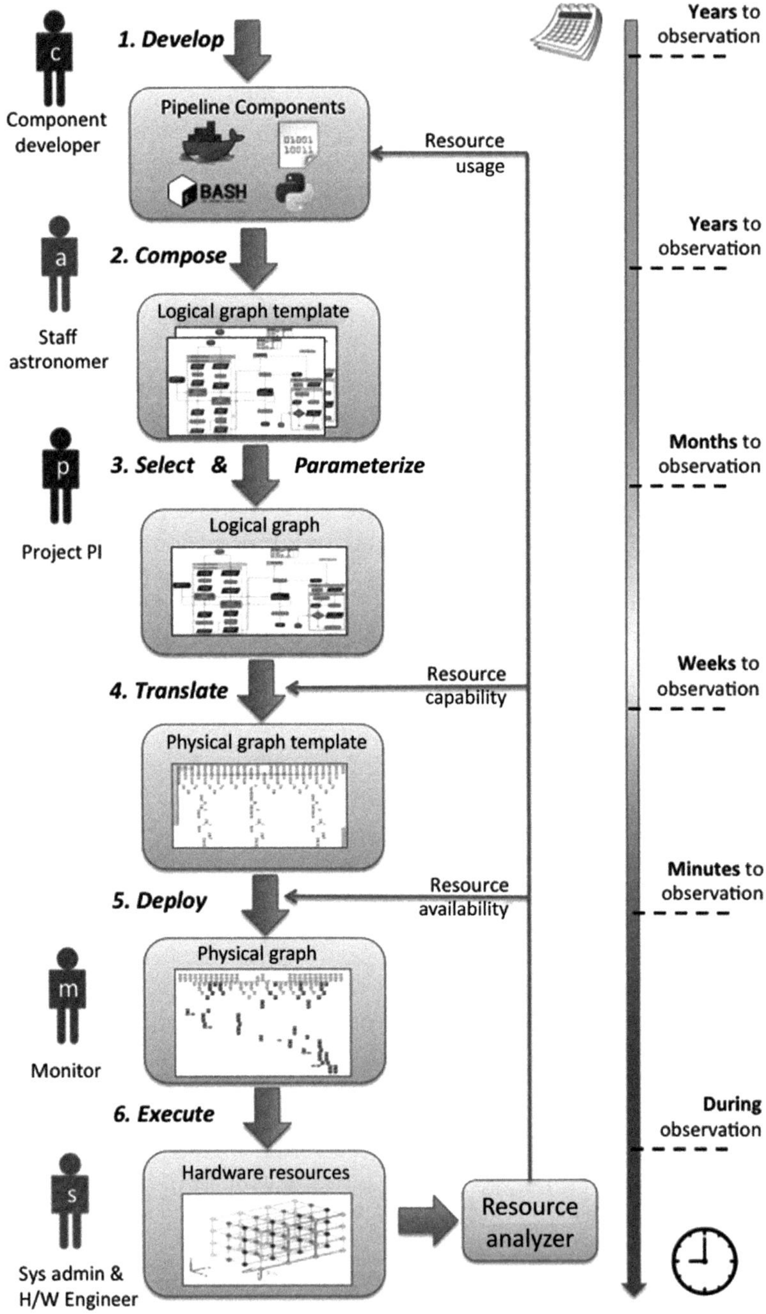

Fig. 9.5 The DALiuGE system and the various stages of the lifecycle of a pipeline in the SKA SDP sense. Reprinted with permission from C. Wu, R. Tobar, K. Vinsen, A. Wicenec, D. Pallot, B. Lao, R. Wang, T. An, M. Boulton, I. Cooper, et al., DALiuGE: a graph execution framework for harnessing the astronomical data deluge, Astron. Comput. 20 (2017) 1–15. Copyright (2017) Elsevier.

It is important to enumerate the stages during designing and implementing a pipeline to clearly distinguish the various issues related, allowing multiple participants to concentrate on what they do best.

Though we define the earlier-mentioned stages in the SKA SDP context, they generally apply to many astronomical processing pipelines being developed, deployed, and executed.

Stage 1. Wrapping the astronomical data processing *Pipeline Component* as ApplicationDrops. Tasks such as Docker images, Python modules, or shell scripts can currently be implemented. That is to say, existing astronomical data processing code can be reused in DALiuGE with few modifications. The *Resource Analyzer* calculates the *Resource Usage* information and provide parameters for the pipeline components. It is necessary to develop and optimize pipeline components for actual data processing needs. While the tasks of computing are stateless, the ApplicationDrops are stateful as they go through the Drop lifecycle (Fig. 9.5)—to initialize, deploy, expire, and delete according to the state at different stages.

Stage 2. A *Logical Graph Template* is designed by astronomers to handle high-level data processing needs (e.g., MS file generation, calibration). The pipeline components of the *Logical Graph Template* are tested and released and resource-oblivious dataflow constructs are used. A Logical Graph Template leaves room for investigators (PIs) to change the procedure to satisfy their research goals. For example, the number of output channels for imaging and subsequent data products can be defined. Generally, a Logical Graph Template specifies the data and component orchestration, which will be offered to the PIs after careful testing and verification. The collection of Logical Graph Template basically describes different data processing needs of the SKA SDP, and which will be released in a fully controlled repository for configuration and version.

Stage 3. When creating a detailed observation, the PI will first select one of the released Logical Graph Templates and specify values of parameters for the observation. This will allow the execution framework to convert the Logical Graph model to a real *Logical Graph*, which is dedicated to the specified observation, in accordance with the observation definition.

Stage 4. The DALiuGE translates a Logical Graph into a *Physical Graph Template*. The *Physical Graph Template* is an unrolled and logical representation of the Logical Graph not tied to specific physical computing resources. This translation is automated and the information collected from the resource analyst *Resource Capability* is used to acquire the cost targets. It should be noted that a single node in a Logical Graph may correspond to several nodes in a Physical Graph Template.

Stage 5. The Physical Graph Template will be deployed when actual data processing is required. The DALiuGE first retrieves the corresponding Physical Graph Template from the graph repository and then generates a *Physical Graph* effectively by associating each node in the Physical Graph Template with an available resource unit, based on the *Resource Availability* (real-time resource usage such as compute node, storage, etc.) from the Resource Analyzer. The DALiuGE deploys the Physical Graph on the *Drop Managers*, a group of daemon processes responsible for the run-time aspect of DALiuGE, minutes before an observation begins. Therefore, *Drops* are created, which are responsible for wrapping and handling both the data and the applications contained in the Physical Graph, and driving the execution of the Physical Graph. The concept of Drops will be discussed in detail in Section 3.8.

Stage 6. The execution stage takes place once the observation finally starts, and the execution cascades down automatically through the edges of the Physical Graph. Some data are persistently stored as science-ready items when all tasks are completed. To continuously profile workload behaviors for each job, the DALiuGE Drop managers can be fitted with plugins to any third-party Resource Analyzer. Such resource usage information can be fed back to the pipeline component development.

Note that for a specified data processing pipeline, the first two stages are executed only once, regardless of how many observations use the same pipeline definition (unless the logic of the pipeline changes that would produce a new Logical Graph Template).

3.2 Develop

How pipeline components are developed to separate the interests of four main stakeholders/roles is described in Fig. 9.6.

First, domain specialists concentrate on algorithms designed to solve a particular domain-specific issue. These algorithms are implemented by the developers of the Pipeline module, which are later converted into ApplicationDrops. DALiuGE has built-in support for wrapping common task types, for example, system executables, bash commands, Docker containers, and so on.

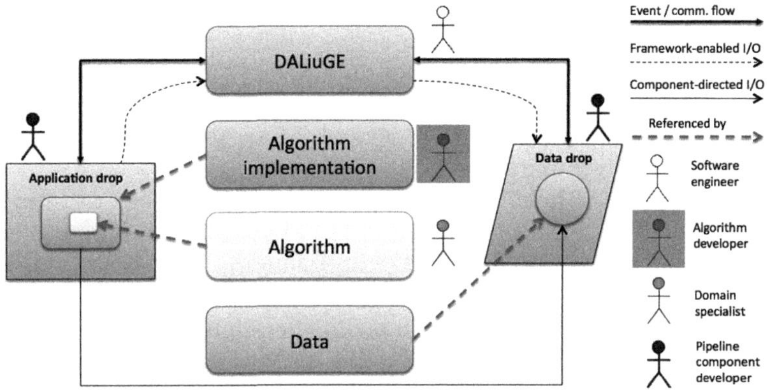

Fig. 9.6 During development process, Drops are associated with algorithms, pipeline components, data, and the DALiuGE graph execution framework, which allows separating concerns. Reprinted with permission from C. Wu, R. Tobar, K. Vinsen, A. Wicenec, D. Pallot, B. Lao, R. Wang, T. An, M. Boulton, I. Cooper, et al., DALiuGE: a graph execution framework for harnessing the astronomical data deluge, Astron. Comput. 20 (2017) 1–15. Copyright (2017) Elsevier.

Developers of pipeline components are also responsible for representing and wrapping data sets in Data Drops, either using a built-in Drop form, expanding one, or writing their own.

3.3 Compose

A Logical Graph Template describes logical operations in a data processing pipeline without regard to observation details and hardware resources.

The building blocks of a Logical Graph Template are called **Constructs**, which are elements exposed to the Logical Graph Template developer through the Logical Graph Template editor. **Data** and **Component** are two basic constructs. Data Drops and Application Drops are instantiated, respectively, from the two constructs. Remember that one Data or Component construct in a Logical Graph Template could result in Physical Graph Template producing several Data or Application Drop instances. Each construct has several related properties, the values of which will be calculated in Section 3.4. The Data and Component constructs, in particular, reveal the properties of *data volume* and *execution time*, respectively, indicating how much data are stored in Data and how long a task will take to execute. Values of these properties can be derived directly from parametric models or calculated from the Resource Analyzer's resource utilization data.

Data and component are linked by inputs or outputs. This relation principle is essential to the development of dataflow-like programs where both tasks and data are graph nodes. A variety of control flow constructs, in addition to the basic constructs, make it possible to create more complex Logical Graph Templates.

They form the structure of the logical graph, and determine the generation of the final Physical Graph Template (shown in Fig. 9.9). Currently, the following flow constructs are supported by the DALiuGE:

- ***Scatter*** represents data parallelism. Data fed into a *Scatter* construct are broken down into several partitions according to the parameter setting, and a *Scatter* construct consumes one of the partitions of Data within the enclosing *Scatter*. The Scatter property `num of copies` controls the number of partitions to be generated. The DALiuGE uses this parameter when creating parallel execution branches.
- ***Gather*** represents a data barrier. Constructs within a *Gather* construct consume the data partition series as a whole. *Gather* has a `num_of_inputs` property that indicates the number of partitions that one instance can handle. The DALiuGE uses this property in-turn to determine how many *Gather* instances to be generated in the physical graph. As shown in Fig. 9.8, *Gather* can be used together with *GroupBy* so that data held in a series of groups are processed together by components enclosed inside the *Gather* component.
- ***GroupBy*** carries out data reordering, also known as the problem of corner turning in radio astronomy [15]. The semantic of `Group By` is similar to that in the SQL statement used in relational databases. It is comparable to the "static" shuffling of MapReduce, where all `Reducers` collected keys are priori known. Fig. 9.7 gives an explanation of the *GroupBy* construct. In this example, a list of two-dimensional (2D) points are first sorted based on the axes order of (x, y), and then resorted based on the axes order of (y, x) after performing *GroupBy*.

  GroupBy and nested *Scatter* constructs are required to be used together so that data Drops (e.g., Calibrated MS file) originally sorted by the outer partition—`Scatter by Timeslice`—are resorted by the inner partition—`Scatter by Channel`. Fig. 9.8 shows an example.
- ***Loop*** indicates iterations. Constructions within a loop are produced repeatedly for a specified number of times. Currently, dynamic branch conditions for Loop are not supported by DALiuGE. A property named `num_of_iterations` can be

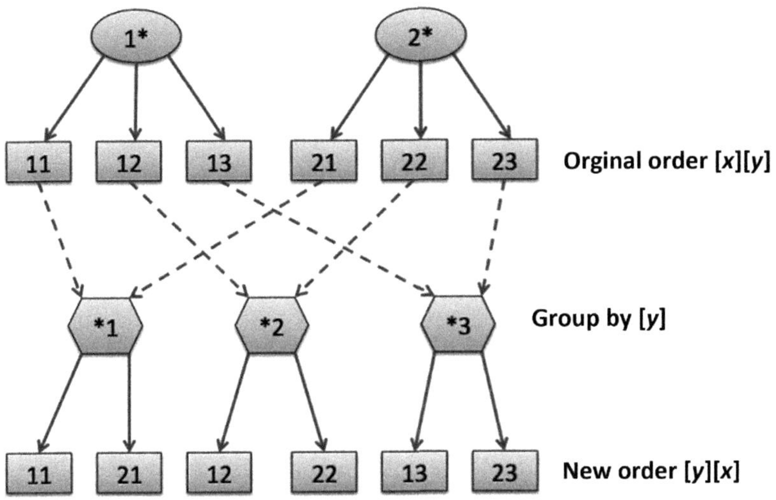

Fig. 9.7 *GroupBy* construct example, resorting a list of 2D points from axes (x, y) to axes (y, x). Reprinted with permission from C. Wu, R. Tobar, K. Vinsen, A. Wicenec, D. Pallot, B. Lao, R. Wang, T. An, M. Boulton, I. Cooper, et al., DALiuGE: a graph execution framework for harnessing the astronomical data deluge, Astron. Comput. 20 (2017) 1–15. Copyright (2017) Elsevier.

assigned by each Loop construct. Fig. 9.8 gives an example of the Loop construct.

To allow astronomers to design Logical Graph Templates and manage a Logical Graph Template repository easily, A Javascript-based Logical Graph Editor Web browser has been developed. Such astronomers would be trained staff astronomers in an operating observatory setting and familiar with the details of the available constructs. At present, the repository is performed as a managed file system directory. Each Logical Graph Template is stored as a JSON-formatted textual file, which can be remotely accessed and modified through the Logical Graph Editor via its RESTful interface.

3.4 Select and parameterize

The following abbreviations are used to simplify the text in the follow-up paragraphs:

Principle investigator	PI
Observation	OBS
Logical Graph Template	LGT
Logical Graph	LGR
Physical Graph Template	PGT
Physical Graph	PGR
Drop Manager	DM

Several months before an observation, the PI of a science project is in charge of generating the Logical Graph for OBS. A suitable

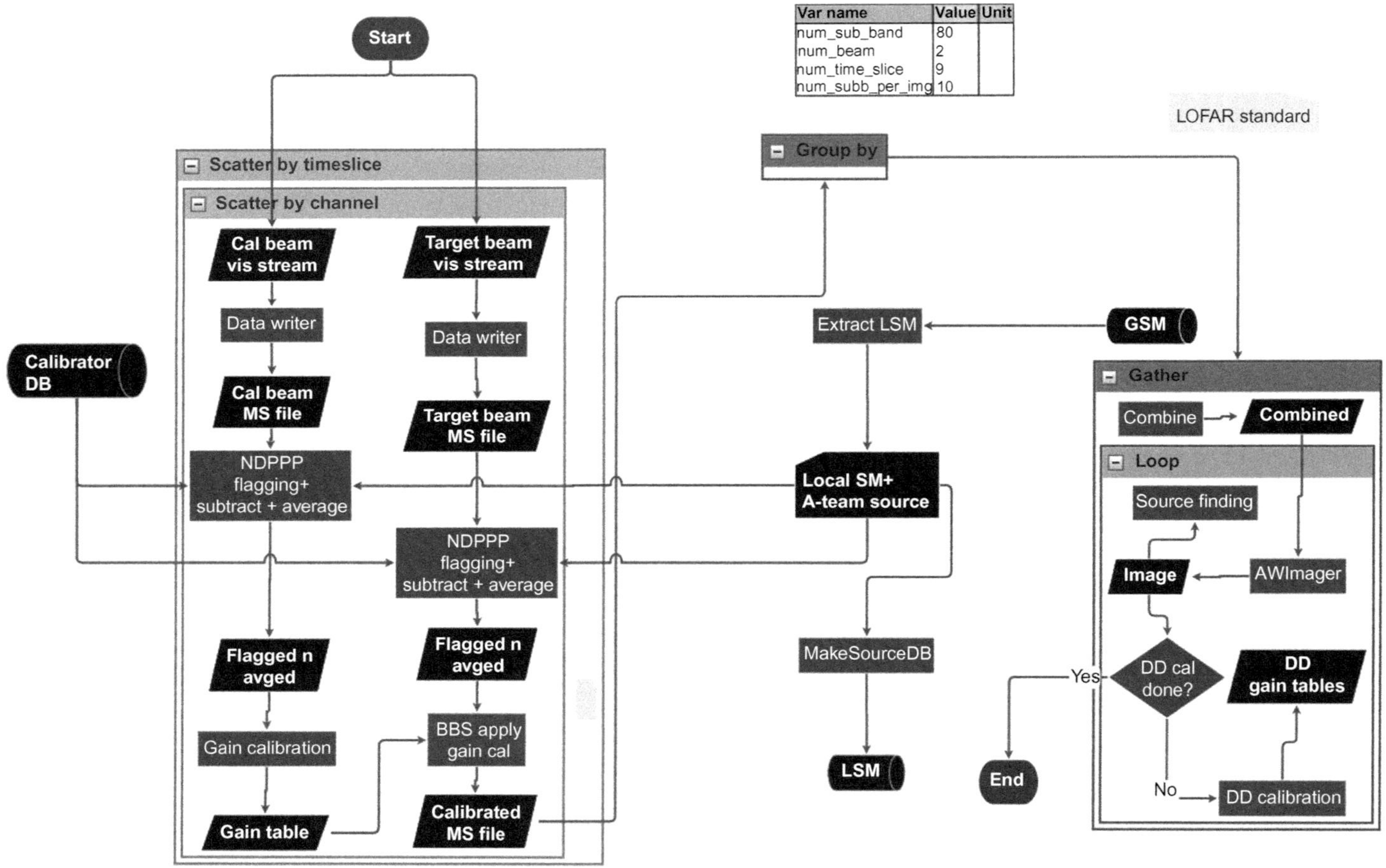

Fig. 9.8 The Logical Graph Template describing the LOFAR telescope [16] imaging pipeline [17]. The original data is split by time and frequency domain using two *Scatter* constructs. A *GroupBy* construct groups the data by frequency and a *Gather* and *Loop* construct for the imaging process. Reprinted with permission from C. Wu, R. Tobar, K. Vinsen, A. Wicenec, D. Pallot, B. Lao, R. Wang, T. An, M. Boulton, I. Cooper, et al., DALiuGE: a graph execution framework for harnessing the astronomical data deluge, Astron. Comput. 20 (2017) 1–15. Copyright (2017) Elsevier.

LGT from the LGT database is first selected and specific parameter values are supplied for all LGT properties according to an observation schedule. The values filled in by the PI plan are the only difference between LGT and LGR.

3.5 Translate

A Logical Graph provides a concise way to express a complex processing pipeline and contains high-level control flow parameters; however, it cannot be directly mapped to resources by the underlying graph execution engine. So, in DALiuGE, a Logical Graph is translated to a Physical Graph Template (PGT) to do this. The translation contains three steps: validation, unrolling construction, and logical partitioning.

- First, the Logical Graph and the constructs included are validated to ensure the translation. For instance, the Logical Graph cycles are not allowed in the current DALiuGE execution framework. This process is similar to checking the syntax of the compiler to ensure the correctness of the processed program structure.
- Second, all necessary Drop specifications are created (may also include Drops that do not appear in the original logical graph) and the logical graph is unrolled. And directed edges of these Drop specifications are established to link them together in the PGT. An example of a PGT in Fig. 9.9 explains how logical graph constructs (in Fig. 9.8) are unrolled to generate a series of Drops according to the semantics of the constructs.
- Finally, the PGT is divided into a series of logical partitions and an execution sequence of the Drops is produced within each partition. Therefore, under certain conditions (e.g., resource footprint, collocation criteria, device locality, etc.), performance requirements such as total data movement, total running time can be met. Under the assumption that hardware resources are homogeneous, two graph partitioning algorithms—*min_time* and *min_res*—are implemented on the DALiuGE currently.

  The *min_time* algorithms produce an optimal partition number so that the total pipeline completion time, which depends both on execution time and the cost of data movement, is minimized. In addition, the number of drops running in parallel in a single partition at any point in time is also guaranteed to be no greater than a degree of parallelism (DoP) threshold.

  The *min_res* algorithms produce a minimized partition number to satisfy the deadline of completion and the threshold

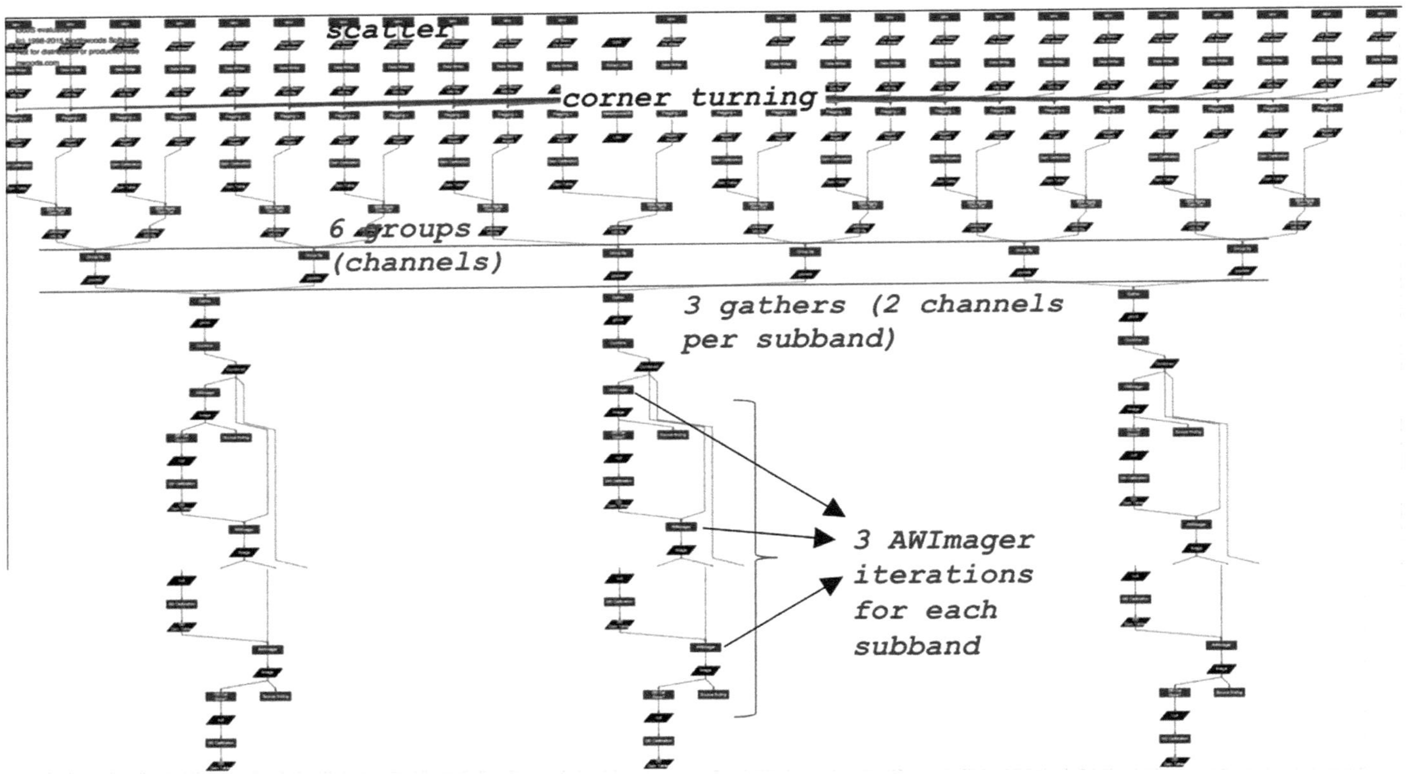

Fig. 9.9 A Physical Graph Template unrolled from the Logical Graph Template in Fig. 9.8. Reprinted with permission from C. Wu, R. Tobar, K. Vinsen, A. Wicenec, D. Pallot, B. Lao, R. Wang, T. An, M. Boulton, I. Cooper, et al., DALiuGE: a graph execution framework for harnessing the astronomical data deluge, Astron. Comput. 20 (2017) 1–15. Copyright (2017) Elsevier.

of DoP constraints. The number of partitions corresponds to the resource footprint because a partition has a restricted DoP. Referring to the method for hardware/software codesign used in embedded systems [18], a look-ahead search strategy and several stochastic local search heuristics such as optimization [19] and particle swarm simulated annealing [20] are used to find a global optimum during graph partitioning.

3.6 Deploy

To execute a PGT, the DALiuGE retrieves it from the graph repository first. Each Drop specification in the PGT is assigned to an available resource. By doing this, the Physical Graph Template is transformed to a **Physical Graph** (PG), which is bound to actually available resource units at the time of execution. Eventually, the DALiuGE submits the PG to the **Drop Managers** (DM) to instantiate and execute.

3.6.1 Resource mapping

In this phase, logical partitions of the PG are mapped to a set of available resources in certain optimal ways such as minimum cost of data movement and load balancing. A two-phase scheduling method [21] is used here to separate the previous graph partitioning step, which is somewhat oblivious of underlying resources, from the resource mapping step, which assigns each Drop onto a physical compute node in the cluster. Real-time information of available resources from the **Resource Analyzer** is required in such placement, which is shown in Fig. 9.5. Similar to the last step, we assume that nodes in the resource pools have the same storage, computing, and interconnect capabilities. The METIS software library [22], whose internally is a multilevel k-way partition algorithm [23], to merge the p PGT partitions into m virtual clusters if $p > m$, where m represents the current available nodes number.[a] If the resources are all homogeneous, physical mapping the m merged clusters to m compute nodes becomes a straightforward round-robin assignment. Once a mapping is established from a cluster c to a compute node n, the corresponding network address will be assigned to all Drops in the cluster.

3.6.2 Drop managers

Drop Managers are a collection of daemon processes for managing and executing PGs. DMs provide external users with a single interface to communicate with the run-time framework, which

[a]http://glaros.dtc.umn.edu/gkhome/metis/metis/overview

allows users to submit and deploy physical graphs and monitor the status of graph execution. DMs are hierarchically organized for considerations of scalability. Currently implemented hierarchy levels comprise three levels; however, the current design is versatile, allowing additional intermediate levels to be introduced in the future if appropriate. There is a **Node Drop Manager** at the bottom of the DM hierarchy for each compute node. Basically, they are responsible for the creation and removal of Drops. Because compute nodes are organized into Data Islands, a **Data Island Drop Manager** operates at the level of Data Island, maintaining a Node Drop Manager list. And, a **Master Drop Manager** is used to manage all Data Island Managers to monitor a single contact point. To represent a given PG execution, drop managers introduce the notion of a **Session**, which are absolutely isolated from each other. This allows the deployment and parallel execution of multiple PGs within a given Drop Manager, providing the suggested available resource by the Resource Analyzer is sufficient to generate valid resource mapping during the translation step. Sessions are defined with a simple life cycle: creation, attaching a PG (complete or a partial), and deployment. It leaves the session in a running status until the execution is complete, and the session can be deleted after completion.

3.6.3 Physical graph deployment

A Drop Manager will prepare the graph deployment on resources it manages once it receives a PG. The deployment procedure, as shown in Fig. 9.10, goes through the DM hierarchy recursively. For Node Drop Managers, the lowest level in the DM hierarchy, this includes validity checking of the PG and creation of the Session and all its drops. While for a Data Island Drop Manager *d*, this procedure separates the PG according to information of the node placement. Hence, there will be no edges between Drops belonging to different computing nodes *n*, and the subgraph will be submitted to the Drop Manager, which is running on *n*. The Data Island Drop Manager records edges from the original PG crossing node boundaries and subsequently communicates them to relevant node managers to create necessary connections between them. Similarly, the island placement information will be used for the Master Drop Manager to split the PG.

3.7 Execute

The Physical Graph root Drops (i.e., those without inputs or producers) will be triggered to execute during the execution phase. The data of root Data Drops are assumed to be present

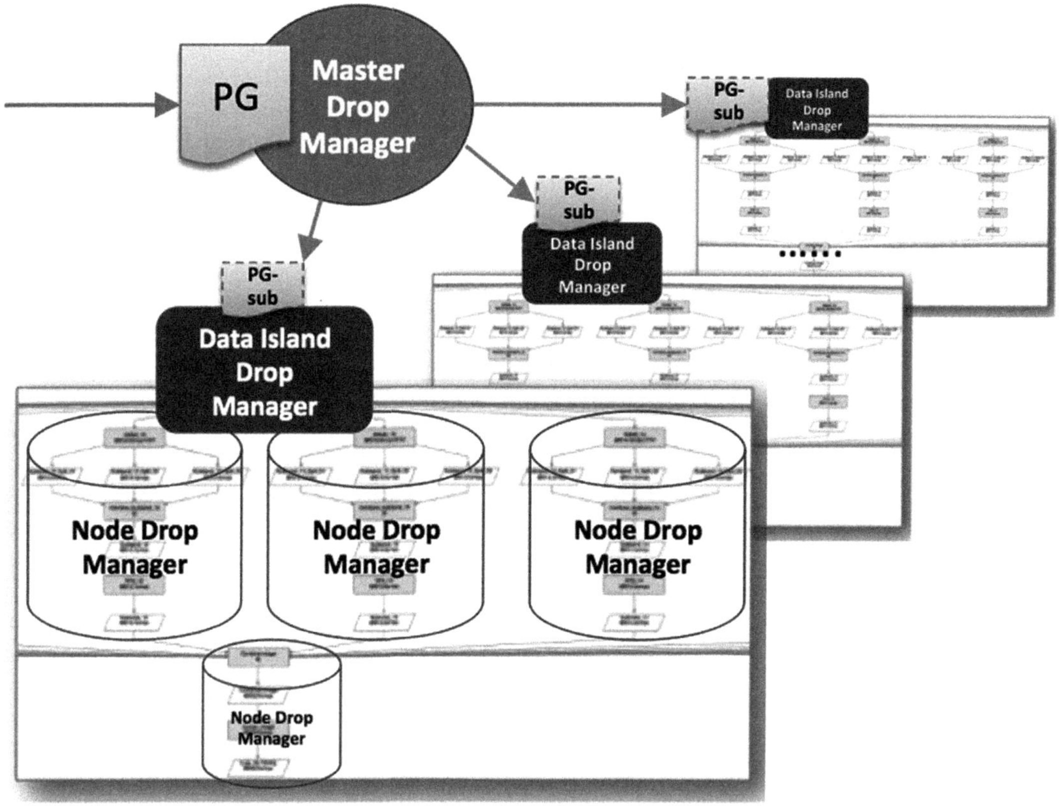

Fig. 9.10 Deployment of physical graph recursively goes through the Drop Manager hierarchy. Reprinted with permission from C. Wu, R. Tobar, K. Vinsen, A. Wicenec, D. Pallot, B. Lao, R. Wang, T. An, M. Boulton, I. Cooper, et al., DALiuGE: a graph execution framework for harnessing the astronomical data deluge, Astron. Comput. 20 (2017) 1–15. Copyright (2017) Elsevier.

and are therefore marked as completed. Once a Data Drop is marked as complete, the incident will be released to all its consumers, who will then determine whether to begin execution on the basis of their configuration. Particularly, applications in batch-style Drops will wait for the completion of all their inputs to begin their execution. Data Drops will correspondingly obtain an event when their producers finish their execution. If all producers of a Drop finish execution, the Drop will mark itself as complete and notify its consumers. The Drops therefore have the ability to conduct their own execution, but not rely on a central organizer.

Failures on Drops are propagated automatically though events in a similar fashion. If any of its producers is marked as ERROR, the Data Drops switch to ERROR status. When an error-tolerant threshold, *t*, inputs move to ERROR, an Application Drop moves to ERROR. Setting t to a value greater than zero allows other execution branches to fail without stopping the main execution branch from advancing if there are appropriate inputs after an error-tolerant collection point is reached.

If there are enough inputs reaching an error-tolerant gathering point, setting t to a value greater than zero allows certain branches of execution to fail without preventing the main execution branch from making progress.

Fig. 9.11 shows a physical graph execution with simulated failures. To produce this execution, we have intentionally let some Application Drops fail (either raising random exceptions or blocking the event flow) at the upstream of the graph execution. Consequently, many downstream Drops have been marked in Red as failures through event propagation. Because the error-tolerant threshold t was set to 50% in this example, the light-yellow Application Drop *A2* is still waiting for one of its inputs *D2* to get ready, although the other input Data Drop *D1* has already entered the ERROR state. The root cause of such waiting is the artificially blocked event flow from the Application Drop *A1*. As a result, many downstream Drops are also in the WAITING state. Eventually, *A2* will also enter ERROR due to timeout, rendering the entire rest of the main branch in the failure mode.

3.8 Drop

In a nutshell, Drops [24] are software objects wrapping a generic payload. Fig. 9.12 shows the simplified UML diagram of the Drop system in DALiuGE. All the various Drop types, described in the next section, are derived from the abstract `Drop` class. The abstract `ContainerDrop` class implements a hierarchical grouping of Drops into a single entity, and currently has two concrete subclasses—`ApplicationDrop` and `DataProductDrop`. The `ApplicationDrop` container allows grouping together everything required to make an actual application stand alone into a single Drop. This is equivalent to the way applications are bundled in Mac OSX [25] or inside a Docker container [26]. The `DataProductDrop` container allows bundling Drops together that belong to a specific SKA data product. In general, such a `DataProductDrop` will consist of thousands of individual Drops, including Data Drops and potentially ApplicationDrops, depending on the definition of what the actual data product entails. For the system, this design allows managing and handling data products as single entities.

The Drop object wrapper associates methods and provenance as well as life cycle and access control data with the payload, thus

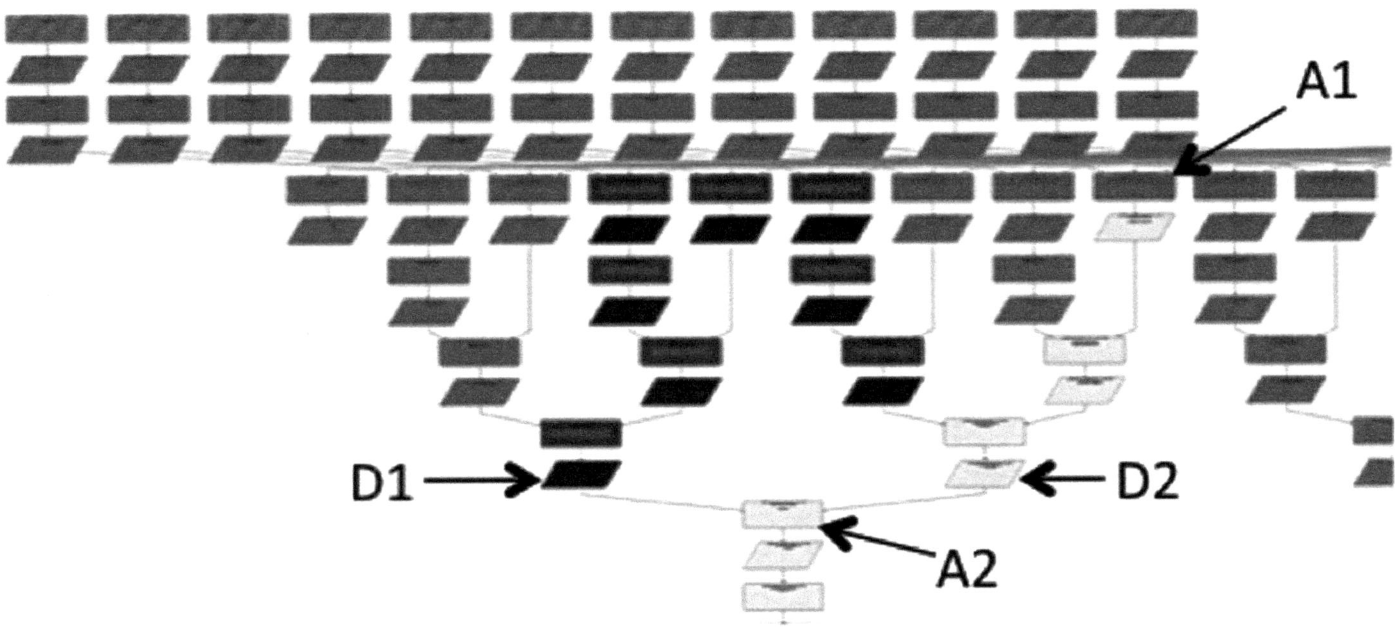

Fig. 9.11 A Physical Graph execution in case of failure. The Application Drop *A2* is configured with $t = 50\%$ and therefore can still continue its execution if the Data Drop *D2* is successfully completed. Reprinted with permission from C. Wu, R. Tobar, K. Vinsen, A. Wicenec, D. Pallot, B. Lao, R. Wang, T. An, M. Boulton, I. Cooper, et al., DALiuGE: a graph execution framework for harnessing the astronomical data deluge, Astron. Comput. 20 (2017) 1–15. Copyright (2017) Elsevier.

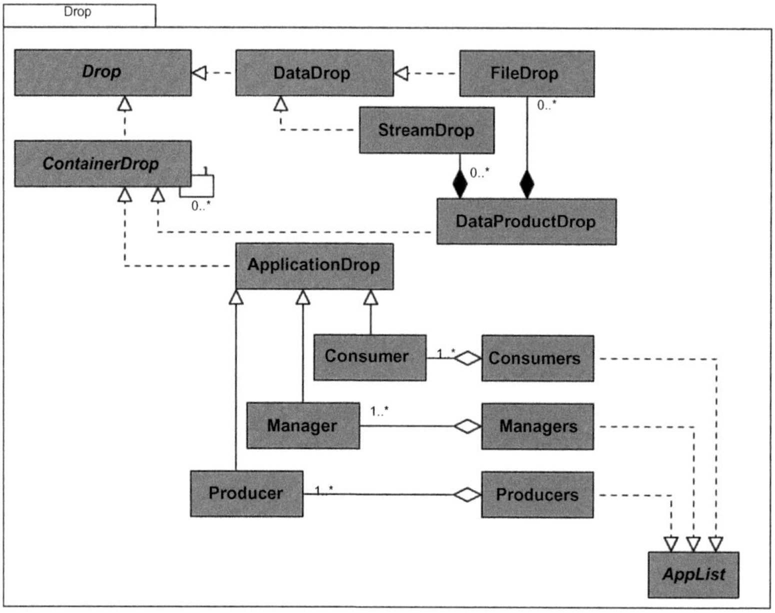

Fig. 9.12 The Drop object diagram in UML notation. This diagram shows the abstract Drop object as well as some of the existing concrete objects. Reprinted with permission from C. Wu, R. Tobar, K. Vinsen, A. Wicenec, D. Pallot, B. Lao, R. Wang, T. An, M. Boulton, I. Cooper, et al., DALiuGE: a graph execution framework for harnessing the astronomical data deluge, Astron. Comput. 20 (2017) 1–15. Copyright (2017) Elsevier.

making data virtually active. In particular, it means that a Data Drop can react and raise events, which then drive the actual execution of the Physical Graph in DALiuGE. While a Drop's payload is stateless (write once and read multiple times), Drops themselves are stateful, which not only allows us to manage Drops through persistent check-pointing, versioning, and recovery after restart, but also enables data sharing among multiple processing pipelines in situations such as reprocessing or commensal observations. Each Drop exposes its payload location via a URL, has a unique identifier, and is instantiated, monitored, and destroyed by a Drop Manager, which subscribes to the events fired by the Drops as they transition through their different states.

Fig. 9.13 shows how Application Drops and Data Drops are connected to each other with the former being either *consumers* or *producers* of the latter (or *inputs* and *outputs* from the opposite point of view, respectively). DALiuGE distinguishes between normal inputs/consumers and their streaming counterparts. A normal consumer application works in a batch-like fashion, running only after all its inputs are marked as completed, which in turn occurs only when their producer applications have finished. Streaming consumers work instead by continuously consuming their input data, which is continuously produced in turn by a producer application.

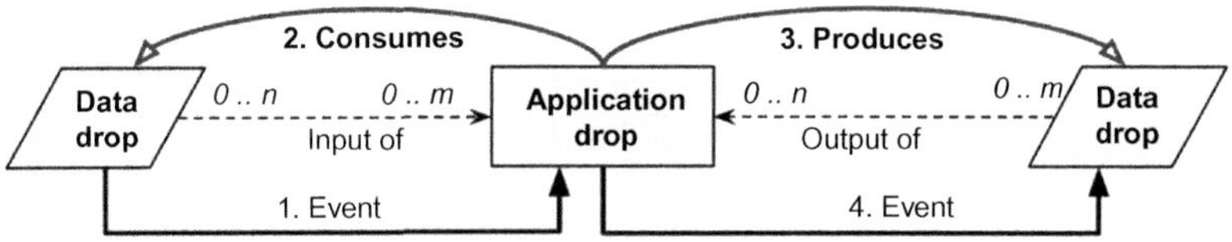

Fig. 9.13 Data drops and application drops are conceptually related and connected to form an event flow/graph. Note that there is a n..m relationship between Data Drops and Application Drops, allowing for scattering, gathering, and reordering applications. Reprinted with permission from C. Wu, R. Tobar, K. Vinsen, A. Wicenec, D. Pallot, B. Lao, R. Wang, T. An, M. Boulton, I. Cooper, et al., DALiuGE: a graph execution framework for harnessing the astronomical data deluge, Astron. Comput. 20 (2017) 1–15. Copyright (2017) Elsevier.

3.8.1 Drop channels

During execution, Drops need two different channels of information exchange. The first is to send events to interested parties, which is a crucial mechanism for the graph execution as explained in Section 3.7. The second is to be able to interact with other Drops and query information about their properties, such as their `dataURL`, `uid`, or other properties. In both situations, Drops residing in the same node (and therefore in the same process as the Node Manager) use direct object invocation to achieve communication. If two Drops reside in two different nodes, we use a publish/subscribe mechanism to fire events between nodes, and a Remote Procedure Call interface to allow Drops to communicate with each other. Note that these channels are merely communication channels, as they do not carry the data payload. The DALiuGE design very cleanly separates communication from bulk data operations, which allows us to change the underlying mechanisms independently, but it also allows binding those two very different traffic types to dedicated networks, if the underlying platform provides that distinction such as the former Fornax supercomputer at the University of Western Australia [27].

3.8.2 Drop I/O

Currently, two options are available for Application Drops to perform I/O operations on Data Drop payloads:

1. DALiuGE Data Drops expose their data payload via a simple I/O abstraction layer. This is designed based on the basic `open`, `read`, `write`, and `close` POSIX calls, which follow a byte stream data model. A pipeline component developer can use these built-in POSIX calls ("Framework-enabled I/O" dotted arrows in Fig. 9.6). Moreover, the abstraction can be extended by refining a set of reusable classes, which allow I/O framework developers to realize more

sophisticated I/O methods or optimizations (e.g., RDMA-based data transfer). In either case, the data are passing through the Drop instance, and the I/O framework will take care of remote I/O access across node boundaries. Neither the pipeline component developers nor the logical graph developers will be exposed to those details. In particular, this enables a fully transparent replacement of both storage methods as well as storage formats to achieve a better throughput.

2. The pipeline component developer can instruct the algorithm implementation to directly perform I/O without going through DALiuGE ("Component-directed" solid arrows in Fig. 9.6). In this case, the Application Drop obtains the `dataURL` (e.g., POSIX file paths, RDMA buffer pointers, etc.) of the Data Drop, and passes it onto the algorithm implementation as a parameter. It is the responsibility of the pipeline component developer to ensure that I/O is occurring in the correct location.

Acknowledgments

The work in this chapter is supported by the National Key Research and Development Program of China (2018YFA0404603) and the Joint Research Fund in Astronomy (U1831204 and U1931141), the National Natural Science Foundation of China (11903009), the Yunnan Key Research and Development Program (2018IA054), and the major scientific research project of Guangdong regular institutions of higher learning (2017KZDXM062).

Thanks for the International Center for Radio Astronomy Research (ICRAR) and the resources provided by the Pawsey Supercomputing Center with funding from the Australian Government and the Government of Western Australia. This work is also supported by Astronomical Big Data Joint Research Center, cofounded by National Astronomical Observatories, Chinese Academy of Sciences and Alibaba Cloud.

References

[1] E.D. Feigelson, G.J. Babu, Modern Statistical Methods for Astronomy: With R Applications, Cambridge University Press, Cambridge, MA, 2012.

[2] The square kilometre array—exploring the universe with the world's largest radio telescope, 2016. https://www.skatelescope.org/.

[3] R. Braun, E. Keane, T. Bourke, J.A. Green, J. Wagg, Advancing astrophysics with the square kilometre array, Proc. Sci. 174 (2015).

[4] G.H. Tan, T.J. Cornwell, P.E. Dewdney, M. Waterson, The square kilometre array baseline design V2.0. 2015 1st URSI Atlantic Radio Science Conference (URSI AT-RASC), May 2015, p. 1, https://doi.org/10.1109/URSI-AT-RASC.2015.7303195.

[5] S. Ratcliffe, F. Graser, B. Carlson, B. Opperman, SKA1 low SDP–CSP interface control document, SKA Project Document Number 100-000000-002 1 (1) (2016).

[6] S. Ratcliffe, F. Graser, B. Carlson, B. Opperman, SKA1 mid SDP–CSP interface control document, SKA Project Document number 300-000000-002 1 (1) (2016).
[7] P.C. Broekema, R.V. van Nieuwpoort, H.E. Bal, The square kilometre array science data processor. Preliminary compute platform design, J. Instrum. 10 (7) (2015) C07004.
[8] I. Raicu, I.T. Foster, Y. Zhao, Many-task computing for grids and supercomputers, 2008 Workshop on Many-Task Computing on Grids and Supercomputers, November, 2008, pp. 1–11, https://doi.org/10.1109/MTAGS.2008.4777912.
[9] I. Raicu, Many-Task Computing: Bridging the Gap Between High-Throughput Computing and High-Performance Computing (Ph.D. thesis), The University of Chicago, 2009.
[10] P. Dewdney, W. Turner, R. Braun, J. Santander-Vela, M. Waterson, G.-H. Tan, SKA1 System BaselineV2 Description, Document number SKA-TEL-SKO-0000308 1 (1) (2015). https://www.skatelescope.org/wp-content/uploads/2014/03/SKA-TEL-SKO-0000308_SKA1_System_Baseline_v2_DescriptionRev01-part-1-signed.pdf.
[11] C. Wu, A. Wicenec, D. Pallot, A. Checcucci, Optimising NGAS for the MWA archive, Exp. Astron. 36 (3) (2013) 679–694.
[12] S.J. Tingay, R. Goeke, J.D. Bowman, D. Emrich, S.M. Ord, D.A. Mitchell, M. F. Morales, T. Booler, B. Crosse, R.B. Wayth, The Murchison widefield array: the square kilometre array precursor at low radio frequencies, Publ. Astron. Soc. Aust. 30 (2013) e007.
[12a] S. Wei, F. Wang, H. Deng, C. Liu, W. Dai, B. Liang, Y. Mei, C. Shi, Y. Liu, J. Wu, OpenCluster: a flexible distributed computing framework for astronomical data processing, Publ. Astron. Soc. Pac. 129 (972) (2016) 024001.
[13] Python, Python Data Structures 2014. https://docs.python.org/2/tutorial/datastructures.html#dictionaries/.
[14] Webpy.org, Web.py Cookbook. 2015. http://webpy.org/cookbook/.
[14a] C. Wu, R. Tobar, K. Vinsen, A. Wicenec, D. Pallot, B. Lao, R. Wang, T. An, M. Boulton, I. Cooper, et al., DALiuGE: a graph execution framework for harnessing the astronomical data deluge, Astron. Comput. 20 (2017) 1–15.
[15] A. Lutomirski, M. Tegmark, N.J. Sanchez, L.C. Stein, W.L. Urry, M. Zaldarriaga, Solving the corner-turning problem for large interferometers, Mon. Not. R. Astron. Soc. 410 (3) (2011) 2075–2080.
[16] M.P. Van Haarlem, M.W. Wise, A.W. Gunst, G. Heald, J.P. McKean, J.W. T. Hessels, A.G. De Bruyn, R. Nijboer, J. Swinbank, R. Fallows, LOFAR: the low-frequency array, Astron. Astrophys. 556 (2013) A2.
[17] G. Heald, J. McKean, R. Pizzo, G. van Diepen, J.E. van Zwieten, R. Van Weeren, D. Rafferty, S. van der Tol, L. Birzan, A. Shulevski, et al., Recent LOFAR imaging pipeline results, in: ISKAF2010 Science Meeting, vol. 112, SISSA Medialab, 2010 p. 057.
[18] G. De Michell, R.K. Gupta, Hardware/software co-design. Proc. IEEE 85 (3) (1997) 349–365, https://doi.org/10.1109/5.558708.
[19] A.R. Jordehi, J. Jasni, Particle swarm optimisation for discrete optimisation problems: a review, Artif. Intell. Rev. 43 (2) (2015) 243–258.
[20] E. Aarts, J. Korst, W. Michiels, Simulated annealing, Search Methodologies-Springer, 2014, pp. 265–285.
[21] J.-C. Liou, M.A. Palis, A comparison of general approaches to multiprocessor scheduling, Parallel Processing Symposium, 1997. Proceedings, 11th International IEEE, 1997, pp. 152–156.
[22] G. Karypis, METIS* a software package for partitioning unstructured graphs, partitioning meshes, and computing fill-reducing orderings of sparse matrices (Version 5.1.0), 2013. http://glaros.dtc.umn.edu/gkhome/fetch/sw/

metis/manual.pdf. University of Minnesota, Department of Computer Science and Engineering, Army HPC Research Center, Minneapolis, MN.
[23] G. Karypis, V. Kumar, Multilevel k-way partitioning scheme for irregular graphs, J. Parallel Distrib. Comput. 48 (1) (1998) 96–129.
[24] A. Wicenec, D. Pallot, R. Tobar, C. Wu, Drop Computing: data driven pipeline processing for the SKA, N.P.F. Lorente, K. Shortridge (Eds.), ADASS XXV, ASP Conf. Ser., ASP, San Francisco, 2016.
[25] Apple, About bundles. 2015. https://developer.apple.com/library/content/documentation/CoreFoundation/Conceptual/CFBundles/AboutBundles/AboutBundles.html.
[26] Docker, Docker–build, ship run, 2016. https://www.docker.com/.
[27] Computerworld, iVEC acquires Fornax supercomputer, 2011. http://www.computerworld.com.au/article/402308/ivec_acquires_fornax_supercomputer/.

10

Application design for execution framework

Ying Mei[a], Rodrigo Tobar[b], Chen Wu[b], Hui Deng[a], Shoulin Wei[c], Feng Wang[a,c]

[a]*Center for Astrophysics, Guangzhou University, Guangzhou Higher Education Mega Center, Guangzhou, China.* [b]*Kunming University of Science and Technology, Chenggong District, Kunming, China.* [c]*International Center for Radio Astronomy Research (ICRAR), The University of Western Australia, Crawley, Perth, WA, Australia*

1 OpenCluster applications design

OpenCluster offers many APIs to allow the client under OpenCluster to quickly develop applications. At a high level, every application of OpenCluster consists of a manager program running the main function of the client, and several worker programs receiving tasks from managers in a cluster. To implement a user-defined manager and worker, the object-oriented approach is adopted (Fig. 10.1).

The base class "Manager" includes two methods:

- getAvailableWorkers(worker_type), which returns available instances of the worker by a given type of worker in a cluster.
- schedule(task), which is an empty method that must be redefined in a derived class by users. In general, users split a task into subtasks, which they dispatch to available workers.

In the same way, the key methods of the class "Worker" are as follows:

- ready(type,host,port). When invoked, it denotes that the worker is ready to accept the command. The first parameter is the type or identity of the worker; the second and third parameters are the host and port to which the worker instance binds.
- perform(task) is an empty method that users must override with detailed data processing logic. The parameter is "task" and the return result is a type of WorkPiece.
- interrupt() interrupts the running worker instance.

Big Data in Astronomy. https://doi.org/10.1016/B978-0-12-819084-5.00011-0

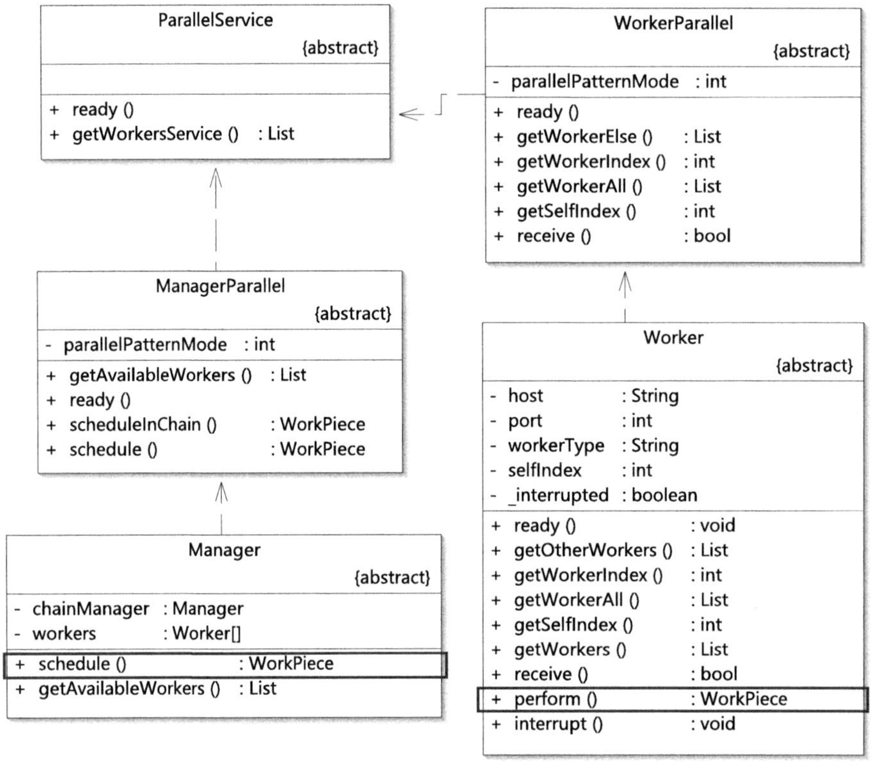

Fig. 10.1 Class hierarchy of "Manager" and "Worker." Reprinted with permission from S. Wei, F. Wang, H. Deng, C. Liu, W. Dai, B. Liang, Y. Mei, C. Shi, Y. Liu, J. Wu,OpenCluster: a flexible distributed computing framework for astronomical data processing, Publ. Astron. Soc. Pac. 129 (972) (2016) 024001. Copyright (2016) IOP Publishing.

It is very easy to develop applications on the OpenCluster framework. The pseudocode (see Tables 10.1 and 10.2) shows how to program a classic example "WordCount," which is a simple application that counts the number of occurrences of each word in a given file.

Class WordCountManager is derived from the base class of Manager. The method "schedule" (lines 4–11) is overridden to split the task, dispatch subtasks to workers in turn, combine a result from the worker into the entire "wordcount" (line 10), and finally return the result. Class WordCountWorker inherits from the base class of Worker. It overrides the perform function (lines 5–13), where it receives the work from WordCountManager. It then opens the file, splits the line into tokens separated by whitespaces, and returns a work piece containing a key-value pair of ⟨⟨word⟩, count⟩. In the main entry (lines 15–16), a worker instance is defined. After invoking the function of "and returns

Table 10.1 Pseudocode for Manager of WordCount in OpenCluster.

```
class WordCountManager(Manager)                                    :
    def__init__(self):
        super(WordCountManager,self).__init__()
    def schedule(self,task):
        workers = self.getAvailableWorkers("WordCount")
        wordcount = {}
    split task into subtasks
        for worker in workers:
            subwordcount = worker.perform(tasks[i])
            wordcount.join(subwordcount)
        return wordcount
if__name__ == "__main__":
    manager = WordCountManager()
    result = manager.schedule"/logs/wordcount.log")
    printresult
```

Reprinted with permission from S. Wei, F. Wang, H. Deng, C. Liu, W. Dai, B. Liang, Y. Mei, C. Shi, Y. Liu, J. Wu,OpenCluster: a flexible distributed computing framework for astronomical data processing, Publ. Astron. Soc. Pac. 129 (972) (2016) 024001. Copyright (2016) IOP Publishing.

Table 10.2 Pseudocode for Worker of WordCount in OpenCluster.

```
class WordCount Worker(Worker):
    def__init__(self, name):
        super(WordCountWorker,self).__init__()
        self.name = name
    def perform (self, work):
        filepath = work.getObj("filepath")
        offset = work.getObj("offset")
        size = work.getObj("size")
        wordcount = {}
        #count the number of words
        wp = WorkPiece(True)
        wp.setObj("word",wordcount)
        return wp
if__name__ == "__main__":
    worker = WordCountWorker("worker1")
    worker.ready("WordCount","*",9280)
```

9Reprinted with permission from S. Wei, F. Wang, H. Deng, C. Liu, W. Dai, B. Liang, Y. Mei, C. Shi, Y. Liu, J. Wu,OpenCluster: a flexible distributed computing framework for astronomical data processing, Publ. Astron. Soc. Pac. 129 (972) (2016) 024001. Copyright (2016) IOP Publishing.

a work piece containing a key-value pair of ⟨⟨word⟩, count⟩," the "WordCount" worker listens at TCP port 9280.

2 MUSER pipeline using OpenCluster

OpenCluster can be adapted to develop different distributed schemes of computing. It migrates an existing project with minimal customization and refinement to a distributed computing platform. OpenCluster users must only understand Python's basic usage and astronomical data processing; they do not need professional skills or experience in distributed computing or multiple computing.

One of the most important parts of MUSER is the data processing pipeline. MUSER's huge data should be processed in real time. MUSER generates 1.92 GB data per minute (low-frequency subarray; MUSER-I) and 3.6 GB/min (high-frequency subarray; MUSER-II) [1]. The sum of raw data can be up to around 4.4 TB per day, as long as 10 h of analysis happens in 1 day.

We have built a pipeline to completely exploit the computing resource that can support multiple run rates at the same time. A MUSER data processing pipeline architecture diagram is shown in Fig. 10.2. The MUSER data processing pipeline allows two modes of data processing at the same time, due to the hardware platform's limitations. One is a mode of batch processing of data that is used for scientific research. The other is data processing in real time, which is used in routine observations for observational

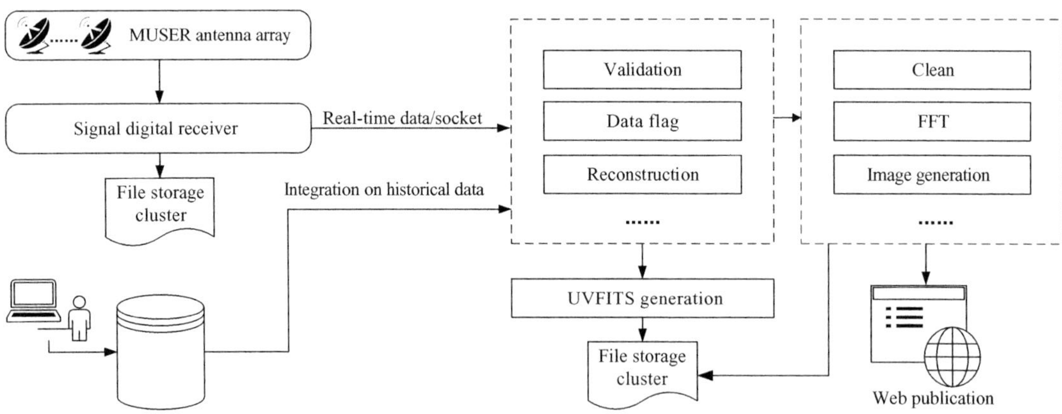

Fig. 10.2 MUSER data processing pipeline. Reprinted with permission from S. Wei, F. Wang, H. Deng, C. Liu, W. Dai, B. Liang, Y. Mei, C. Shi, Y. Liu, J. Wu,OpenCluster: a flexible distributed computing framework for astronomical data processing, Publ. Astron. Soc. Pac. 129 (972) (2016) 024001. Copyright (2016) IOP Publishing.

monitoring. In order to ensure routine observation, the functions of real-time data processing must be guaranteed. If the computer system has spare resources, other batch data processing activities can be performed.

We are therefore providing two examples of developing data processing pipelines using OpenCluster with multiple simultaneous tasks.

The observational data acquired from the digital receiver contain the two parts: (1) Full raw data, which will be stored in the file storage cluster. (2) Sampled raw data (a quick-look display of data quality assurance raw data) that will be sent as a TCP/IP stream on a regular schedule. When astronomers submit a special request for integration via a web page at a certain time, the data processing system is responsible for providing batch computing for UVFITS files or images. In the meantime, the system receives the sampled data in real time and performs the same web publishing computing tasks. Focused on the two computing modes, we present in this section two examples of OpenCluster's ability to run vast amounts of astronomical data processing using a few commands. The examples are implemented in MUSER's real environment.

2.1 Data file format transformation

The file format transformation feature is necessary for scientific research and cooperation to be followed up. The raw observational data of MUSER, however, cannot be exchanged directly because the raw data do not have corresponding observation information, such as observation targets and target positions. To implement OpenCluster's function to support the conversion of raw data to a UVFITS file, we first designed a standalone function called generateUVFITS that can open a specified file, read related raw data over time, and finally generate UVFITS files in the specified output folder. To integrate generateUVFITS into the OpenCluster platform, we simply derived three classes from their base classes:

1. UVFITSWorker, which is derived from the "Worker" class and wrapped generateUVFITS in the perform() method;
2. UVFITSManager, which is derived from the "Manager" class. The schedule() method is implemented to schedule UVFITSWorker;
3. UVFITSTaskDispatcher, which is a daemon program that periodically polls a task queue to determine if it has a new task to submit.

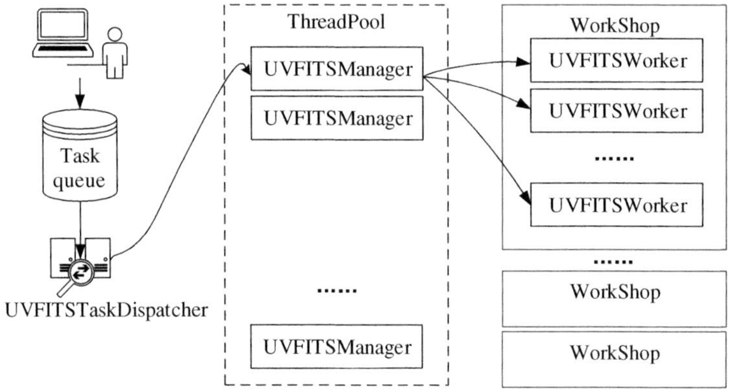

Fig. 10.3 Flow diagram of data file format transformation. Reprinted with permission from S. Wei, F. Wang, H. Deng, C. Liu, W. Dai, B. Liang, Y. Mei, C. Shi, Y. Liu, J. Wu,OpenCluster: a flexible distributed computing framework for astronomical data processing, Publ. Astron. Soc. Pac. 129 (972) (2016) 024001. Copyright (2016) IOP Publishing.

In the implementation of UVFITSTaskDispatcher, we created a UVFITSManager thread pool (the optimum pool size should be equal to the number of CPU cores) to improve the performance of computing. This task is stored in a task queue when a new batch computing task is submitted. If a thread is available in the UVFITSManager thread pool, UVFITSTaskDispatcher will launch a UVFITSManager instance from the thread pool. Then it starts running a new thread. The corresponding task is divided into a series of subtasks when the UVFITSManager thread is started. The subtask is then handled by the corresponding UVFITSWorker, respectively and concurrently. When the UVFITSManager thread is getting started, the corresponding task is divided into a series of subtasks. Then, the subtask is processed by the corresponding UVFITSWorker, respectively and concurrently. The flow diagram of data file format transformation is shown in Fig. 10.3.

2.2 Real-time imaging and monitoring

After performing a series of processes such as gridding, flagging, and CLEAN, MUSER can ultimately produce high-resolution images. Among them, the deconvolution for dirty images is a highly time-intensive calculation procedure that must be executed on machines with GPUs.

OpenCluster can model a complex pipeline for data processing that can use MUSER's hybrid hardware resources. Observational data monitoring, for instance, is a feature essential for routine observation. A dedicated computer sends the monitoring system raw observational data every 5 min. The data and images should be processed as quickly as possible by the monitoring system. OpenCluster offers an implementation of a TCP socket server that users can reuse to extend user-defined handling of data in a pooling thread model.

1. MUSERSocketServer is an asynchronous TCP socket server. After new data are received, a new MUSERImageManager instance is invoked to process the new data;
2. MUSERImageManager, which is derived from the Manager class and rewritten as a schedule() method. The task is split into subtasks, and it arranges subtasks for MUSERImageWorker instances;
3. MUSERImageWorker, which is derived from the Worker class, is deployed on machines with GPUs. The business logics are implemented in the *perform()* method to fetch subtasks, perform computing issues (i.e., CLEAN, FFT, and so on), and return results.

It creates a new MUSERImageManager instance when MUSERSocketServer receives raw observational data and stores the instance in a thread pool queue. The instance of MUSERImageManager is presented from the queue and executed when resources are available to run. MUSERImageManager splits one observational data item into multiple slices according to the band and frequency. It dispatches each slice to multiple MUSERImageWorker instances to, respectively, generate images. The flow diagram of data file format transformation is shown in Fig. 10.4.

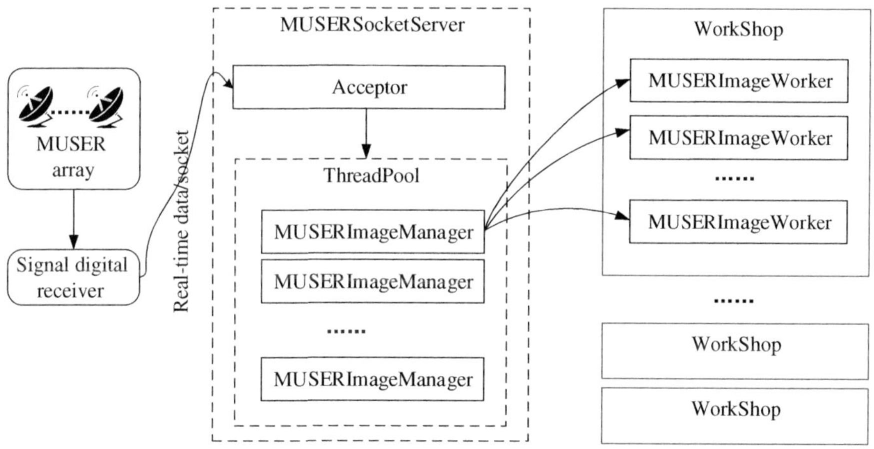

Fig. 10.4 Flow diagram of real-time imaging. Reprinted with permission from S. Wei, F. Wang, H. Deng, C. Liu, W. Dai, B. Liang, Y. Mei, C. Shi, Y. Liu, J. Wu,OpenCluster: a flexible distributed computing framework for astronomical data processing, Publ. Astron. Soc. Pac. 129 (972) (2016) 024001. Copyright (2016) IOP Publishing.

3 Design CHILES on AWS using DALiuGE

The Cosmos HI Large Extragalactic Survey (CHILES) [2] is deployed for the Karl G. Jansky Very Large Array (VLA), a 27 antenna array. The upgraded front-end (wide-band L-band receivers) and back-end (the WIDAR correlator) [3] provide full function for spectral line observing between ~940 and ~1430 MHz in the sky (15 spectral windows of 32 MHz, each giving a total of 480 MHz in each session). The array configuration for CHILES is VLA-B, which has 11 km baselines and a typical beam size of ~5 × 7 arcsec at 1.4 GHz, assuming natural weighting.

The observations are dithered in frequency to smooth out the edges of the spectral windows. The antennas observe about 0.5 degrees (~2000 arcsec) across the sky at the phase center.

In our previous work on CHILES data reduction [4], we designed three different hard code scripts for computing environments, respectively. Therefore, we want to take a challenge that drives the execution of the "new" pipeline with five pipeline components running under DALiuGE.

1. Splitting the measurement set for an individual day into measurement sets containing 4 MHz chunks. The current implementation has 42 days worth of data and this will grow with the release of the next semester's data.
2. Subtracting the local sky model and moving the corrected fluxes into the "corrected" column of the measurement set on each of the 4 MHz chunks across all days.
3. Cleaning by taking all the day measurement sets for a 4 MHz band and apply clean to them all to produce a single cleaned file for that 4 MHz band.
4. Converting each 4 MHz measurement set to JPEG2000.
5. Concatenating the final cleaned 4 MHz files into a single measurement set.

3.1 Setup

There were three simple steps to set up AWS for this exercise. Step 1 selects the suitable AWS EC2 instance to meet the task's requirements regarding CPU, memory, disc, and I/O bandwidth. Step 2 runs a Python script to begin the required number of DALiuGE Drop Manager servers, ensures disks are enabled and the correct software is installed, and starts a data island manager to orchestrate the physical graph deployment to the different Drop Managers. Step 3 runs a Python script that calculates what needs to be done and sends the physical graph to the manager of the data island.

To speed up the startup times of the EC2 instances, we created an Amazon Machine Image (AMI) with a recent version of DALiuGE preinstalled and a Python virtual environment precreated and ready to be used. We pulled the latest version of DALiuGE from GitHub at instance-creation time and installed it in the precreated virtual environment to ensure that we worked with the latest version.

The computing components were primarily installed using Docker. A number of tasks of CasaPy [5] and SkuareView [6] process the CHILES data. We wrapped the CasaPy and SkuareView code inside Docker images and uploaded them to DockerHub.

DALiuGE has features of supporting built-in Docker, which could integrate the processing components into the pipeline graph easily.

When an Application Drop is triggered to "run," it first invokes a Python program written by boto3 [7] to create Ed2 cluster instances. After the EC2 instances are created, the corresponding Drop Managers are launched. Meantime, the corresponding Physical Graph is calculated, and once the Drop Managers are up and ready, the Physical Graph is deployed onto DALiuGE and executed. The Physical Graph consists of a special Application Drop that shuts down each of the EC2 instances automatically so as to guarantee no pay for servers without computing tasks. To reduce costs, we call for Amazon EC2 Spot Instances instead of on-demand nodes. Depending on the load characteristics of the pipeline, the Python script always looks for the cheapest spot price in the regions specified for the instance type required.

The configuration of each EC2 instance is loaded by submitting a YAML file at creation time. The AWS and Docker files are created subsequently. A Bash file is applied to configure the disks. The DALiuGE Drop Manager will start after pulling the latest version of docker containers, DALiuGE and the CHILES pipeline from Github. We used three AWS storage options during different stages of the pipeline—Elastic Block Store (EBS), Ephemeral Storage, and Simple Storage Service (S3).

3.2 Results and costs

The initial measurement sets for semester 1 of the CHILES project amount to approximately 15 TB of data.[a] Splitting the original measurement sets into 4 MHz chunks creates 120 subdirectories that each contain 42 measurement sets requiring 9.1 TB. The

[a]Data transfer from non-AWS sources into AWS EC2 instances is free of charge.

model subtraction requires 9.1 TB. The clean step creates 120 4 GB files for a total of 480 GB of data. The final concatenation produces a single file of 480 GB.

The `splitting` process requires either an i2.2xlarge or an i2.4xlarge EC2 instance, depending on the size of the measurement set on a particular day. If the measurement set is greater than 500 GB, we require the greater power (and more expensive) i2.4xlarge instance. The `model subtraction` requires an i2.2xlarge EC2 instance. It takes approximately 18 h to perform the 5040 model subtractions if given 20 nodes. The spot price for an i2.2xlarge is usually of the order of US$0.20/h, which gives an indicative cost of US$72. The `clean` process requires an i2.4xlarge EC2 instance because of the increased load on the CPU and memory. It takes approximately 20 h to perform the 120 cleans if given 20 nodes. The spot price for an i2.4xlarge is usually of the order of US$0.35/h, which gives an indicative cost of US$140. The `conversion` to the JPEG2000 images is performed using a single i2.2xlarge instance and only takes a few hours. The final `concatenation` has only been performed once as the resultant image is too large to be downloaded quickly and for systems to process. It took more than 48 h using a i2.4xlarge instance.

4 The migration of SAGECal/MPI to DALiuGe

A remarkable characteristic of radio data is data-intensive computation, and distributed computing has become a popular approach to tackle mass radio data processing. DALiuGE [8] is a scalable and high-performance distributed execution system designed to process huge astronomical datasets of the magnitude needed by the Phase 1 (SKA1) Square Kilometer Array. However, most existing radio astronomical algorithms support parallel computing based on the message-passing interface (MPI).

Therefore, the integration of existing MPI algorithms into the distributed computing framework becomes a meaningful issue. An important objective of this initiative is to determine the feasibility of using DALiuGE to drive the implementation of existing distributed calibration solutions. In addition, a deeper understanding of the features of the SAGECal workload is envisaged.

4.1 About SAGECal

SAGECal [9] is a technique of self-calibration using the Expectation Maximization (EM) algorithm [10] to achieve maximum likelihood estimation of the instrument and sky parameters.

While the plain EM algorithm was used to estimate parameters for nonlinear, overlapping signals (e.g., Ref. [11]), SAGECal uses an improved EM algorithm known as the SAGE algorithm [12] to:

- speed up the convergence (compared to the plain EM algorithm);
- reduce the computational complexity (compared to least-squares estimation methods) [13]; and
- improve the quality of calibration (compared to the Peeling calibration method).

To deal with large data volumes produced by next-generation radio interferometers such as the SKA, the Distributed SAGECal algorithm [14] exploits data parallelism across the frequency axis to perform distributed calibration. This is reformulated as a minimization problem with an additional constraint based on the "smoothness prior." Specifically, a given frequency f must share, with all other frequencies, the same set of F **frequency-independent** (hidden) variables Z, which, when linearly combined with F **frequency-dependent** polynomial terms B_f, impose the smoothness over f to its Jones matrix J_f. In Fig. 10.5, the core concept of Distributed SAGECal is demonstrated. The agent is assigned a frequency and works independently, without interfering with others, to minimize a cost function $g'i(J)$. The fusion center enforces the smoothness consensus of all agents on Z during each iteration by solving a problem of consensus optimization by using the consensus alternating direction method of multipliers (C-ADMM) [15]. A more detailed description of the algorithm should be found in the literature [14].

Fig. 10.5 also shows that each agent only needs to access a small partition of the large datasets and can be executed in parallel with only relatively minor communication overheads to the fusion center (see Section 4.3). Therefore, the Distributed SAGE-Cal algorithm is somewhat "insensitive" to the data size, working

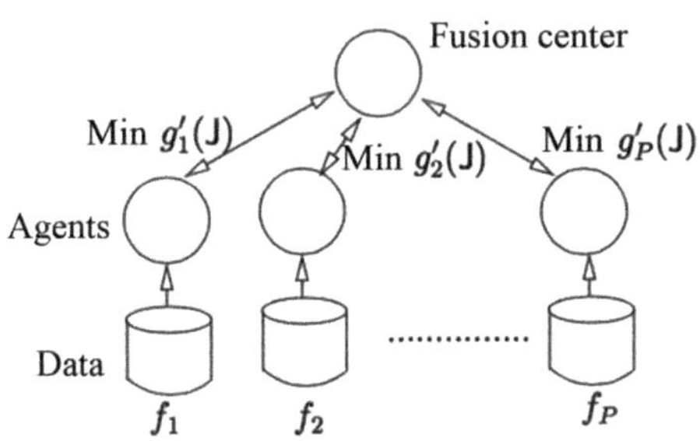

Fig. 10.5 The basic idea of Distributed SAGECal. Modified from S. Yatawatta, Distributed radio interferometric calibration, Mon. Not. R. Astron. Soc. 449 (4) (2015) 4506–4514.

as effectively for large datasets as for small ones. By intentionally varying the total P number of agents, one can achieve an optimal balance between the completion time and resource footprint (i.e., degree of data parallelism).

4.2 Code analysis

An analysis of the source code[b] in the MPI-based Distributed SAGECal implementation has generated the UML sequence diagram shown in Fig. 10.6. This presents detailed data exchanges between the *Master* and *Slaves*. It is essential to deeply understand the interactions for the integration of Distributed SAGECal into DALiuGE.

The *Master* process represents the "fusion center" and several *Slave* processes represent the "agents," as depicted in Fig. 10.5. The interactions between *Master* and *Slaves* are structured within a nested loop. The outer loop, shown as *Timeslots Iteration*, iterates over data chunks, each of which consists of a fixed number *tilesz* of time samples. The inner loop contains *ADMM Iterations*, the number *Nadmm* of which is a predefined parameter. Within a particular ADMM iteration n, the minimization of $g_i^{\prime}(J)$ is performed by a *Slave fi* (i.e., inside the `sagefit_vis`), which sends current the Lagrange multiplier $(Y_{fi})^n$ and the newly estimated Jones matrix $(J_{fi})^{n+1}$ to *Master* upon completion. After receiving this information from all *Slaves*, the *Master* computes an updated global consensus $(Z)^{n+1}$, and broadcasts it back to all the *Slaves*. The *Slave fi* then updates its local Lagrange multiplier from $(Y_{fi})^n$ to $(Y_{fi})^{n+1}$ based on the "goodness of fit" of the smoothness constraint $[(J_{fi})^{n+1} - B_{fi}(Z)^{n+1}]$. Interactions inside the *Timeslots Iteration* mainly serve the purpose of housekeeping for preparing and preprocessing input data chunks and writing the solutions to the I/O system.

As part of the MPI-based application, both *Master* and *Slaves* are long running processes that explicitly maintain internal states, send/receive messages across multiple iterations, and manage data synchronization. Section 4.4 discusses a method to incorporate such "stateful" MPI applications directly into the DALiuGE execution framework. Section 4.7 explores an alternative to decomposing such long-running MPI processes into many fine-grained "stateless" Application and Data Drops. This decomposition will lead to a dataflow-style application that shifts the burden of state management completely to execution frameworks such as DALiuGE. Before giving detailed descriptions of the solutions,

[b] git://git.code.sf.net/p/sagecal/code.

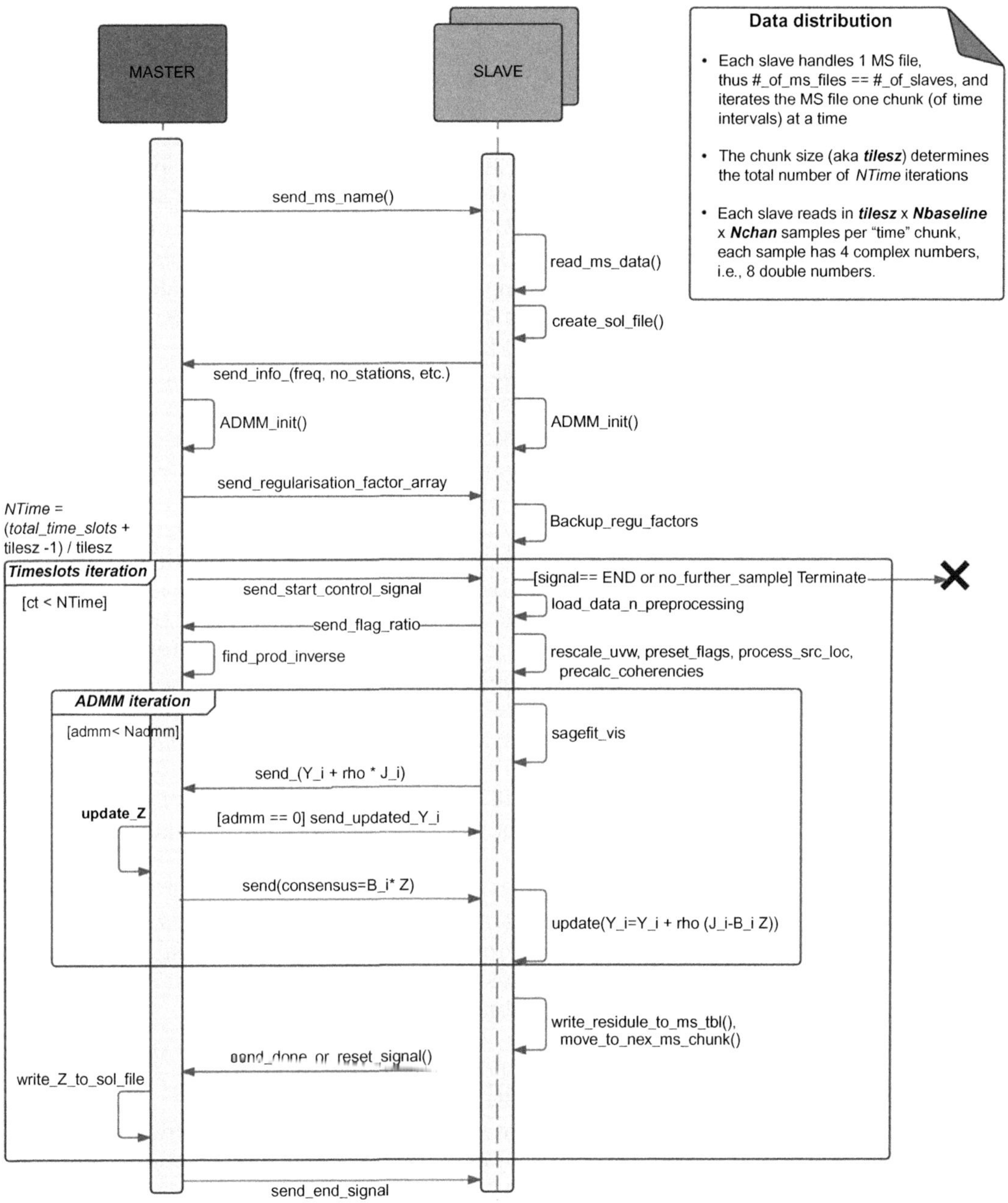

Fig. 10.6 A UML sequence diagram showing message exchanges defined in [14].

Section 4.3 first presents a more detailed run-time analysis of the workload characteristics of the Distributed SAGECal algorithm.

4.3 Workload characterization

Distributed SAGECal was profiled using four processors with one MPI task on each node of Magnus (Cray XC40 architecture) with the Allinea Map profiling tool. Each Magnus node has a main memory of 64 GB. Each MPI task of the executable `sagecal-mpi` is capable of spawning six threads, whereas the multithreading has been implemented in the pthreads library. The tile size simulated was `-t 2` and ADMM iterations were set to `-A 5`. Two thread-numbered simulations are profiled as n = 1 and n = 4. Compared to the one run with one thread, which completed in 842 s, a simulation with four pthread threads took 326 s. Fig. 10.7 shows the runtime work characterization of the Distributed SAGECal for the above state run.

The executable `sagecal-mpi` costs approximately 70% on node local activity, which includes thread management. Approximately 29% of the time is spent in MPI work, which includes MPI environment setup and communication. The message size is of the order of 0.1 MB/s, that is, small messages, and the communication cost is dominated by the blocking `MPI_Recv` call.

Contrary to this, CPU memory access accounts for 90.7% of CPU time, whereas in real floating point calculations, there is not much use of CPU. The pattern of CPU memory access occurs on salves, not on the MPI master mission. Finally, the use of memory per MPI task is modest compared to the main memory per node available.

Because `sagecal-mpi` requires a strong scaling test on the MS file per MPI task only if the number of pthreads is limited (the code itself limits a maximum of six). A weak scaling test, which can reveal how the overhead MPI communication scales with the size of the problem, may be helpful.

4.4 Drop wrapping

One of the alternatives that we researched to integrate Distributed SAGECal into DALiuGE was the wrapping of the application into one or more Application Drops.

This was achieved by adding a new Application Drop class to DALiuGE to integrate any existing MPI applications seamlessly. We named this new class `MPIApp`. At the creation time, it receives the MPI application (the *target*) to run its parameters and the number of processes to be used as parameters (further

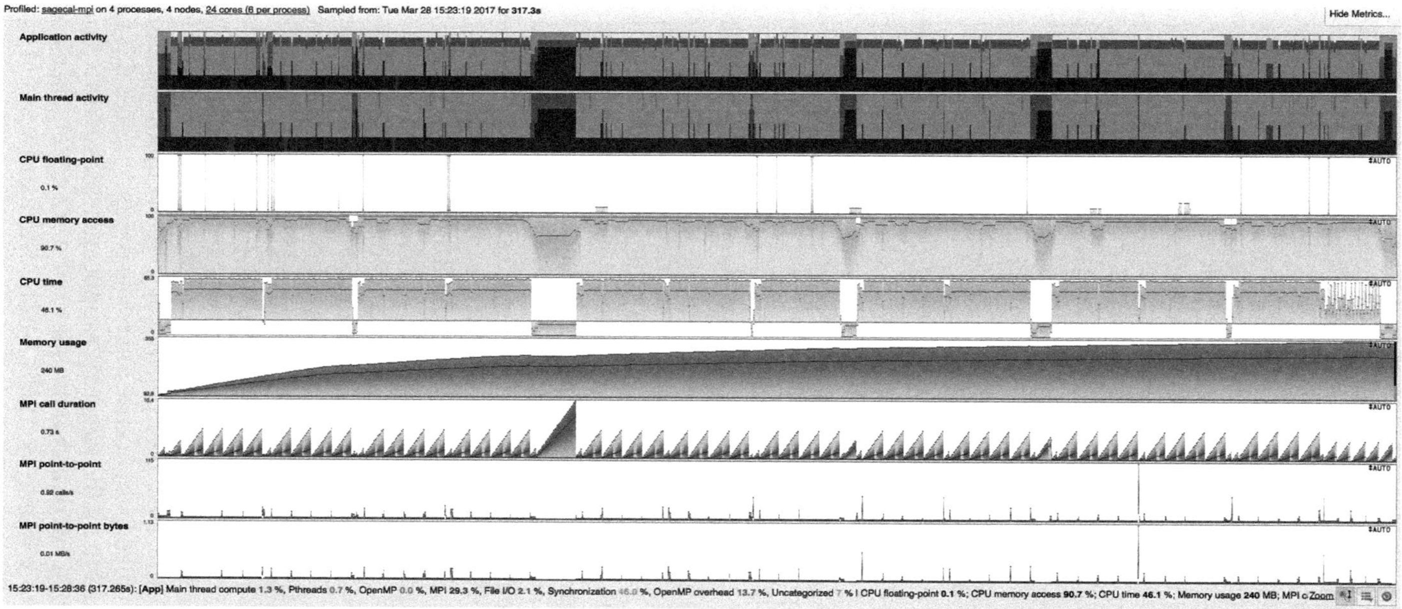

Fig. 10.7 Allinea Map profile of SAGECal-MPI run with four MPI tasks on four nodes with four pthreads per MPI task on each node sharing 64 GB per node. The `x-axis` is the time at which samples were taken by the profiler, whereas `y-axis` is the time reading from ranks at a particular time stamp. The *darker shading of color* depicts clustering of MPI ranks around the value of the `y-axis` for every `x`. Application activity means the cumulative activity on all the MPI tasks whereas main thread activity is the profile of the master thread on each MPI task. Of the total 317.3 sampled seconds, the CPU operation takes 46.1% and the MPI communication between MPI tasks costs 29.3%. 46.0% time is lumped as synchronization which is attributed to pthread activity, that is, create, join, destroy, and anything in between. Of the total sampled time, 2.1% is spent in file I/O.

parameters can be added in the future if needed). When fired, the MPIApp uses the MPI_Comm_spawn function (part of the MPI 2 standard) to start the target MPI application with a given number of processes. MPI_Comm_spawn not only creates an intercommunicator between the parent and children process groups for them to communicate, but also ensures that the children processes that form the target MPI application have an MPI_COMM_WORLD communicator different from that of the parent, providing complete isolation between the two groups of processes. Finally, MPI_Comm_spawn allows passing down arbitrary, implementation-dependent information and instructions to the runtime system on how to start the new processes.

4.5 Dynamic MPI process control

The MPIApp will wait until the execution of the target MPI program is complete in order to comply with the Drop lifecycle. This can only be ensured through the use of the newly created intercommunicator to synchronize child and parent execution. We opted to use a simple MPI Barrier for this experiment, although other types of collective calls could be used if appropriate (e.g., to collect information about the exit status of children in the parent process).

It was easy to add the barrier to the MPIApp side. It was more difficult to apply it to the children's systems. To apply this, we studied two different methods. In the first one, MPIApp explicitly spawns the intended MPI request. Consequently, to receive the parent intercommunicator and add the required barrier, the target request must be slightly modified. This involves making two extra calls: one to MPI Comm get parent and one to MPI Barrier. Distributed SAGECal is a simple application, so it was easy to incorporate this support.

In the second method, MPIApp spawns a DALiuGE internal Python script. This script first obtains the parent intercommunicator, then forks and executes the target MPI application. After the MPI application exits, the script finally calls the barrier on the parent intercommunicator. When using this method, the existing MPI application must **not** be aware that it has a parent communicator, or if it is aware, it must not use it. Most MPI applications, including Distributed SAGECal, do not use this communicator. This requirement exists because internally the MPIApp uses this communicator to wait for the target MPI application processes to finish and collect their exit data (stdout, stderr, and exit code). This second method is more fragile as it relies on nonstandard ways of interacting with MPI applications, using a fork and an execv syscall (something OpenMPI actually recommends against), and eventually working with the parent intercommunicator after the target MPI application has called `MPI_Finalize`.

Finally, the target MPI application (in our case, Distributed SAGECal) should be function correctly and completely. In the case of Distributed SAGECal, we needed to optimize it slightly in order to properly call `MPI_Abort` instead of using a combination of `MPI_Finalize` and `exit`. The latter left some processes blocked because they were not part of the collective action of aborting the whole program.

4.6 Process location

Depending on the scheduling system in use, an important consideration is that care needs to be taken to leave enough unused resources so that the dynamically spawned MPI processes can be scheduled and run. In the experiments, SLURM 17.02.0 in combination with OpenMPI 2.0 are installed. We utilize four computing nodes, and realized that two tasks would be run on each one (using `sbatch -N 4 -ntasks-per-node 2`), allowing for a total of eight MPI processes to run. The queued script then launched DALiuGE using only two MPI processes (`mpirun -n 2`) to run only two Drop Managers. This left six free slots for subsequent scheduling of dynamic tasks, which we used to run Distributed SAGECal with six MPI processes.

Fig. 10.8 presents the effect observed in the system as we spread the load of the new processes. The left side of the picture corresponds to an execution of Distributed SAGECal from within DALiuGE using six dynamically spawned MPI processes running in one node only (`pleiades14`). In such a case, only the one node can see its CPU load and memory increase while the rest remain relatively static. The right side shows the same execution, but indicates OpenMPI to use two nodes (`pleiades13` and `pleiades14`). In this case, as expected, the use of load and memory is shared between the two nodes. This approach can therefore run MPI processes from within DALiuGE while giving us control over the process placement.

Different scheduling systems behave differently though, and accordingly some MPI implementations do not even implement `MPI_Comm_Spawn` API. Other MPI implementations might also not offer the flexibility found in OpenMPI to place the new MPI processes. Therefore, care must be taken when choosing the system.

4.7 Dataflow and graphs

To migrate MPI-based Distributed SAGECal into dataflow-style applications whose execution is driven by DALiuGE, two approaches were investigated. The first one is at a much coarser

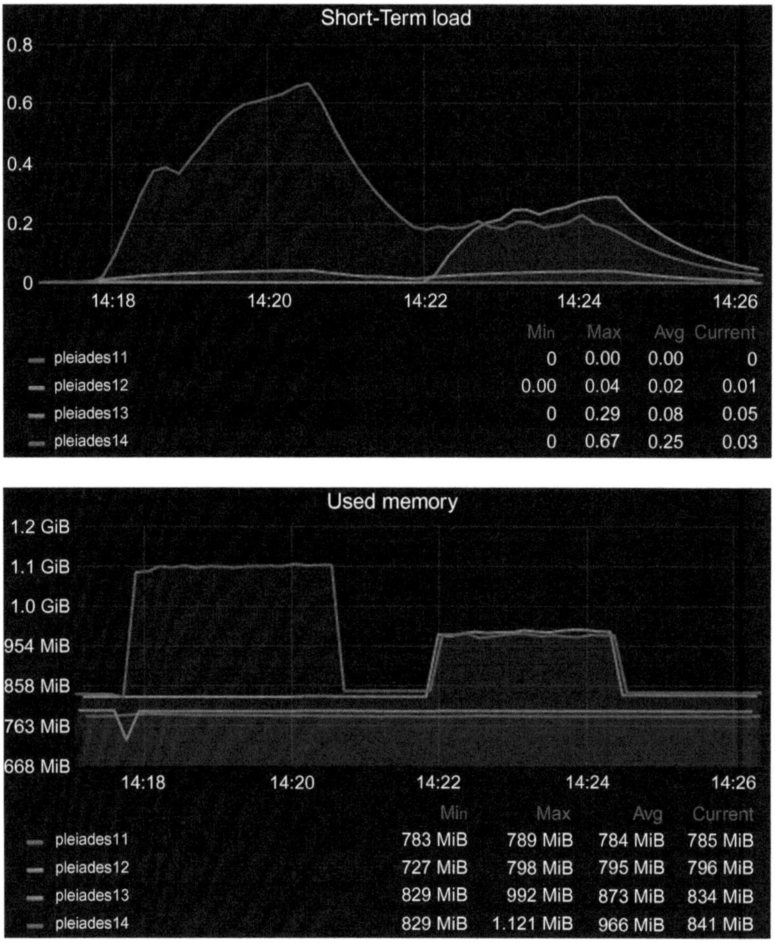

Fig. 10.8 Short-term CPU load (*top*) and used memory (*bottom*) per node with all six MPI children processes concentrated in one node (*left*) and distributed in two nodes (*right*). Both resources show that the dynamically scheduled MPI processes are indeed being run in the requested nodes. Short-term CPU load is measured as an average over 1 min, which explains the smoother lines.

level of granularity, and uses `MPIApp` discussed in Section 4.4 to wrap the entire Distributed SAGECal algorithm. The second approach decomposes the existing algorithm into much more fine-grained Drops, each of which performs "stateless" computation.

4.7.1 Coarse-grained graphs

The octagon shape in the logical graph (Fig. 10.9) represents an `MPIApp`, which has a Drop property `num_of_procs` set to 5. During the graph translation stage [8], five `MPIApp` Drops are developed. One of them represents the Master process (not shown in Fig. 10.9), and the other four represent the Slave processes. Each `MPIApp` Drop has its own input and output files. For the

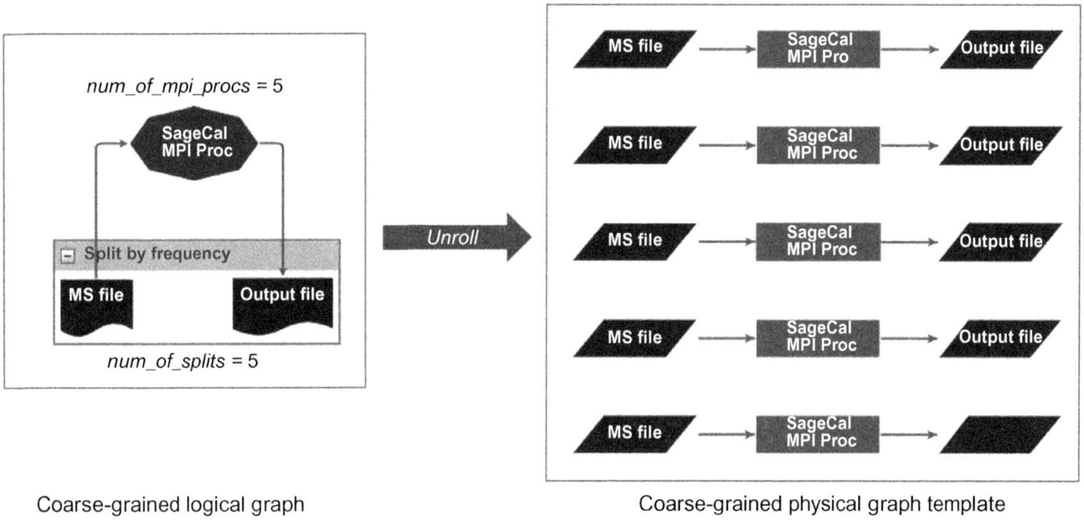

Fig. 10.9 Wrap the entire Distributed SAGECal using `MPIApp` Drops in the logical graph, which is then unrolled into the physical graph template with five MPI processes and split MeasurementSet files and output files.

Master, the input is the metadata file describing dataset locations, and the output is the result file. For Slaves, the inputs are MeasurementSet files, and the outputs are residual files.

The message exchanges (exhibited in Fig. 10.6) are not modeled in detail within coarse-grained graphs (logical or physical) and are completely managed internally by the MPI processes as before. Therefore, this option requires little modification or augmentation of the existing Distributed SAGECal implementation for DALiuGE integration. Additionally, the `MPIApp` Drop wrapper will monitor the execution progress of its enclosing MPI process during DALiuGE graph execution, ensuring that each application is executed in the node assigned at the scheduling time (using the mechanism described in Section 4.6), and trigger events across edges of the graph to downstream Drops for cascaded execution.

Because no modification is made to the underlying MPI-based communication and I/O handling, this option also requires a global file system as a "typical" high-performance computing application.

4.7.2 Fine-grained graphs

To generate fine-grained graphs, the first step is to decompose the MPI application into smaller Drop constructs, each of which represents either stateless computation (Application Drop) or persistent/transient datasets (Data Drop). The motivation for such a decomposition is to convert the complex message interactions shown in Fig. 10.6 into a Directed-Acyclic Graph (DAG) dataflow. This is because a dataflow-based execution framework does not require any explicit constraints (e.g., synchronization and its associated state management) beyond what is already inherent in the data dependencies (along the "flow"). Accordingly, an execution naturally exploits the full parallelism without incurring any application-level overheads at run time. Concretely, for Distributed SAGECal, this means the algorithm developers no longer deal with message controls via `MPI_Send`, `MPI_Recv`, `MPI_Barrier`, etc. Instead, they focus solely on the computational aspects of the algorithm implementation, and let the execution framework worry about state management and the order or timing of operations stipulated in the dataflow definition. DALiuGE provides two built-in Drops (`BashShellApp` and `DylibDrop`), which can encapsulate C/C++ codes to allow rapid porting of SAGECal.

4.7.3 Implementation with BashShellApp

The logical graph of SAGECal in DALiuGE in Fig. 10.10 includes around 30 Drop constructs implemented with `BashShellApp` and `FileDrop`, which are the result of the decomposition of the original MPI-based Distributed SAGECal application. The general rule for creating these Drops is based on the communications between the Master and Slaves and between the application and the underlying I/O system (in this case, the file system). The *Scatter* construct—FREQUENCY SPLIT PER SLAVE—on the left represents multiple instances of the Slave process. The two nested *Loop* constructs inside each Slave are "stepwise locked" with their counterparts in the *Master* instance. This means only those Drops that are tagged with the same iteration number can be connected via event/data flow between the *Master* and the *Slave*. Readers should refer to [8] for details on logical graph constructs such as *Scatter* and *Loop*. It should be noted that the implementation of such a decomposition requires considerable modifications to the existing MPI wrapper of the Distributed SAGECal algorithm[c] —refactoring the two monolithic functions (`sagecal_master` and `sagecal_slave`) into a number of more self-contained C++ classes/functions and wrapping them with `BashShellApp` and `FileDrop` rather than `MPIApp` used

[c]Although it does not require any changes to the algorithm implementation.

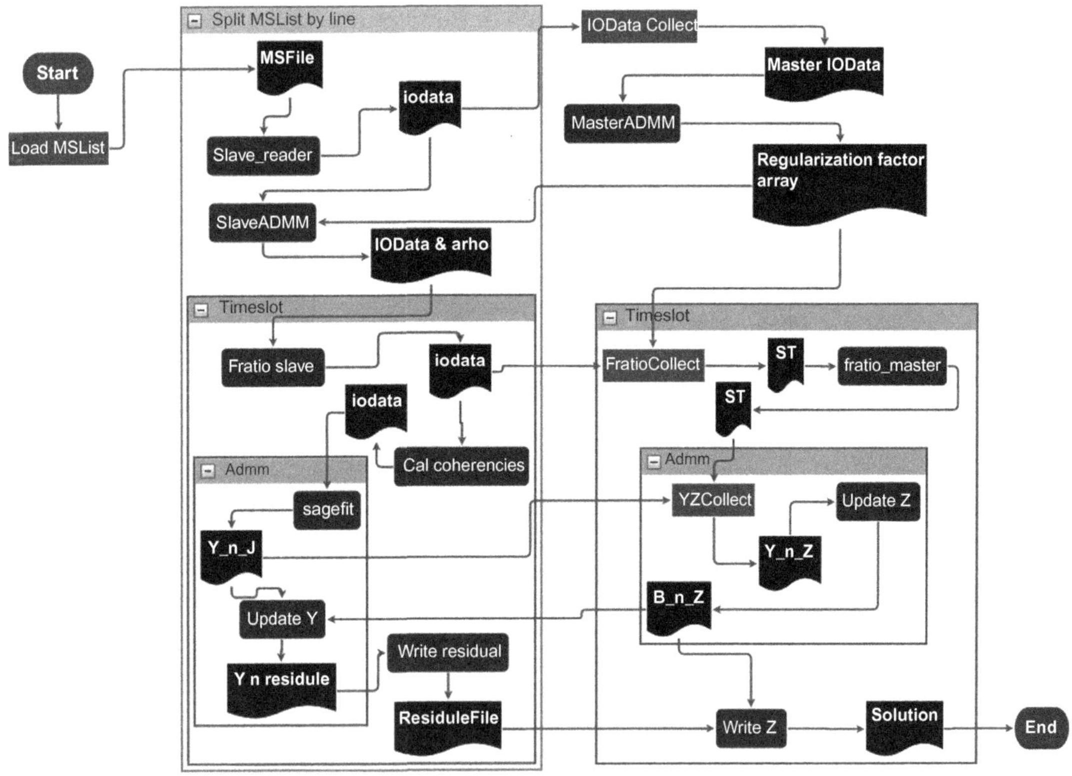

Fig. 10.10 Logical graph of the Distributed SAGECal algorithm with BashShellApp and FileDrop.

in Section 4.4. In fact, `BashShellApp` is provided as a wrapper of the actual executable program. Therefore, in each BashShellDrop the user needs to define an entry for executable programs and related arguments. While these changes are not directly related to the execution framework, and hence beyond the scope of this investigation at this stage, we have performed similar decompositions [16] in order to incorporate other imaging software components, such as the ARL [17], into DALiuGE this way.

At this stage, the logical graph of the Distributed SAGECal can still be "unrolled" into the physical graph template. Although it cannot be further translated into an executable physical graph due to the lack of complete realization of the decomposed software components mentioned earlier, it can clearly visualize the complete dataflow as a DAG.

We have finished the entire implementation of those fine-grained component Drops with `BashShellApp` and `FileDrop`; such component-level details are completely opaque to the execution framework anyway. All codes can be found at https://github.com/astroitlab/sagecal-daliuge. More importantly, Distributed SAGE-Cal can use DALiuGE's scheduling capability to run these physical graphs on distributed resources (rather than on a single node) in an optimal way in terms of total resource footprint, data movement volume, and execution time. Readers are requested to refer to [8] for more technical details about DALiuGE.

4.7.4 Implementation with DynlibApp

The previous version of DALiuGE only supports the revoking of C/C++ programs with `BashShellApp` and transmit variables with `FileDrop`. This implies that there are more processes for the serialization and deserialization of variables in the file style. If the transmission of variables can be performed between memories, it can obviously improve the performance. Now, the new DynLibApp was provided in the latest version of DALiuGE. `DynlibApp` allows C/C++ programs to read and write variables in `MemoryDrop`. As shown in Fig. 10.11, compared to the `BashShellApp` version, `BashShellApp` was replaced by `DynlibApp`. Meanwhile, we used `MemoryDrop` to replace all `FileDrop`.

By using CTYPES, which is a foreign function library for Python, `DynlibApp` supports the manipulation of Input and Output DataDrops in C/C++ codes. It means that a C/C++ program can invoke the methods (functions) of a Python MemoryDrop to receive and send the data. Thus, the difficulty of porting 24 applications would be significantly decreased. According to the requirements of CTYPES, one property is considered to store the location of shared libraries for the `DynlibApp`. All the C/C++ programs should be compiled as ".so" files. In each program, two methods, `init()` and `run()`, must be implemented. When the Drop is executed, `init()` would be invoked first for the environment and variable initialization. The concrete algorithm implementation would be coded in `run()`. The two methods are declared as follows[d] :

```
int init(dlg_app_info *app, const char ***params);
int run(dlg_app_info *app);
```

[d]The detailed specifications are available in *dlg_app.h* at https://github.com/ICRAR/daliuge

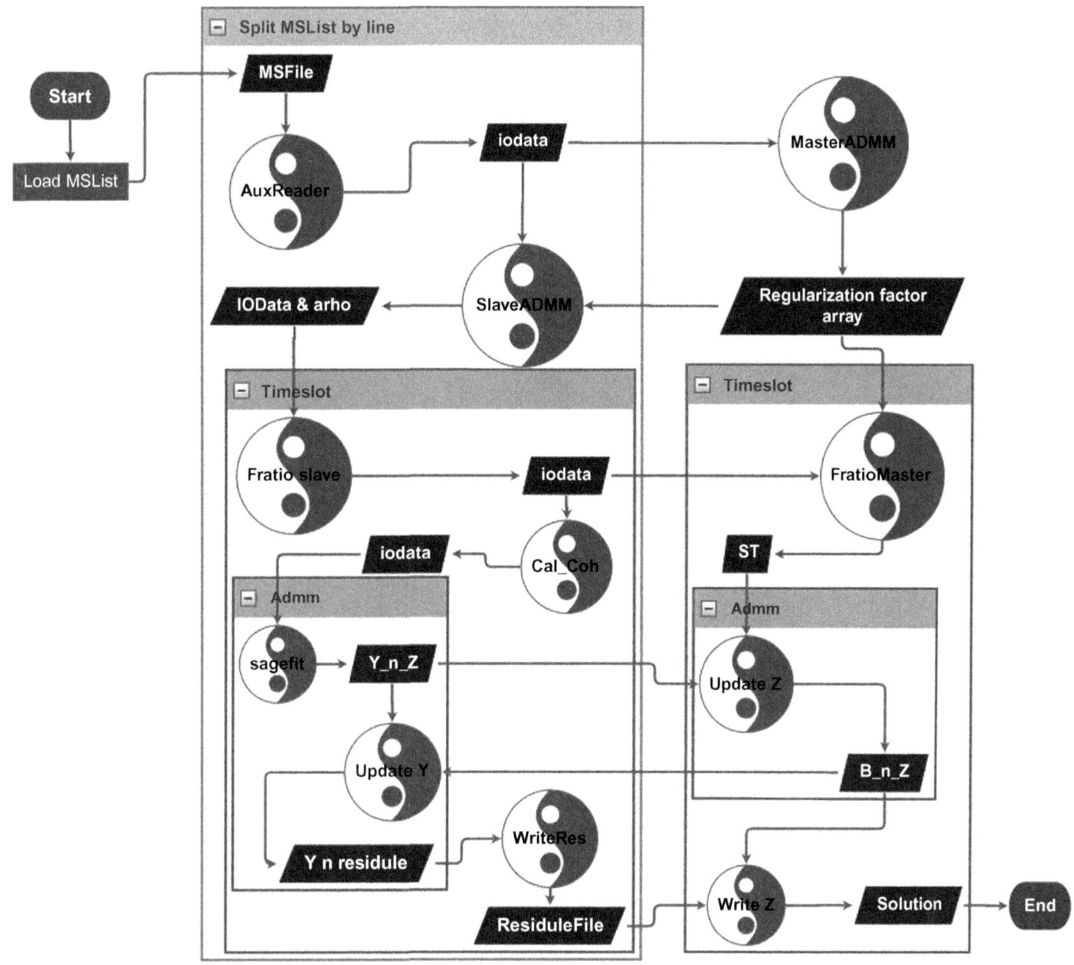

Fig. 10.11 Logical graph of the Distributed SAGECal algorithm with DynlibApp and MemoryDrop.

For porting SAGECal to DALiuGE, as shown in Fig. 10.11, it contains 11 different `DynlibApp` Drops. Each `DynlibApp` Drop needs to specify a dynamic link library file (".so" files in Linux). We created two classes, Master and Slave, and compiled them into two ".so" files to simplify the structure of the program and reduce the difficulty of transplantation. According to the arguments in `dlg_app_info`, Master and Slave could invoke different functions.

We also have completed the coding changes. All codes have been released at https://github.com/astroitlab/sagecal-daliuge-dynlib.

Acknowledgments

The work in this chapter is supported by the National Key Research and Development Program of China (2018YFA0404603) and the Joint Research Fund in Astronomy (U1831204 and U1931141), the National Natural Science Foundation of China (11903009), and the Yunnan Key Research and Development Program (2018IA054), the major scientific research project of Guangdong regular institutions of higher learning (2017KZDXM062).

Thanks for the International Center for Radio Astronomy Research (ICRAR) and the resources provided by the Pawsey Supercomputing Center with funding from the Australian Government and the Government of Western Australia. This work is also supported by Astronomical Big Data Joint Research Center, cofounded by National Astronomical Observatories, Chinese Academy of Sciences and Alibaba Cloud.

References

[1] F. Wang, Y. Mei, W. Wang, K.F. Ji, Distributed data-processing pipeline for Mingantu Ultrawide Spectral Radioheliograph, Publ. Astron. Soc. Pac. 127 (950) (2015) 383–396.

[2] X. Fernandez, J.H. van Gorkom, E. Momjian, Chiles Team, The COSMOS HI large extragalactic survey (CHILES): probing HI across cosmic time, in: American Astronomical Society Meeting Abstracts, American Astronomical Society Meeting Abstracts, vol. 225, 2015, p. 427.03.

[3] E. Momjian, R. Perley, The impact of the new thermal gap receiver assembly on the sensitivity of the EVLA at L-band (1–2 GHz), 2012, http://library.nrao.edu/public/memos/evla/EVLAM_165.pdf.

[4] R. Dodson, K. Vinsen, C. Wu, A. Popping, M. Meyer, A. Wicenec, P. Quinn, J. van Gorkom, E. Momjian, Imaging SKA-scale data in three different computing environments, Astron. Comput. 14 (2016) 8–22.

[5] J.P. McMullin, B. Waters, D. Schiebel, W. Young, K. Golap, CASA architecture and applications, Astronomical Data Analysis Software and Systems XVI, vol. 376, 2007, p. 127.

[6] V.V. Kitaeff, A. Cannon, A. Wicenec, D. Taubman, Astronomical imagery: considerations for a contemporary approach with JPEG2000. Astron. Comput. 12 (2015) 229–239, https://doi.org/10.1016/j.ascom.2014.06.002.

[7] AWS, Boto 3 Documentation, 2016. http://boto3.readthedocs.io/en/latest/.

[8] C. Wu, R. Tobar, K. Vinsen, A. Wicenec, D. Pallot, B. Lao, R. Wang, T. An, M. Boulton, I. Cooper, DALiuGE: a graph execution framework for harnessing the astronomical data deluge, Astron. Comput. 20 (2017) 1–15.

[9] S. Yatawatta, S. Zaroubi, G. De Bruyn, L. Koopmans, J. Noordam, Radio interferomietric calibration using the sage algorithm, in: Digital Signal Processing Workshop and 5th IEEE Signal Processing Education Workshop, 2009. DSP/SPE 2009. IEEE 13th, IEEE, 2009, pp. 150–155.

[10] A.P. Dempster, N.M. Laird, D.B. Rubin, Maximum likelihood from incomplete data via the EM algorithm, J. R. Stat. Soc. B (Methodol.) 39 (1) (1977) 1–22.

[11] M. Feder, E. Weinstein, Parameter estimation of superimposed signals using the EM algorithm, IEEE Trans. Acoust. Speech Signal Process. 36 (4) (1988) 477–489.

[12] J.A. Fessler, A.O. Hero, Space-alternating generalized expectation-maximization algorithm, IEEE Trans. Signal Process. 42 (10) (1994) 2664–2677.

[13] D.W. Marquardt, An algorithm for least-squares estimation of nonlinear parameters, J. Soc. Ind. Appl. Math. 11 (2) (1963) 431–441.
[14] S. Yatawatta, Distributed radio interferometric calibration, Mon. Not. R. Astron. Soc. 449 (4) (2015) 4506–4514.
[15] S. Boyd, N. Parikh, E. Chu, B. Peleato, J. Eckstein, Distributed optimization and statistical learning via the alternating direction method of multipliers, Found. Trends Mach. Learn. 3 (1) (2011) 1–122.
[16] R. Tobar, Pipelines running on DALiuGE using the Algorithm Reference Library, 2017. https://github.com/ICRAR/daliuge-pips/tree/master/arl_int.
[17] T. Cornwell, P. Wortmann, Algorithm Reference Library, 2017. https://github.com/SKA-ScienceDataProcessor/algorithm-reference-library.

Further reading

S. Wei, F. Wang, H. Deng, C. Liu, W. Dai, B. Liang, Y. Mei, C. Shi, Y. Liu, J. Wu, OpenCluster: a flexible distributed computing framework for astronomical data processing, Publ. Astron. Soc. Pac. 129 (972) (2016) 024001.

11

Heterogeneous computing platform for backend computing tasks

Yongxin Zhu[a,b,c], Tian Huang[d,e], Junjie Hou[b], Sen Du[b], Shijin Song[b]

*[a]Shanghai Advanced Research Institute, Chinese Academy of Sciences, Shanghai, China. [b]School of Microelectronics, Shanghai Jiao Tong University, Shanghai, China. [c]University of Chinese Academy of Sciences, Beijing, China. [d]Astrophysics Group, Cavendish Lab, Cambridge University, Cambridge, United Kingdom. [e]Institute of High Performance Computing, Agency for Science, Technology and Research (A*STAR), Singapore, Singapore*

1 Introduction

Several large software packages designed for the calibration and the management of interferometric datasets—e.g., CASA [1], AIPS [2], MIRIAD [3], and GILDAS [4]—generally propose routing schemes implemented on a CPU-based single computer architecture.

However, faced with the constraints of I/O constraints or computing constraints, modern computer architectures do not provide the best possible solution because they are only suitable for general-purpose computing scenarios. At this point, the choice of a single architecture obviously cannot solve the problem, so relatively speaking, the comprehensive utilization of different characteristics of computer architecture can be regarded as a feasible solution for large-scale computing [5].

In this chapter, we will introduce a variety of computing architectures and platforms, which are dedicated to back-end computing tasks in radio astronomy. Back-end computing tasks refer to the analysis of the visibility saved in long-term storage. The analysis results of the core algorithm and performance in the specific case can give the audience a general understanding of the advantages and disadvantages of the architecture and platform.

Big Data in Astronomy. https://doi.org/10.1016/B978-0-12-819084-5.00008-0

Some case studies also give a brief introduction to the radio telescope project to help the audience understand the background. It is because radio telescope projects usually have different requirements for back-end computing tasks, while back-end computing tasks usually do not share a common programming model.

2 Computing architecture and platform

A decade ago, the first multicore CPU came out, sacrificing the single-core peak performance to achieve parallelism. For many core architectures, this concept has been brought to the extreme. In addition to introducing more cores, dedicated logic designed for specific tasks can also be included , such as leveraging additional areas in exchange for higher power efficiency or throughput/latency [6]. The use of heterogeneity can further improve performance and/or power efficiency. It could be done at the node level by deploying multiple types of devices (such as CPU and GPU), or in a single device (such as ARM big. LITTLE [7]). The other heterogeneity comes in the form of additional reconfigurable FPGAs, for example, Molen polymorphic processor [8] and the systems of Convey Computer [9].

Most of the improvements in terms of power efficiency and throughput recently are on the basis of these methodologies. For instance, GPUs adopt the multicore mode with thousands of small cores, which are often utilized in heterogeneous settings. CPUs can run faster with the help of wider and more specialized instructions [10] or small accelerators [11].

2.1 Graphical processing unit

A graphical processing unit (GPU), equipped with a large number of smaller logical cores, i.e., arithmetic logic units (ALUs), control units, and memory cache, is basically designed to process a set of simpler and more similar computations in parallel.

GPUs are very popular in astronomical research. Although the algorithm itself may have bottlenecks, GPU is relatively easy for programming compared with other computing platforms. At present, some basic calculations such as FFT and matrix multiplication have been widely concerned and successfully implemented on GPU, which performs faster and consumes less power than their counterpart on CPU.

"NVIDIA TITAN RTX is the fastest PC graphics card ever built. It's powered by the award-winning Turing architecture, bringing 130 Tensor TFLOPs of performance, 576 tensor cores, and 24 GB of ultra-fast GDDR6 memory to your PC" (Figs. 11.1 and 11.2; Tables 11.1 and 11.2).

Fig. 11.1 NVIDIA TITAN product. Cited from: NVIDIA Corporation. NVIDIA TITAN Product. Accessed: 2019. [Online]. Available: https://developer.nvidia.com/cuda-gpus.

Fig. 11.2 Nvidia Titan RTX GPU. Cited from: NVIDIA Corporation. Nvidia Titan RTX GPU. Accessed: 2019. [Online]. Available: https://www.nvidia.cn/content/dam/en-zz/Solutions/titan/documents/titan-rtx-for-creators-us-nvidia-1011126-r6-web.pdf.

Table 11.1 GeForce and TITAN products.

GPU	Compute capability
NVIDIA TITAN RTX	7.5
Geforce RTX 2080 Ti	7.5
Geforce RTX 2080	7.5
Geforce RTX 2070	7.5
Geforce RTX 2060	7.5
NVIDIA TITAN V	7.0
NVIDIA TITAN Xp	6.1
NVIDIA TITAN X	6.1
Geforce GTX 1080 Ti	6.1
Geforce GTX 1080	6.1

Cited from: NVIDIA Corporation. GeForce and TITAN Products. Accessed: 2019. [Online]. Available: https://developer.nvidia.com/cuda-gpus.

Table 11.2 Specifications of NVIDIA TITAN RTX.

GPU memory	24 GB GDDR6
Memory interface	384-bit
Memory bandwidth	Up to 672 GB/s
NVIDIA CUDA cores	4608
NVIDIA tensor cores	576
NVIDIA RT cores	72
Single-precision performance	16.3 TFLOPS
Tensor performance	130 TFLOPS
NVIDIA NVLink	Connects 2 TITAN RTX GPUs
NVIDIA NVLink bandwidth	100 GB/s (bidirectional)
System interface	PCI Express 3.0 × 16
Power consumption	280 W
Thermal solution	Active
Form factor	4.4″ H × 10.5″ L, Dual slot, full height
Display connectors	3 × DisplayPort, 1 × HDMI, 1 × USB Type-C
Max simultaneous displays	4 × 4096 × 2160 @ 120 Hz, 4 × 5120 × 2880 @ 60 Hz, 2 × 7680 × 4320 @ 60 Hz
Encode/Decode engines	1 × encode, 1 × decode
VR ready	Yes
Graphics APIs	Microsoft DirectX 12 API3, Vulkan API4, OpenGL 4.64
Compute APIs	CUDA, DirectCompute, OpenCLTM

Cited from: NVIDIA Corporation. Specifications of NVIDIA TITAN RTX. Accessed: 2019. [Online]. Available: https://www.nvidia.cn/content/dam/en-zz/Solutions/titan/documents/titan-rtx-for-creators-us-nvidia-1011126-r6-web.pdf.

"The 2nd Gen Ryzen Threadripper processors enable blazing fast renders, exports and encodes. With up to 32 cores and 64 threads, the 2nd Gen Ryzen Threadripper is a beast for digital content creators that saves you time" (Fig. 11.3 and Table 11.3).

A combination of CPU and GPU also plays an important role in the field of radio astronomy. The CPU excels in flexibility, whereas GPU is known for its large-scale parallel computing capabilities with a regular pattern of memory access. Therefore, the key to the success of heterogeneous computing architecture is to build a proper collaborative relationship between them.

2.1.1 GALARIO

The radio interferometer needs to undergo its own Fourier transform and discrete sampling calculations before generating a large amount of complex visibility. With the improvement of angular resolution and the data rate of modern radio

Fig. 11.3 AMD Ryzen Threadripper 2990WX. Cited from: AMD Corporation. AMD Ryzen Threadripper 2990WX. Accessed: 2019. [Online]. Available: https://www.amd.com/en/products/cpu/amd-ryzen-threadripper-2990wx.

Table 11.3 Specifications of AMD Ryzen "Threadripper" 2990WX.

System memory frequency	2933 MHz
System memory type	DDR4
Memory channels	4
CPU cores	32
Threads	64
Base clock	3 GHz
Max boost clock	Up to 4.2 GHz
Total L1 cache	3 MB
Total L2 cache	16 MB
Total L3 cache	64 MB
Unlocked	Yes
CMOS	12 nm
Package	sTR4
PCI express version	PCIe 3.0
Thermal solution (PIB)	Not included
Default TDP/TDP	250 W
Max temps	68°C
OS support	Windows 10—64 Bit Edition, RHEL ×86 64 Bit, Ubuntu ×86 64 Bit
Supported technologies	Zen Core Architecture, AMD SenseMI Technology, AMD Ryzen Master Utility

Cited from: AMD Corporation. Specifications of AMD Ryzen "Threadripper" 2990WX. Accessed: 2019. [Online]. Available: https://www.amd.com/en/products/cpu/amd-ryzen-threadripper-2990wx.

interferometers recently, comparing the results predicted by the model with the observational complex visibility dataset requires more computational costs since the matrix grows larger and the number of visibility points explodes.

Matrix operations such as the Fourier transform, quadrant swaps, and interpolation are often applied to calculate comprehensive visibility from the model brightness. These computational operations, though time consuming, are utilized in every probabilistic assessment. In the general workflow, in order to fit the model to the real data, it is often necessary to call the probability evaluation thousands of times.

To minimize the time required for probabilistic assessment, modern GPUs are applied to GALARIO [12] to improve the computing power of the overall system. As a result, GALARIO achieved a huge improvement in performance compared to a python project with the same functionality, that is, the execution speed of GALARIO on the GPU increased by 150 times and on the CPU improved by 90 times. More generally, this kind of acceleration schemes can be deployable on any GPU specification, even desktop-class GPUs.

2.1.2 Precision

In recent years, IEEE 754-2008 half-precision floating-point numbers, a compact data representation method, have been regarded as an emerging hotspot in GPU architecture. With the help of this method, the memory bandwidth can be reduced, and the number of accommodated computing units in the dead zone can be increased, which finally speed up the program. The report of Seznec et al. [13] shows that half-precision floating-point numbers do achieve a certain degree of acceleration on 2D convolution, however, it also indicates the nonnegligible precision loss of 2D convolution. Thus, a conclusion can be drawn that in order to achieve a good compromise between acceleration and calculation error, half storage space, and single computation are usually used.

2.2 Many integrated core

The demand for high-performance computing (HPC) power in radio astronomy has been growing with the goal of achieving an exaflop performance by 2020. To meet this huge computing demand, novel CPU and accelerator architectures, such as GPUs and FPGAs, are being developed and tested. Previously, Intel proposed an objective to reach exascale computing power by 2018 on the basis of Intel's multicore architecture and the IntelMany Integrated Core (Intel MIC) architecture.

Intel Xeon Phi is a series of ×86 multicore processors based on the MIC architecture. It is suitable for the applications on supercomputers, servers, and high-end workstations. Standard programming languages and APIs like OpenMP are allowed in its architecture.

"The second generation Intel Xeon PhiTM processor has up to 72 cores with each core supporting four threads giving a total of 288 threads in a socket. The cores are organized as Tiles. Each Tile contains 2 cores, 4 vector processing units (VPU), L2 Cache and CHA. So, there are total of 36 Tiles in the processor CPU" (Fig. 11.4 and Table 11.4).

The MIC has been applied to the square kilometre array (SKA) project. The SKA, one of the 21st century's multiwavelength facilities, is substantially a telescope in centimeters and meters. As part of an international development project, SKA initially designed to explore hydrogen arrays. Researchers later discovered

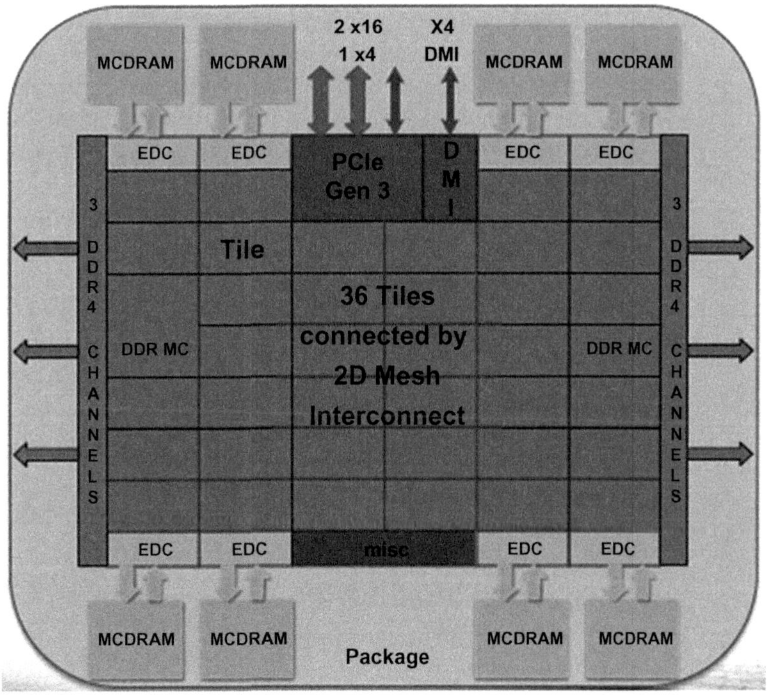

Fig. 11.4 Overview of Intel Xeon Phi x200 processor. Cited from: Intel Corporation. Overview of Intel Xeon Phi x200 processor. Accessed: 2019. [Online]. Available: https://ark.intel.com/content/www/us/en/ark/products/95830/intel-xeon-phi-processor-7290-16gb-1-50-ghz-72-core.html.

Table 11.4 Specifications of Intel Xeon Phi Processor 7290 (16GB, 1.50 GHz, 72 core).

CPU cores	72
Threads	72
Processor base frequency	1.5 GHz
Max turbo frequency	1.7 GHz
Cache	36 MB L2 Cache
TDP	245 W
VID voltage range	0.550–1.125 V

Cited from: Intel Corporation. Specifications of Intel Xeon Phi Processor 7290. Accessed: 2019. [Online]. Available: https://ark.intel.com/content/www/us/en/ark/products/95830/intel-xeon-phi-processor-7290-16gb-1-50-ghz-72-core.html.

that this extremely sensitive telescope could detect the 21-cm-high H line in the Milky Way-like galaxy. Since then, it has become clear that such telescopes can be used in a wider range of scientific projects. The international community has also proposed important research programs for the SKA, including scientific issues in astronomy, physics, and astrobiology.

NAG Ltd. proposed the technologies to transplant and optimize the radio astronomy convolution grid code on the platforms of Xeon, Xeon Phi, and NVIDIA GPUs. As an important part of the radio astronomy data processing pipeline, convolution gridding is implemented as follows: selecting a random sequence of visibilities, convolving them with a series of kernels, and assembling the results into the grid. Involved in this project, four sets of visibilities were fully utilized: one integrated, and the other three extracted from the simulation of similar SKA configurations.

There are several attempts to implement radio astronomy applications on MIC [14–16], which are related to front-end computing tasks. Since these tasks all occur before the signals enter the long-term storage for imaging or source detection, they are outside the scope of this chapter.

2.3 Cell broadband engine

Prior to the advent of the Cell BE, hardware designers focused on improving the system performance of general processors from three aspects of clock frequency, superscalar problem, and super-pipelined design. In this case, maintaining the original memory speed may result in adverse results such as increased memory latencies and even more complex logic intrusions. Moreover, execution pipelines and memory bandwidth are often not fully utilized by complex cores, which lack the capability to handle numerous concurrent accesses to memory. Eventually, the area utilization of the chip is extremely low, and the power consumption of the system increases disproportionately with the system performance [17].

The Cell BE is designed to improve performance/area and performance/power ratios by leveraging powerful yet simple cores, which make more efficient use of area and lower power consumption. These cores can work independently or collaboratively with the assistance of high data bandwidth interconnection. Further, the efficiency of the memory bandwidth utilization is greatly improved due to numerous concurrent memory accesses from the cores (Fig. 11.5).

The case study of implementing and optimizing the most time-consuming radio astronomical imaging kernel (i.e., data grid and de grid) on the Cell BE. processor was proposed by the author of [18]. These two kernels are dependent of convolutional resampling, which is the basic processing module to determine the workload of image synthesis. The optimization of these kernels directly affects the performance and hardware requirements of any modern large radio telescope [19]. On the basis of brief

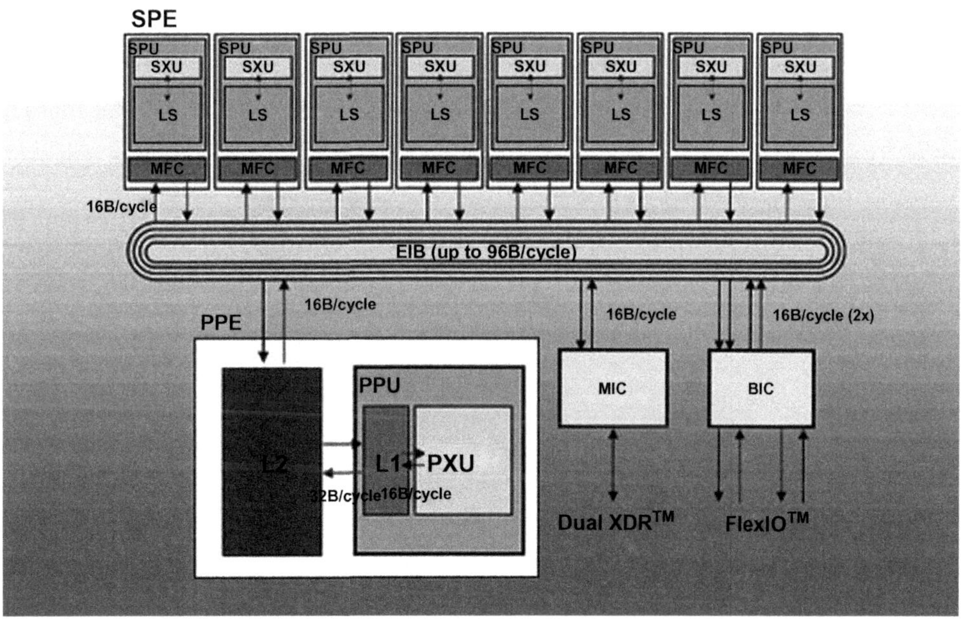

Fig. 11.5 The architecture of cell broadband engine. Cited from: IBM Corporation. The Architecture of Cell Broadband Engine. Accessed: 2019. [Online]. Available: https://www.ibm.com/ibm/history/ibm100/us/en/icons/cellengine/.

analysis, we utilize the master-workers example to model parallel applications. Although there are some challenges in the implementation and subsequent optimization, especially in terms of appropriate load balancing and data distribution, the performance obtained proves that this example is a good choice for processing data-intensive applications on Cell BE processors.

2.4 ASIC/FPGA

After the digital signal is preprocessed, the ASIC structure is not a universal choice for radio astronomy algorithms. Part of the reason is that the GPU provides a more friendly development environment and does not involve much hardware design skills. However, with the help of reasonable design methods, ASIC architectures are generally superior in power consumption. Modern radio telescopes such as the SKA [20] require a high-power/energy-efficient processing architecture to process terabytes of raw data per second. For example, based on the modern technology, double-precision with multiply-add operations will consume 7.2 MW of power in the medium-frequency array of SKA [20]. Researchers currently use analog or field-programmable gate arrays (FPGAs) (Fig. 11.6) to explore energy-efficient radio astronomy algorithms.

Considering the energy-efficiency, [21] presented a new heterogeneous architecture for least squares (LS) acceleration. Combining optimized-precision (accurate) and reduced-precision (approximate) computational cores can offer proper mass output

Fig. 11.6 Xilinx Virtex UltraScale+ FPGA VCU1525. Cited from: Xilinx Corporation. Xilinx Virtex UltraScale+ FPGA VCU1525. Accessed: 2019. [Online]. Available: https://www.xilinx.com/products/boards-and-kits/vcu1525-a.html#hardware.

while reducing energy consumption compared to a precise optimized architecture. The inherent error resilience of an iterative workload can utilize an approximate computational core to handle the initial iteration of the LS algorithm. According to a case study of radio astronomy calibration, up to 24% (and 29%) of energy can be reduced compared to the precise method of biased (unbiased) design.

3 Algorithm benchmarking

In this section, we review the backend computing tasks that benefit from the heterogeneous computing platform.

To fully exploit the potential of heterogeneous computing platforms, researchers need to analyze and understand the bottleneck of the tasks. The most common issue is the irregular memory access pattern that brings down the efficiency of algorithms. We will give a brief review on how researchers tackle this issue and improve the performance.

3.1 Gridding

Gridding is an important algorithm in the radio astronomy image process. When inverse fast Fourier transform (IFFT) algorithms are applied, convolution gridding transforms the nonuniform sampled UV-plane (visibility plane) to a uniform sampled one. The data flow of UV imaging is shown in Fig. 11.7. Fringe visibilities of celestial objects can be measured by interferometers. Visibility measurements are convolved with an appropriately chosen convolution function $C(u, v)$, however, which partly have the same size support. Then, the resultant UV-plane is then sampled uniformly with the use of an IFFT. The result is an image on the lm-plane (intensity distribution plane) that is corrected by a simple element-by-element convolution. The output is a dirty image of the intensity distribution, which is the true intensity of the nonuniform and incomplete UV coverage. In the opposite step, the model visibility on continuous coordinates is calculated from the discrete model, which is called degridding. The flow chart of GRIDDING and DEGRIDDING is shown in Figs. 11.8 and 11.9, respectively.

In order to improve imaging performance using FFT transforms, the measured visibility must be gridded into a two-dimensional array. To smooth and interpolate the data, the visualization data are convolved with an appropriately selected grid convolution function (GCF) before being resampled using different weighting methods. In practice, this precise approach

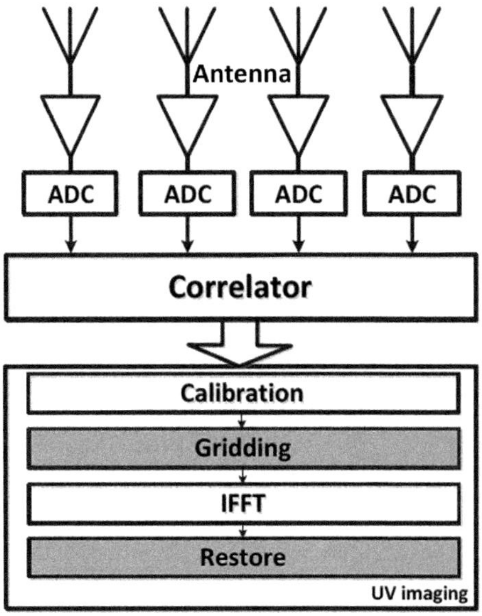

Fig. 11.7 Data flow of UV imaging. Cited from: Wu Q. Design and Optimization of High Efficiency Hardware Accelerator for the Key Algorithm of SKA Science Data Processing[D]. Shanghai, China: Shanghai Jiao Tong University, 2019.

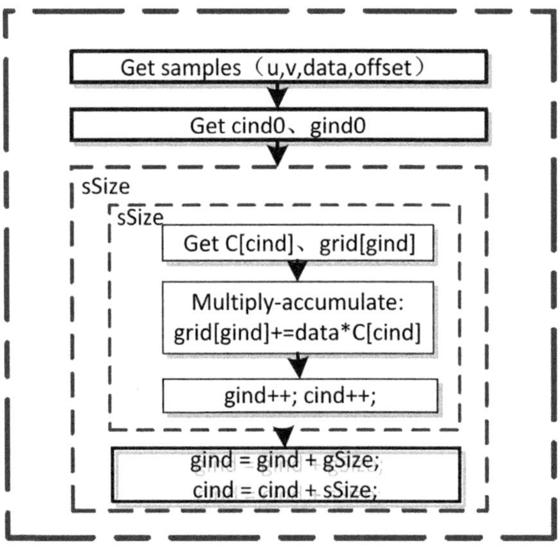

Fig. 11.8 The flow chart of GRIDDING. Cited from: Wu Q. Design and Optimization of High Efficiency Hardware Accelerator for the Key Algorithm of SKA Science Data Processing[D]. Shanghai, China: Shanghai Jiao Tong University, 2019.

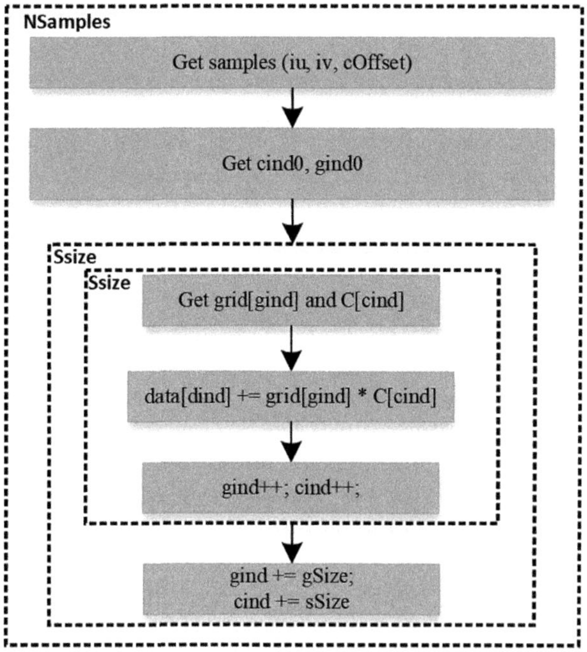

Fig. 11.9 The flow chart of DEGRIDDING. Cited from: Hou J, Zhu Y, Du S, et al. FPGA-based Scale-out Prototyping of Degridding Algorithm for Accelerating Square Kilometer Array Telescope Data Processing[J]. IEEE Access, 2020.

can greatly affect the macroscopic properties of the resulting image, such as aliasing from edges and uniform spacing of the grid (Schwab 1984). Many functions, such as pillbox functions, exponential functions, sine functions, and pillbox functions, can be considered as convolution functions. The MIRIAD software uses a spheroidal function [3]. According to Schwab's choice of the optimal gridding method, the gridding function is taken as the optimal GCF.

The naive implementation method of grid computing intensity is very low [22], and carried out many atomic add operation. This is shown in Refs. [23, 24] for both CPUs and GPUs. Due to bandwidth constraints, both platforms achieved about 5% of their peak performance. An implementation on the cell [18] achieves a peak performance of about 25%, but only after considerable optimization of memory bandwidth usage. The GPU research carried out in Ref. [25] tried to eliminate the atomic addition instruction as much as possible, and the GPU reached 25% of its peak performance. Even with almost no atomic additions (0.23% of all grid updates), they still account for 26% of the time. For gridding, a simple kernel will work well on most architectures. Bottlenecks are memory bandwidth and atomic additions.

The computational process of gridding and degridding is essentially two-dimensional convolution. For the gridding algorithm, each (u, v) spatial sample point is convolved with a two-dimensional window. Then, convolution result is mapped to a regular two-dimensional grid. Because of the randomness of the sampling points in the (u, v) space, the computational features of the gridding and degridding algorithms are memory-intensive. In other words, memory access performance is the bottleneck of these algorithms.

Romein [25] proposed the GPU's work assignment strategy, which effectively convolved the radio telescope data onto the grid. It is one of the first practical algorithms for GPU-accelerated gridding. While being different from existed work allocation strategies, this strategy keeps a low number of device memory accesses without incurring overhead for sorting or searching through telescope data. Performance measurements show that this strategy is an order of magnitude faster than existing accelerator-based grid generators. Despite being the most advanced technology, it usually only spends about 25% of the GPU's computing power on actual convolution operations. Storage systems have bottlenecks, but the computing overhead associated with address computing is also significant.

For radio interferometry, Muscat [26] has developed a new high-performance imaging synthesis tool which is called malta-imager. Malta-imager takes use of CUDA-compatible NVIDIA GPU to compose images via w-projection. Besides the method that the backbone deals with numerical calculations, they also develop a new framework called the general array framework (GAFW) to handle different hardware, such as GPUs and FPGAs. This feature is only implemented for CUDA-compatible GPUs to achieve the main purpose. The malta-imager is built on this framework and can perform all GPU-based calculations. The test cases presented in their paper shows that this imaging tool can synthesize images nearly 100 times faster than CPU-based conventional imaging tools. The main goal of their work has been fully achieved due to the successful development of high-performance imaging tools.

Merry [27] attempts to reduce the memory communication overhead so that more flops can be available for convolution calculation. The main contribution is that a modification of the algorithm is made in which each thread of execution processes multiple elements of the grid. This is a standard transformation called threading coarsening, but she has adapted the method to this specific problem. This allows some overhead to be amortized over multiple grid elements, which improves the overall performance.

According to the current running state of MUSER, Mei et al. [28] realized a high-performance imaging pipeline based on GPU technology [29]. The imaging process, including gridding and deconvolution, is performed on the GPU platform. Preliminary imaging results demonstrate the feasibility of this approach. They focus on phase correction in the current status and the detection of the solar disk and sky brightness based on the Gaussian mixture model. As such, it automatically determines the number of iterations of the cleaning algorithm. What's more, the performance of GPU-based Steer clean is much better than the commonly used CASA software [30].

Gridding is applied to astronomical data to transform data from irregular sampling space to regular grid space [31]. This can lead to an abnormal memory access mode, which can cause difficulties for CPU-only or GPU-only computing platform. In order to solve these problems, [30] proposed a CPU-GPU hybrid convolution-based meshing algorithm. First, an efficient lookup table based on HEALPix was introduced in the preordering step. GPU-accelerated collection is then achieved through proper data placement and thread coarsening. Experiments show that the solution has a significant improvement in performance and can be applied to the actual data pipeline of large interferometers and single discs.

Zhao and Jing [32] had an attempt on implementing gridding algorithm on a many-integrated-core architectural processor for SKA project. They proposed an efficient data partitioning scheme to the gridding software calculation kernel. For parallel software design, the data partitioning scheme determines the parallelization scheme. Zhu and Tong [33] presented a scheme based on two-dimensional data partitioning. Its purpose is to improve the locality of data to improve memory access efficiency.

According to the data partitioning scheme, the parallelization scheme of the grid is determined. Since the gridding software is an MPI parallel software, the computational tasks of each MPI process are the same, so the parallelization design in Ref. [32] focuses on the multithread parallelization design within a single MPI process. This step is implemented using OpenMP. After that, combined with the characteristics of the MIC platform, two optimization methods are mainly used. For the first one, the main working array of Gridding needs to be optimized on MIC chipset's multilevel cache to improve its memory access performance. The second method is to combine the vectorization features of MIC chipset to vectorize the gridding software.

Zhao et al. also pay attention to the problem of memory access and memory-write conflict. In order to further improve the memory access efficiency of gridding, a fast sorting algorithm for gridding was developed. By sorting the data, the jump step size of its memory access is reduced, and the performance of the gridding is further improved. In order to solve the memory-write conflict problem of gridding caused by the parallelization process, a method of using a temporary buffer is proposed. This method not only completely avoids write conflicts logically, but also does not require synchronous operations.

3.2 Deconvolution

Due to the irregular sub-Nyquist sampling of the Fourier domain, the use of earth rotation to synthesize radio telescope imaging is an ill-posed problem. Due to the high side lobe of the array response, this subsampling will lead to aliasing in the image. To solve this problem, the influence of the instrument response needs to be removed from the image (a process called deconvolution) and compensate for inaccuracies in the array response (called self-calibration, but it has many similarities to blind deconvolution). It is important to keep in mind that improved imaging capabilities are the results of better equipment combined with new imaging technologies. Each generation of radio telescopes requires a lot of hardware development [34].

Due to this subsampling, there are multiple possible image matching measurements. The simplest imaging method (i.e., generating an image) treats all unmeasurable visibilities as zero. This led to the so-called dirty image and suffered from artifacts. Due to the all-zero hypothesis, the point spread function (PSF) convolves efficiently with the sky image, and it is the measuring results of the interferometer, which measures the point source at the center of the phase.

The most deconvolution algorithms subtract PSF from the location of the source (usually the brightest point of the dirty image) to eliminate the convolution. Deconvolution selects the image from the possible images closest to the actual sky. Under the background of low intensity noise, many images will consist of only several small point light sources, which will result in sparse images. Thus, finding sparse images, which match the measured results is a good method of deconvolution.

3.2.1 Clean

The CLEAN algorithm is a widely used deconvolution approach in the field of radio synthesis imaging, which represents the image with a large number of point sources in the view of an otherwise empty field. The procedure of CLEAN is used to find the strengths and locations of these point sources.

The Högbom CLEAN algorithm is raised by Högbom, J. A. in 1974 and it attends to find the brightest pixel (it is called a CLEAN component) in a designated area of the image (it is called the CLEAN window, and it contains all the real emission of the image) [35]. Unfortunately, there are errors or noise in the calibration errors and previous CLEAN iterations, and this indicates that negative components will be obtained as well. To clean up the negative ingredients is usually very effective. When the peak pixel is found, the dirty image at the brightest pixel subtracts a small portion of the dirty beam (it is called a loop gain, typically about 0.1 or less). The subtracted image is called the residual image. The repeat of the subtraction and search loop will stop until the side lobes fall below the noise level. If you only consider a single point source, it's easy to see how CLEAN works. For extended sources, we treat emission as a gathering of point sources. In addition to point sources, the flux density units of dirty images are not very useful. But convolution of CLEAN components does make sense for the unit Jy per clean beam area, and it can be converted to Jy per unit area, since we can calculate the equivalent region of the cleaned beam.

Zhao and Jing [32] proposed an optimized version of CLEAN, which made use of the most advanced many integrated core (MIC) architecture and high-bandwidth memory. Several optimizations were implemented, including vectorization optimization implementation, thread load balancing implementation, and cache optimization implementation. Vectorization is implemented through compile options, which makes full use of the AVX512 instruction set on KNL. This method improves the performance of the program through high vectorization. Then, through OpenMP static balancing and scheduling, the load between threads can be achieved by dividing the data of the image. At last, with the consideration of the much higher bandwidth of MCDRAM compared with DDR4 memory, MCDRAM is used for storing the data which requires frequent access by utilizing the Memkind Library.

3.2.2 Richardson-Lucy method

The Richardson-Lucy method [36, 37] was the most widely used deconvolution method in the area of astronomy. This method has advantages of preserving the number of counts and the nonnegativity of the original object. However, it is known that Richardson-Lucy is an inefficient method. In most cases, especially for low noise levels, the cost of an acceptable solution is hundreds of iterations. Aiming at the problem of constrained minimization of continuous differentiable convex functions, a general optimization method, proportional gradient projection method (SGP) [38] is proposed.

In Ref. [39], the characteristics of SGP are illustrated to the astronomical community, and such methods are implemented to solve the problem of multiimage deconvolution and boundary effect correction. The first problem is basic. One example is reconstructing images of future large binocular telescope (LBT) interferometers, which can be named LINC-NIRVANA [40–42], The second problem is more related with single image deconvolution as well as multiimage deconvolution. All implementations of such algorithms are in interactive data language (IDL) and the codes are open access and mostly free. Prato et al. [39] proposed a method of implementing a GPU.

The SGP version of RL greatly improves efficiency. It allows for the significant acceleration of all RL-type algorithms. In general, acceleration depends heavily on the particular object that needs to be reconstructed. When the desired object is given, acceleration is more dependent on the noise level. When the GPU version is implemented, consistent acceleration results are obtained when compared to the report of [43]. Since the computational core of the RL algorithm is FFT, and the calculation basis of the SGP algorithm is stepping, etc., the acceleration of the RL algorithm is faster than that of the SGP algorithm. However, the gain relative to the RL is very significant. Sometimes, it is possible to allow us to deconvolute 2048 × 2048 images in a matter of seconds.

3.2.3 MEM

The class of image-restoration algorithms is important when these algorithms are used to produce satisfactory images. Images obtained in this way are consistent with the measured visibility within the noise level. Besides, the level of image quality can be maximized to a certain degree. Among them, the maximum entropy method (MEM) has attracted the most interest in the area of radio astronomy. In Ref. [44], it is found that one of the best entropy forms is:

$$F = -\sum_{k} I_k \ln \frac{I_k}{M_k e}$$

This form can be widely used for general purposes. In the formula, I_k is the image to be deconvolved. And M_k is a default image that has incorporated some prior information. In general, the candidate default image can be the target image, which has a low resolution. Besides, the model has to be measured through a χ^2 statistic to guarantee that the intensity of the model and the visibility data are consistent. This measurement can be made through the calculation of the mean-squared difference between the measurements, V_k, and the corresponding values of the model, V_k'. The concrete representation is

$$\chi^2 = \sum_k \frac{|V_k - V'_k|^2}{\sigma_k{}^2}$$

This solution is an iterative process that is similar to CLEAN [45].

For the large-scale image synthesis problem, the GPU implementation of MEM was developed by Carcamo et al. [46]. The GPU implementation is applied to Atacama Large Millimeter/Submillimeter Wave Array (ALMA), which is an international millimeter/submillimeter interferometer built in the Atacama Desert in northern Chile [47]. ALMA will generate rather massive data, which requires a massively parallel and unfailingly high-throughput approach. However, today's multicore computers are not possible to solve such a big date problem. In contrast, GPUs have evolved from specific uses to scientific and general-purpose applications. The various functions and APIs of GPUs provide convenience for the scientific community to leverage their computing power without complex configurations [48]. Recently, a GPU implementation [49] was proposed to solve the problem of radio observation Bayesian inference [50]. In this method, Bayesian inference is used to sample the parameter set, which represents the sky model. Furthermore, the idea that visibilities best match with the model is proposed. Their models can handle points and Gaussian sources. In order to solve the Bayesian model of the maximum entropy prior in a direct way, a nonlinear optimization problem was adopted by this team. The power of GPU computing was verified through some tests of several ALMA datasets. Their computing performance was also compared with multithreaded solutions.

Carcamo's research showed that high-performance GPU implementations using MEM can reconstruct ALMA images. The calculations show that GPU achieves 524 times faster image reconstruction than the equivalent 20-core multithreaded solution. As the reconstruction time grows slowly relative to the size of the image and the amount of visibility, GPU seems to have more scalability. This is rather important and meaningful because ALMA and the new radio telescope SKA will produce more amounts of observation data. In the future, the current framework can be updated to leverage the energy of more advanced GPUs and use a multi-GPU approach to assigning data and make calculations.

In 2018, Carcamo's team continued to work on GPU implementations to solve the problem of maximum entropy image reconstruction in interferometry [51]. The main idea of this solution is to process single-channel data based on a single GPU processing and process multispectral data based on multi-GPU processing. A host algorithm in this implementation is responsible for managing the main iteration loop and simultaneously orchestrating the kernel calls. Small kernels are determined with the consideration of improving data locality and minimizing thread differences. For reducing the time cost data movement between the host and the device, most of the data is stored in the device's memory, besides, the algorithm shows the intrinsic nature of a SIMT paradigm, which makes this algorithm proper for an effective GPU implementation.

The resulted implementation achieved approximately 1681 × acceleration on the smallest dataset. Obviously, the rebuild of CO(6-5) only needs about 2.1 min on GPU, while the same task takes about 2.5 days on a single CPU. As for the task of Long baseline HL Tau band 6 with spectral window 0 and 1042 image size, it requires about 58 min using this implementation. But for full-band and multispectral datasets such as SDP 8.1 and HL Tau, it is still challenging to handle nongrid MEM reconstruction with acceptable processing time. For example, 200 min are required per iteration when processing a complete dataset with 6 HL Tau reconstructions. As such, this long time shows that the cost is still expensive when conventional use of hardware is considered.

From the analysis of the MEM image synthesis results, the method has removed the background noise of the system. Though a better resolution is obtained, the intensity is lower at the point source. In addition, the MEM has a smaller signal variation in the flat areas and a smoother fit on the extended structure than CLEAN. In a way, MEM users can benefit from this method because it allows users to identify a finer morphology of the object with not too much loss of signal intensity.

3.2.4 NNLS

Nonnegative least squares (NNLS) were originally studied by Briggs [52] in the context of radio astronomy imaging. The NNLS problem is a class of constrained least squares that do not allow the coefficient to be negative. Lawson and Hanson (1974) raised the NNLS algorithm, it provides a solution to the form AX = B of a matrix equation, where A denotes a dirty beam and B denotes a dirty image in a radio astronomy application. In the form AX = B, the intensity X does not contain negative values, the algorithm provides a least-mean-squares solution on this problem.

In Ref. [53], the authors evaluated the NNLS algorithm on the synthesized dataset with the same configuration as the LOFAR telescope. van Haarlem et al. [54] LOFAR, a low-frequency array, is a radio telescope designed and built by ASTRON that opens the lowest-frequency radio regime for a wide range of astrophysics studies. Sardarabadi et al. [53] used a large number of documents related to solving the least squares problem, and developed a parallel NNLS active-set solving method, and explained the relationship of its method and CLEAN. The active-set method involves selecting a column in the measurement matrix corresponding to the pixel of the maximum strength image on the remaining image repeatedly, and adjusting the measured values based on the pixel of the selected image to obtain the same technique as CLEAN. This process continues until the residual image achieves a noise-based threshold.

Hardware-accelerated NNLS has not yet attracted widespread attention in the radio astronomy community. However, in other application areas, there has been a large amount of literature on using GPU, FPGA, and other computing architectures to accelerate NNLS.

For example, Luo and Duraiswami [55] reduce the computational and memory access costs by reordering the columns in the decomposition, parallelizing a version of the original active-set iteration algorithm derived from Lawson and Hanson on a multicore architecture. Mejía-Roa et al. [56] proposed NMF-mGPU, which is an efficient and easy-to-use NMF algorithm implementation that takes advantage of the high computational performance provided by GPUs. Koitka and Friedrich [57] proposed nmfgpu4R, which provides nonnegative matrix factorization (NMF) calculations on the compute unified device architecture (CUDA) platform in the R environment.

3.2.5 Sparse regularization

MUFFIN [58] is a variant of the sparse regularization algorithm used to solve different problems of multiwavelength reconstruction of large-scale data cubes. It sets up the SKA telescope project [59] as a target application. The advantage of the MUFFIN algorithm is the use of spatio-spectral separation to implement it in a parallel architecture w.r.t the frequencies. This upstream consideration is the primary use of parallel computing problems. Since the most demanding steps of the algorithm are separable w.r.t. the wavelengths, Bouhaik obtains a parallel implementation on a multinode system (such as multiple GPU), where each slave node processes one wavelength. Another important part of the algorithm is the use of isotropic undecimated wavelet transform (IUWT) to achieve spatial regularization. This wavelet transform is accelerated on GPU systems because it requires computational power and is suitable for graphics architecture.

3.3 Compressed sensing

According to Nyquist-Shannon sampling theory, in order to reconstruct an image perfectly, at least twice the measured value of each point in the image is required. However, a priori knowledge of the image (sparse in the pixel dictionary) can be used to increase the measurements required number [45].

Compressed sensing theory [60] states that if a linear combination of measured signals can be calculated, the probability of refactoring the s-sparse signal (signal only nonzero value) or s-weakly sparse signal (the signal s values are dominates, and other values close to zero) is very high, assuming that each measurement of linear coefficient is orthogonal (a measure of inner product of different coefficient is close to zero) [61].

The measurements of radio interferometry are the points on the two-dimensional Fourier plane. Therefore, the coefficients are columns of a two-dimensional Fourier transform matrix. Fortunately, they are the Cartesian constraints [60]. It means that if the image is sparse (the number of nonzero pixels is set to be less than half as many as the number of measurements), there is a great possibility to reconstruct the image of the actual sky by interferometry.

Orthogonal matching pursuit (OMP) is a compressed sensing algorithm that minimizes differences in an image and optimizes the process by recursively selecting a single pixel (the difference between the transformation and measurement of the image) and then weighting the pixels. This way they have the best approximation. Schwardt [62] found that this approach produced better results (in terms of sparsity, contrast, and execution time) than other mainstream compressed sensing technologies.

The algorithm flow of OMP is shown in Algorithm 11.1. Like MP, OMP algorithm also operates until the best element is selected. Because we are going to weight this component in each subsequent iteration, its index is stored in an (initially empty) vector D. For each selected element, α_i, its contribution to y is determined by the ith row of Θ. Therefore, matrix A is set to contain only columns of Θ_D, and the index of Θ_D is in set D. Θ_D perfectly represents the value of each selected element, so the solution vector of $\arg\min \|y - \Theta_D\alpha\|_2$ minimizes the possible residuals of the selected element.

Algorithm 11.1

$r \leftarrow y$

$D \leftarrow \phi$;

while r is significant **do**

  $p \leftarrow \Theta_r^{\mathrm{H}}$

  $i \leftarrow \arg\max |p_j|$ where $j \in D$

  $D \leftarrow D \cup \{i\}$

  $\alpha \leftarrow \arg\min \|y - \Theta_D\alpha\|_2$ ◁ Only the selected components α_j, $j \in D$ are determined by this problem. Define $\alpha_j = 0$ for $j \notin D$.

  $r \leftarrow y - \Theta\alpha$

Belle et al. [45] implemented an OMP algorithm for radio interferometric deconvolution. The algorithm is implemented on the GPU. The OMP implementation in the report is significantly better than CASA's CLEAN in source discovery. Of the 100 synthetic sky models we used in our tests, a whopping 82% matched the light source, while CLEAN matched only 61%. In addition, the average intensity of the OMP residual image is 2.9 times lower than that of the CLEAN. In addition, the GPU implementation is 83 times faster than a single-threaded CPU implementation and 23 times faster than a four-thread implementation.

Research in other application areas indicated that OMP is an algorithm that has good parallelism to be exploited. For example, Kulkarni and Mohsenin [63] proposed a platform independent CS reconstruction architecture is proposed by Kulkarni and Mohsenin [63], it can reduce the reconstruction time. The architecture can be implemented on a variety of platforms, like general-purpose CPUs, GPUs, a Virtex-7 FPGA, and a domain-specific many core.

3.4 Source extraction

Source extraction is a step in the process of astronomical image processing that calculates and reports the properties and characteristics of astronomical points of interest or sources. There are many software packages that automate the complex radio source identification and extraction process.

Duchamp was used as a benchmark for base timing and output verification because it represents the latest software suite for three-dimensional source extraction in radio astronomy [64]. Through optimization and algorithm redesign, Duchamp's source code extraction process is accelerated by Resnick et al., allowing the routine to map well to the GPU architecture.

Resnick and his team treated Duchamp as a naive CPU implementation that iteratively optimizes these components in two phases to reduce computation time while maintaining accuracy. The first phase requires optimal CPU rewrite of the source detection routines, as well as general framework development. The second phase uses NVIDIA's CUDA to assign computationally expensive routines to the GPU.

Five different datasets were used in the experiments. The datasets were generated by Kurt Van Heyden of the University of Cape Town's Department of Astronomy using the SKA Simulated Skies (S3) Tool provided by the Oxford University SKADS project [65]. This work is not complete on the GPU, but there are loads of research in the graphics environment, which can adapt to radio astronomy [66–68].

4 Telescopes and applications

4.1 SKA

The SKA will be the vastest radio telescope array in the world once completed by integrating the sampled data from a huge number of small antenna nodes in the array to emulate a giant antenna. The SKA project will construct thousands of dishes and up to a million low-frequency antennas, and this will enable astronomers to monitor the sky in unprecedented detail and survey the entire sky much faster than any system currently in existence.

There are 13 member countries in the SKA project, and about 100 organizations from about 20 countries are involved in its design and development. The world's leading scientists and engineers are designing and developing a system that requires supercomputers to be faster than any existing system, and network technology that will generate more data traffic than the entire Internet [69].

4.2 SETI

SETI (search for extraterrestrial intelligence) is a field of scientific research aiming at detecting intelligent life outside the Earth [70]. The University of California, Berkeley's SETI team conducted research based on these ideas. They found that there may already be thousands of computers available to use. Most computers sit on most of the day and the toasters fly around the screen. But nothing has been done, and the startup power is wasted. This is where SETI @ home (and you!) comes together. The SETI @home project hopes to borrow your computer when you are not using it to help us "... find new life and new civilizations." We'll do this using a screensaver that takes a lot of data from us over the internet, analyzes the data, and reports the results to us. When you need a computer, our screensaver disappears immediately, and the analysis continues only after you have finished working.

For small radio astronomy projects, the capability to build or purchase such computing power is unaffordable, which is exactly the tradeoff they need to make. They can use a smaller computer instead of a large computer to work, but it will cost more time in this approach. It seems impossible that so much data could be analyzed in time. Fortunately, it is possible to break up the data analysis task easily into subtasks that are little pieces. And, these pieces can be processed separately and in parallel with no mutual intervention. In addition, the amount of sky that can be seen from Arecibo is finite. As such, the entire sky as seen from the telescope will be scanned three times in the next 2 years. Generally speaking, this will be enough for the project. New telescopes, new experiments, and new approaches will come forth to SETI by the time that the sky is scanned three times.

4.3 MUSER

MingantU SpEctral Radioheliograph MUSER (MingantU SpEctral Radioheliograph) [29] is a solar-dedicated interferometric array located in Mingantu town, Inner Mongolia of China. It enables to image the sun in a wide frequency range (0.4–2 GHz) with high spatial resolution, high temporal resolution, and high-frequency resolution. It was completed after the name was changed from CSRH (China Spectrum Solar Radio). It was renamed according to CSRH (Chinese Spectral Radioheliograph) after its accomplishment [29].

There are some critical temporal constraints of the telescope that need to be respected, which requires these reconstruction algorithms to be implemented on a high-performance parallel architecture in an accelerated and optimized way. Therefore, the main objective is the explorations of different parallel architectures. Then, the implementation of MUFFIN algorithm on a multi-GPU server using OpenCL parallel programming model is accelerated.

In order to meet the requirements of MUSER processing, a high-performance distributed data processing pipeline (DDPP) is built on the Open Cluster infrastructure to handle various tasks for CSRH observational data, including data storage, archiving, preprocessing, image reconstruction, deconvolution, and real-time monitoring. Computing performance can meet the requirements of MUSER data processing in multicomputer parallel and GPU hybrid systems.

4.4 Parkes observatory

The high time resolution universe (HTRU) survey is carried out at a radio telescope observatory, Parkes Observatory. This observatory is located 20 km north of the town of Parkes, New South Wales, Australia. As for the survey, an all-sky search for transient and periodic points sources is adopted using the 64-m dish, in order to discover new pulsars and fast radio transients [71].

The current pulsar and transient search process is composed of several steps. First, filter bank data (that is, the time sequence of each frequency channel) are retrieved from the backend. These data are stored using 2 bits per sample, which gives a data rate of around 4 MB/s per beam. After that, they are sent from the telescope down to the Swinburne supercomputer in Melbourne via a fiber optic link. The transmission distance is about 700 km. On the computer side, these data are written to disks for optionally immediate processing. Alternatively, they could be also written to tapes for longer-term storage and further processing or reprocessing. Generally, a distributed cluster of multicore central processing units (CPUs) are used for the processing. It takes more than 30 min for the current transient search pipeline to process a 10-min observation when the data are in memory [72].

4.5 Radio transient detection

Barsdell et al. [72] developed a transient-detection pipeline based on the real-time operation of the Parkes radio telescope data. The High Time Resolution Universe (HTRU) measurements are currently being performed at the Parkes Observatory, using a 64-m dish antenna for an all-sky search of transient and periodic point sources. Its primary goals are to discover new pulsars and fast radio transients. Since 2008, it has discovered more than 65 previously unknown pulsars [73, 74].

The current pulsar and transient search process begins with the acquisition of filterbank data. (i.e., time series for each frequency channel), the time series is stored from the backend using

2 bits/sample to a data rate of about 4 MB/s per beam, connected via fiber optics, and sent to the Swinburne supercomputer in Melbourne (~ 700 km). There, it is written to disk for immediate processing (if possible) and written to tape for long-term storage and future (re)processing. The processing itself is then executed on a distributed cluster of multicore central processing units (CPUs). Once the data are in memory, the current transient search pipeline will take more than 30 min to process the 10-min observation.

Although the survey data so far need to be processed offline due to the limitation of CPU computing power, ideally, the pipeline should be executed in real-time using observation data. The goal of Barsdell's team is to implement transient detection pipelines on the GPU and take advantage of the boost in processing power to achieve real-time processing rates.

The transient detection part of the software pipeline, starting with the filterbank data, ultimately generates a list of candidate events, including five processes: RFI mitigation, incoherent dedispersion, baseline/red-noise removal, matched filtering, and peak finding. As pointed out by [75], the computationally intensive nature of the algorithm makes it ideal for GPU architectures. Through the GPU implementation of the algorithm, they can implement a transient-detection pipeline based on the real-time operation of the Parks radio telescope data.

4.6 Solar flare detection

The solar flare is an intense, abrupt release of energy that occurs in the sun as the magnetic field changes due to the presence of flux or the movement of sunspots.

Poli et al. [76] provide a unique TNS for detecting solar flare events in solar images obtained using radio astronomy techniques. This study targets the Brazilian Decimeter Array telescope. The Brazilian Decimetric Array (BDA) is a 26-element radio telescope interferometer located in the municipality of Cachoeira Paulista in the state of Sao Paulo, Brazil [77]. The BDA is open to the entire scientific community for the study of solar and nonsolar and space weather phenomena.

Poli proposed an extensible universal framework for processing signals captured by radio interferometers. He introduced and discussed the implementation of a solar flare detection system called SOL-FLARE, which is suitable for the proposed GPU-CUDA framework, which helps to automatically detect solar flare events in solar images.

5 Conclusion

This chapter introduces the heterogeneous computing platform for backend computing tasks. Computing platforms like CPU+GPU, Many-core, and distributed platforms are popular among radio astronomers as their programming environments are more friendly to the end users and do not require much expertise in computer architecture. Backend computing tasks in image synthesis and deconvolution attract extensive attention from the research community and are adapted onto various computing platforms. One of the common challenges is the irregular pattern of memory, which reduces the efficiency of parallel computing architectures. One of the solutions is to reorder the memory access flexible processors, like CPU, so that the computing tasks can fully exploit the power of parallel computing architecture, like GPU. Another possible solution is to rely on the advances of computing architecture and wait for the emergence of memory with higher bandwidth and lower latency.

References

[1] J.P. McMullin, et al., CASA architecture and applications, in: Astronomical Data Analysis Software and Systems XVI, vol. 376, 2007.

[2] E.W. Greisen, AIPS, the VLA, and the VLBA, in: Information Handling in Astronomy-Historical Vistas, Springer, Dordrecht, 2003, , pp. 109–125.

[3] R.J. Sault, P.J. Teuben, M.C.H. Wright, A retrospective view of MIRIAD, in: Astronomical Data Analysis Software and Systems IV, vol. 77, 1995.

[4] G. Team, GILDAS: Grenoble Image and Line Data Analysis Software, Astrophysics Source Code Library, 2013.

[5] J.M. Ford, P. Demorest, S. Ransom, Heterogeneous real-time computing in radio astronomy, in: Software and Cyberinfrastructure for Astronomy, vol. 7740, International Society for Optics and Photonics, 2010.

[6] G. Venkatesh, J. Sampson, N. Goulding, S. Garcia, V. Bryksin, J. Lugo-Martinez, S. Swanson, M.B. Taylor, Conservation cores: reducing the energy of mature computations, SIGARCH Comput. Archit. News 38 (2010) 205–218.

[7] ARM: big.little, http://www.arm.com/products/processors/technologies/biglittleprocessing.php.

[8] S. Vassiliadis, S. Wong, G. Gaydadjiev, K. Bertels, G. Kuzmanov, E. Panainte, The MOLEN polymorphic processor, IEEE Trans. Comput. 53 (2004) 1363–1375.

[9] Convey: Convey Computer Website, http://www.conveycomputer.com.

[10] Intel: Intel SSE and AVX Extensions, http://software.intel.com/en-us/intel-isa-extensions.

[11] Intel: Intel Random Number Generator, http://software.intel.com/sites/default/files/m/d/4/1/d/8/441_Intel_R__DRNG_Software_Implementation_Guide_final_Aug7.pdf.

[12] M. Tazzari, F. Beaujean, L. Testi, GALARIO: a GPU accelerated library for analysing radio interferometer observations, Mon. Not. R. Astron. Soc. 476 (4) (2018) 4527–4542.

[13] M. Seznec, et al., A study on convolution using half-precision floating-point numbers on gpu for radio astronomy deconvolution, in: 2018 IEEE International Workshop on Signal Processing Systems (SiPS), IEEE, 2018.
[14] J.W. Romein, A comparison of accelerator architectures for radio-astronomical signal-processing algorithms, in: 2016 45th International Conference on Parallel Processing (ICPP), IEEE, 2016.
[15] A. Sclocco, et al., Real-time dedispersion for fast radio transient surveys, using auto tuning on many-core accelerators, Astron. Comput. 14 (2016) 1–7.
[16] K. Adámek, J. Novotný, W. Armour, The Implementation of a Real-Time Polyphase Filter, arXiv preprint arXiv:1411.3656, 2014.
[17] T. Chen, et al., Cell broadband engine architecture and its first implementation—a performance view, IBM J. Res. Dev. 51 (5) (2007) 559–572.
[18] A.L. Varbanescu, A.S. van Amesfoort, T. Cornwell, G. van Diepen, R. van Nieuwpoort, B.G. Elmegreen, H. Sips, Building high-resolution sky images using the cell/B.E, Sci. Program. 17 (2009) 113–134.
[19] T.J. Cornwell, SKA and EVLA computing costs for wide field imaging, in: The Square Kilometre Array: An Engineering Perspective, Springer, Dordrecht, 2005, , pp. 329–343.
[20] R. Jongerius, S. Wijnholds, R. Nijboer, H. Corporaal, An end-to-end computing model for the square kilometre array, Computer 47 (9) (2014) 48–54, https://doi.org/10.1109/MC.2014.235.
[21] G.A. Gillani, A. Krapukhin, A.B.J. Kokkeler, Energy-efficient approximate least squares accelerator: a case study of radio astronomy calibration processing, in: Proceedings of the 16th ACM International Conference on Computing Frontiers, ACM, 2019.
[22] A.S. van Amesfoort, A.L. Varbanescu, H.J. Sips, R.V. van Nieuwpoort, Evaluating multi-core platforms for HPC data-intensive kernels, in: Proceedings of the 6th ACM Conference on Computing Frontiers, CF 2009, ACM, New York, 2009, , pp. 207–216.
[23] B. Humphreys, T. Cornwell, SKA Memo 132: Analysis of Convolutional Resampling Algorithm Performance, 2011.
[24] B. Humphreys, T. Cornwell, Analysis of Convolutional Resampling Algorithm Performance, http://www.skatelescope.org/uploaded/59116_132_Memo_Humphreys.pdf, 2011.
[25] J.W. Romein, An efficient work-distribution strategy for gridding radio-telescope data on GPUs, in: Proceedings of the 26th ACM International Conference on Supercomputing, ICS 2012, ACM, New York, 2012, pp. 321–330.
[26] D. Muscat, High-Performance Image Synthesis For Radio Interferometry, arXiv preprint arXiv:1403.4209, 2014.
[27] B. Merry, Faster GPU-based convolutional gridding via thread coarsening, Astron. Comput. 16 (2016) 140–145.
[28] Y. Mei, et al., GPU-Based High-performance Imaging for Mingantu Spectral RadioHeliograph, arXiv preprint arXiv:1712.06913, 2017.
[29] Y. Yan, et al., On mingantu spectral radioheliograph-MUSER: introduction and observations, in: 2017 IEEE Radio and Antenna Days of the Indian Ocean (RADIO), IEEE, 2017.
[30] Q. Luo, et al., HyGrid: a CPU-GPU hybrid convolution-based gridding algorithm in radio astronomy, in: International Conference on Algorithms and Architectures for Parallel Processing, Springer, Cham, 2018.
[31] P. Léna, D. Rouan, F. Lebrun, F. Mignard, D. Pelat, S. Lyle, Observational Astrophysics. Astronomy and Astrophysics Library, third ed., Springer, Heidelberg, 2012. https://doi.org/10.1007/978-3-642-21815-6.
[32] X. Zhao, J. Jing, SDP Memo 093: Initial Scale out Prototyping Based on Supercomputing Platform in CPU+MIC Architecture, , 2018.

[33] Y. Zhu, S.J. Tong, SDP Memo 082: Summarising Initial Scale-Out Prototyping Efforts, 2018.

[34] R. Levanda, A. Leshem, Image Formation in Synthetic Aperture Radio Telescopes, arXiv preprint arXiv:1009.0460, 2010.

[35] J.A. Högbom, Aperture synthesis with a non-regular distribution of interferometer baselines, Astron. Astrophys. Suppl. Ser. 15 (1974) 417.

[36] W.H. Richardson, Bayesian-based iterative method of image restoration, J. Opt. Soc. Am. 62 (1) (1972) 55–59.

[37] L.B. Lucy, An iterative technique for the rectification of observed distributions, Astron. J. 79 (1974) 745.

[38] S. Bonettini, R. Zanella, L. Zanni, A scaled gradient projection method for constrained image deblurring, Inverse Prob. 25 (1) (2008). 015002.

[39] M. Prato, et al., Efficient deconvolution methods for astronomical imaging: algorithms and IDL-GPU codes, Astron. Astrophys. 539 (2012) A133.

[40] T. Herbst, et al., LINC-NIRVANA: a Fizeau beam combiner for the large binocular telescope, in: Interferometry for Optical Astronomy II, vol. 4838, International Society for Optics and Photonics, 2003.

[41] T.M. Herbst, et al., The LINC-NIRVANA interferometric imager for the large binocular telescope, in: Ground-based Instrumentation for Astronomy, vol. 5492, International Society for Optics and Photonics, 2004.

[42] J.M. Hill, P. Salinari, Large binocular telescope project, in: Telescope Structures, Enclosures, Controls, Assembly/Integration/Validation, and Commissioning, vol. 4004, International Society for Optics and Photonics, 2000.

[43] V. Ruggiero, et al., Iterative regularization algorithms for constrained image deblurring on graphics processors, J. Glob. Optim. 48 (1) (2010) 145–157.

[44] R. Narayan, R. Nityananda, Indirect imaging, in: J.A. Roberts (Ed.), Proc. IAU/URSI Symp, Cambridge University Press, London, 1984, p. 267.

[45] J. Van Belle, R. Armstrong, J. Gain, Accelerated deconvolution of radio interferometric images using orthogonal matching pursuit and graphics hardware, J. Astron. Instrum. 6 (04) (2017) 1750009.

[46] M. Cárcamo, et al., High performance GPU Bayesian image synthesis, in: 2015 IEEE International Symposium on Signal Processing and Information Technology (ISSPIT), IEEE, 2015.

[47] A.B. Peck, A.J. Beasley, High resolution sub-millimeter imaging with ALMA, in: Journal of Physics: Conference Series, vol. 131, IOP Publishing, 2008. No. 1.

[48] J.D. Owens, et al., A survey of general-purpose computation on graphics hardware, in: Computer Graphics Forum, vol. 26(1), Blackwell Publishing Ltd, Oxford, UK, 2007.

[49] M. Lochner, et al., Bayesian inference for radio observations, Mon. Not. R. Astron. Soc. 450 (2) (2015) 1308–1319.

[50] S.J. Perkins, et al., Montblanc: GPU accelerated radio interferometer measurement equations in support of bayesian inference for radio observations, Astron. Comput. 12 (2015) 73–85.

[51] M. Cárcamo, et al., Multi-GPU maximum entropy image synthesis for radio astronomy, Astron. Comput. 22 (2018) 16–27.

[52] D.S. Briggs, High Fidelity Deconvolution Of Moderately Resolved Sources, Diss. New Mexico Institute of Mining and Technology, 1995.

[53] A.M. Sardarabadi, A. Leshem, A.-J. van der Veen, Radio astronomical image formation using constrained least squares and Krylov subspaces, Astron. Astrophys. 588 (2016) A95.

[54] M.P. van Haarlem, et al., LOFAR: the low-frequency array, Astron. Astrophys. 556 (2013) A2.

[55] Y. Luo, R. Duraiswami, Efficient parallel non-negative least squares on the GPU, SIAM J. Sci. Comput. (2010). Copper Mountain Special Edition.

[56] E. Mejía-Roa, et al., NMF-mGPU: non-negative matrix factorization on multi-GPU systems, BMC Bioinf. 16 (1) (2015) 43.

[57] S. Koitka, C.M. Friedrich, nmfgpu4R: GPU-accelerated computation of the non-negative matrix factorization (NMF) using CUDA capable hardware, R J. 8 (2) (2016) 382–392.

[58] Y.S. Bouhaik, G.A.C. Nicolas, Parallelization on a Compute-Intensive Server (Multi-Gpu and/or Fpga) of Large Data Processing Algorithms for the Big Radio Astronomy Antenna Network SKA, 2017.

[59] M. Huynh, J. Lazio, An Overview of the Square Kilometre Array, arXiv preprint arXiv:1311.4288, 2013.

[60] E.J. Candès, Compressive sampling, in: Proceedings of the International Congress of Mathematicians, vol. 3, 2006.

[61] S. Kunis, H. Rauhut, Random sampling of sparse trigonometric polynomials. II. Orthogonal matching pursuit versus basis pursuit, Found. Comput. Math. 8 (6) (2008) 737–763.

[62] L.C. Schwardt, Compressed sensing imaging with the KAT-7 array, in: 2012 International Conference on Electromagnetics in Advanced Applications, IEEE, 2012.

[63] A. Kulkarni, T. Mohsenin, Accelerating compressive sensing reconstruction OMP algorithm with CPU, GPU, FPGA and domain specific many-core, in: 2015 IEEE International Symposium on Circuits and Systems (ISCAS), IEEE, 2015.

[64] M.T. Whiting, Duchamp: A 3D source finder for spectral-line data, Mon. Not. R. Astron. Soc. 421 (4) (2012) 3242–3256.

[65] SKA Simulated Skies (S3) Tools. Retrieved 16 July, 2010 From the University of Oxford: http://www.lra.ens.fr/~levrier/Recherche/S3/.

[66] O. Kalentev, et al., Connected component labeling on a 2D grid using CUDA, J. Parallel Distrib. Comput. 71 (4) (2011) 615–620.

[67] P. Chen, et al., Block-run-based connected component labelling algorithm for GPGPU using shared memory, Electron. Lett. 47 (24) (2011) 1309–1311.

[68] P. Ciarach, et al., Real-time FPGA implementation of connected component labelling for a 4K video stream, in: International Symposium on Applied Reconfigurable Computing, Springer, Cham, 2019.

[69] SKAO, The SKA Project, Accessed: 2019. [Online]. Available, https://www.skatelescope.org/the-ska-project/.

[70] E.J. Korpela, et al., Status of the UC-Berkeley SETI efforts, in: Instruments, Methods, and Missions for Astrobiology XIV, vol. 8152, International Society for Optics and Photonics, 2011.

[71] G. Hobbs, et al., The Parkes observatory pulsar data archive, Publ. Astron. Soc. Aust. 28 (3) (2011) 202–214.

[72] B.R. Barsdell, et al., Spotting Radio Transients With the Help of GPUs, arXiv preprint arXiv:1112.0065, 2011.

[73] M.J. Keith, et al., The high time resolution universe pulsar survey. IV. Discovery and polarimetry of millisecond pulsars, Mon. Not. R. Astron. Soc. 419 (2) (2011) 1752–1765.

[74] L. Staveley-Smith, et al., The parkes 21 cm multibeam receiver, Publ. Astron. Soc. Aust. 13 (3) (1996) 243–248.

[75] B.R. Barsdell, D.G. Barnes, C.J. Fluke, Swinburne Research Bank, 2010.

[76] G. Poli, et al., Solar flare detection system based on tolerance near sets in a GPU-CUDA framework, Knowl.-Based Syst. 70 (2014) 345–360.

[77] H.S. Sawant, et al., Brazilian decimetric array (Phase-I), Sol. Phys. 242 (1–2) (2007) 213–220.

Further reading

Y. Arjoune, et al., Compressive sensing: performance comparison of sparse recovery algorithms, in: 2017 IEEE 7th Annual Computing and Communication Workshop and Conference (CCWC), IEEE, 2017.

H. Bal, et al., A medium-scale distributed system for computer science research: infrastructure for the long term, Computer 49 (5) (2016) 54–63.

F. Baron, B. Kloppenborg, J. Monnier, Toward 5D Image Reconstruction for Optical Interferometry, International Society for Optics and Photonics, vol. 8445, Amsterdam, Netherlands, 2012.

T. Blumensath, M.E. Davies, Iterative hard thresholding for compressed sensing, Appl. Comput. Harmon. Anal. 27 (3) (2009) 265–274.

J. du Toit, Convolution Gridding on CPU, GPU and KNL. SDP Memo 036, 2017) http://ska-sdp.org/sites/default/files/attachments/sdp_memo_36.pdf.

R. Edgar, et al., Enabling a high throughput real time data pipeline for a large radio telescope array with GPUs, Comput. Phys. Commun. 181 (10) (2010) 1707–1714.

P. Fernique, D. Durand, T. Boch, A. Oberto, F. Pineau, HEALpix based crosscorrelation in astronomy, in: Astronomical Data Analysis Software and Systems XXII, Astronomical Society of the Pacific Conference Series, vol. 475, 2013, p. 135.

K.M. G′orski, et al., HEALPix: a framework for high-resolution discretization and fast analysis of data distributed on the sphere, Astrophys. J. 622 (2) (2005) 759.

N.M. Gürel, et al., Compressive Sensing with Low Precision Data Representation: Theory and Applications, arXiv preprint arXiv:1802.04907, 2018.

Z. Hong, C. Yu, J. Wang, J. Xiao, C. Cui, J. Sun, Aquadexim: highly efficient in-memory indexing and querying of astronomy time series images, Exp. Astron. 42 (3) (2016) 387–405.

R. Inta, D.J. Bowman, S.M. Scott, The chimera: an off-the-shelf CPU/GPGPU/FPGA hybrid computing platform, Int. J. Reconfig. Comput. 2012 (2012) 2.

R. Khanna, A. Kyrillidis, IHT dies hard: Provable Accelerated Iterative Hard Thresholding, arXiv preprint arXiv:1712.09379, 2017.

M.G. Labate, et al., SKA1-LOW: design and scientific objectives, in: 2017 XXXIInd General Assembly and Scientific Symposium of the International Union of Radio Science (URSI GASS), IEEE, 2017.

C.L. Lawson, R.J. Hanson, Solving Least Squares Problems, vol. 15, Siam, 1995.

A. Leshem, A.-J. Van der Veen, Radio-astronomical imaging in the presence of strong radio interference, IEEE Trans. Inf. Theory 46 (5) (2000) 1730–1747.

Y. Mao, et al., DAG constrained scheduling prototype for an astronomy exa-scale HPC application, in: 2016 IEEE 18th International Conference on High Performance Computing and Communications; IEEE 14th International Conference on Smart City; IEEE 2nd International Conference on Data Science and Systems (HPCC/SmartCity/DSS), IEEE, 2016.

A.K. Maurya, A.K. Tripathi, On benchmarking task scheduling algorithms for heterogeneous computing systems, J. Supercomput. 74 (7) (2018) 3039–3070.

I. Marti-Vidal, et al., UVMULTIFIT: a versatile tool for fitting astronomical radio interferometric data, Astron. Astrophys. 563 (2014) A136.

M. McCool, J. Reinders, A. Robison, Structured Parallel Programming: Patterns for Efficient Computation, first ed., Morgan Kaufmann Publishers Inc., San Francisco, 2012.

J.M. Myre, et al., TNT-NN: a fast active set method for solving large non-negative least squares problems, Proc. Comput. Sci. 108 (2017) 755–764.

G. Resnick, M. Kuttel, P. Marais, Gpu Accelerated Source Extraction in Radio Astronomy: A CUDA Implementation, Department of Computer Science, University of Cape Town, Tech. Rep., 2010.

J. Sum, C.S. Leung, R.C.C. Cheung, T.Y. Ho, HEALPIX DCT technique for compressing PCA-based illumination adjustable images, Neural Comput. Applic. 22 (7) (2013) 1291–1300.

E. Vermij, et al., Exascale radio astronomy: can we ride the technology wave? in: International Supercomputing Conference, Springer, Cham, 2014.

12

High-performance computing for astronomical big data

Yongxin Zhu[a,b], Haihang You[e], Junjie Hou[b], Yu Zheng[b], Tian Huang[c,d], Yuefeng Song[b]

*[a]Shanghai Advanced Research Institute, Chinese Academy of Sciences, Beijing, China. [b]School of Microelectronics, Shanghai Jiao Tong University, Shanghai, China. [c]Astrophysics Group, Cavendish Lab, Cambridge University, Cambridge, United Kingdom. [d]Institute of High Performance Computing, Agency for Science, Technology and Research (A*STAR), Singapore, Singapore. [e]Institute of Computing Technologies, Chinese Academy of Sciences, Beijing, China*

1 Introduction

The SKA (square kilometer array) will become the world's largest radio telescope [1]. It will consist of a low-frequency aperture array with up to 256,000 dipoles, a midfrequency dish array equipped with phased array feeds, and a midfrequency dish array equipped with cooled high-sensitivity single-pixel feeds [2]. It achieves the objective of simulating a giant antenna by integrating the sample data from small antennas, and will eventually produce images and other data over wavelengths from 4.3 m (70 MHz) to 1 cm (30 GHz) [3]. With high spatial resolution and high-sensitivity features, the SKA is built to put forward the understanding of the most important phenomena in the Universe, including some pertaining to the birth and eventual death of the Universe itself.

One of the most critical parts of the SKA project is science data processing (SDP). The function of SDP is to offer the computing platform to process, integrate, and analyze these data on time. This is specifically saying that each collected datum needs to go through signal amplification, ADC, data correlation, UV imaging processing, and other operations as shown in Fig. 12.1. The SKA has a large number of small antennas, and the number of operations grows with the square of the number of elements. As a result, the data rates of the SKA will be enormous and will increase with each generation, so will the processing requirements [4].

Due to the limited storage space, the SKA needs to process massive data in real time, which makes the SKA scientific data

Big Data in Astronomy. https://doi.org/10.1016/B978-0-12-819084-5.00002-X

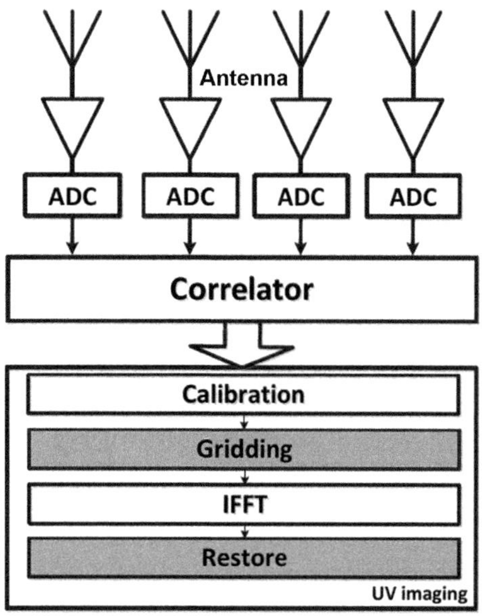

Fig. 12.1 Data flow of SKA-SDP.

processing become a bottleneck of the computational performance. To tackle the challenge, the high performance execution frameworks can be used to enhance the performance of SKA-SDP.

To handle data volumes in such an unprecedented scale, various execution frameworks were proposed. However, these execution frameworks were defined originally for big data processing. It was still confusing if they can be applied on high-performance computing platforms. In this chapter, execution framework experiments on supercomputers will be described to illustrate the feasibility of execution framework.

In addition to the execution framework, algorithm reference library (ARL) is also critical to astronomical data processing. ARL is a reference algorithm library of the imaging processing developed for SKA, which include data model, pipeline, sky components such as a plurality of modules and calibration, imaging, FFT converter, and other algorithms of radio astronomy data processing. ARL is written in Numpy-based Python language. The advantage is that it is easy to understand many complex processes and is easy to use, but the performance of ARL is far from meeting the needs of astronomical data processing. To this end, the optimized version of ARL needs to be developed to enhance the performance of astronomical data processing.

2 Execution framework and prototype test

2.1 High performance execution frameworks

2.1.1 Parallel programming models

Most HPC applications today use both shared-memory (e.g., OpenMP) and distributed memory (e.g., MPI) programming models to achieve a desirable level of parallelism in the field of high-performance computing. Both programming models require developers to exploit potential parallelism, control process/thread synchronization, and manage various overheads associated with parallelism such as race conditions.

OpenMP [5] is an application program interface (API), jointly defined by a group of major computer hardware and software vendors. OpenMP provides a portable, scalable model for developers of shared memory parallel applications. The API supports C/C++ and Fortran on a wide variety of architectures. OpenMP is designed for multiprocessor/core, shared memory machines, and it largely limited to single-node parallelism. Typically, the number of processing elements (cores) on a node determines how much parallelism can be implemented.

Message passing interface (MPI) [6] is a standardized and portable message-passing standard designed by a group of researchers from academia and industry to function on a wide variety of parallel computing architectures. The standard defines the syntax and semantics of a core of library routines useful to a wide range of users writing portable message-passing programs in C, C++, and Fortran. There are several well-tested and efficient implementations of MPI, many of which are open source or in the public domain. These fostered the development of a parallel software industry, and encouraged development of portable and scalable large-scale parallel applications. MPI is not sanctioned by any major standards body; nevertheless, it has become a de facto standard for communication among processes that model a parallel program running on a distributed memory system. Actual distributed memory supercomputers such as computer clusters often run such programs.

2.1.2 Distributed computing frameworks

Due to the rapid increase of data size in recent years, industrial data-intensive applications raise the challenge of processing these huge amount of data in the data center with high performance. MapReduce model [7] provides an efficient data flow engine to improve the performance of data processing in cluster environment. The popular MapReduce-based frameworks such as Apache Spark [8], Apache Hadoop [9] are applied in many distributed computing scenarios. Spark is an open-source distributed framework for big data processing, and it supports in-memory computation of RDDs and provides a provision of reusability, fault tolerance, real-time stream processing [10]. Hadoop is mainly used as the distributed file system.

Apache Spark is an open-source cluster computing framework for big data processing. The first system was developed by the group of the University of California, Berkeley, and it gained incubator status in Apache very quickly in June 2013 [11]. Spark overtook MapReduce and Hadoop, and emerged as the next generation big data processing engine. Spark supports in-memory computing and enables very fast data query operation compared to Hadoop, which is the disk- based engine [10]. Spark maintains the fault tolerance and linear scalability of MapReduce, and it extends a data-sharing abstraction called RDD (resilient distributed dataset). Besides, Spark provides rich API in Scala, Java, Python, so it is easier and convenient to program. Spark also provides many high-level tools such as Spark SQL, MLlib library for machine learning, GraphX for graph processing, Spark Streaming for stream processing, and so on [12].

Apache Hadoop is an open-source software framework used for distributed storage and processing of dataset of big data using the MapReduce programming model. It consists of computer clusters built from commodity hardware. All the modules in Hadoop are designed with a fundamental assumption that hardware failures are common occurrences and should be automatically handled by the framework [13].

The core of Apache Hadoop consists of a storage part, known as Hadoop Distributed File System (HDFS), and a processing part, which is a MapReduce programming model. Hadoop splits files into large blocks and distributes them across nodes in a cluster. It then transfers packaged code into nodes to process the data in parallel. This approach takes advantage of data locality, where nodes manipulate the data they have access to. This allows the dataset to be processed faster and more efficiently than it would be in a more conventional supercomputer architecture that relies on a parallel file system where computation and data are distributed via high-speed networking [14].

2.1.3 Dataflow computation model

The dataflow computation model was initially proposed [15–17] to express programs as directed acyclic graphs (DAG), where the vertices are the stateless computational tasks that compose the program, and edges connect the output of one task with the input of another. Traveling through the edges are data tokens, which are continuously processed by the nodes as they arrive. Unlike OpenMP or MPI, this paradigm does not explicitly place any control or constraints on the order or timing of operations beyond what is inherent in the data dependencies among compute tasks, and therefore exploits the full inherent parallelism of a program. Putting the concept of dataflow into practice still requires control flow operations and data storage in order to make it practical and useful (e.g., the MIT dataflow architecture [18]).

More recently, graph-based models have been used for building scalable data processing pipelines. DALiuGE [19] is a graph-based execution framework, specifically designed to support very large-scale processing graphs for the reduction of interferometric radio astronomy datasets. DALiuGE has already been used for processing large astronomical datasets in existing radio astronomy projects. It originated from a prototyping activity as part of the SDP Consortium and it aimed to prototype the execution framework of the proposed SDP architecture.

TensorFlow is an open-source software library for high-performance numerical computation. Its flexible architecture allows easy deployment of computation across a variety of platforms (CPUs, GPUs, TPUs), and from desktops to clusters of servers to mobile and edge devices. At its core, TensorFlow represents both computations and states (that the computation operates on) as dataflow-like graphs, and it maps graphs onto various hardware platforms such as laptops, GPU clusters, Android/iOS devices, etc.

2.2 DALiuGE scalabilty test on supercomputers

In the classic process of astronomical data processing, astronomy scientists usually adopt static workflow supported by popular high-performance-computing platform technologies.

For example, researchers in Cambridge University developed a Performance Prototype Platform (P3-ALaSKA) OpenStack prototype platform for SKA community. They managed to provide platform services to SKA community for evaluation, testing, and various prototyping activities. P3-ALaSKA prototype platform is based on OpenStack control, which is able to provision hardware as bare metal cloud resources, including compute provisioning resources, storage provisioning resources, and network infrastructure resources. The resource provisioning decisions are made according to static workflow defined by astronomy scientists as users who submit their static workflow as tasks to be scheduled.

However, the assumption of predefined static workflow for classic high-performance computing may not hold for uncertain data processing pipelines for huge telescopes such as SKA since the scale of SKA project is huge enough to incur dynamic workflow developed and redeveloped by multiple international research institutes due to unexpected system operational faults, changes in deployment requirements, different logging level as well as various compiling switches. These uncertainties naturally arose from huge number of dimensions in vast SKA telescope design space that is explored by international researchers in SKA community across the world.

To tackle the issues with existing platform tools, the Data Activated Liu (Flow) Graph Execution Engine—DALiuGE was developed by the International Centre for Radio Astronomy Research (ICRAR) in Western Australia University for the SDP Consortium of the SKA community. DALiuGE was designed to be a distributed data management platform and a scalable pipeline execution environment, supporting continuous, time and power bounded, data-intensive processing for producing SKA science-ready products.

After ICRAR released DALiuGE as an open source project, DALiuGE was tested by SKA community on supercomputers in the Netherlands and China with technical help from ICRAR Australia. Their activities will be described in following subsections to verify if DALiuGE can execute various pipelines in a scalable means and to demonstrate the capability of DAliuGE to execute a pipeline in a complete form under the control of DALiuGe.

2.2.1 DALiuGE prototyping test by ASTRON

ASTRON (Netherlands Institute for Radio Astronomy) evaluated the feasibility of the DALiuGE execution framework by prototyping DALiuGE with the pipeline of their low-frequency array (LOFAR) [20, 21].

Hardware

ASTRON researchers completed their experiments on ASTRON's nodes of Distributed ASCI Supercomputer 5 (DAS-5). DAS-5 was the 5th generation in a series of supercomputers designed by the Advanced School for Computing and Imaging (ASCI) [21]. DAS-5 was released in 2015 to harness wide variety of accelerators, larger memories, larger disks, and software-defined networking (SDN). DAS-5 is a medium-scale distributed supercomputer consisting of 6 clusters, 3252 cores in eight-core Xeon E5-2630v3 CPUs, FDR Infiniband as local area network, and light paths as wide area network, which supports 100 Gbit/s and SDN in the form of network virtualization. Besides the varieties in hardware accelerators, e.g., NVIDIA GPUs, AMD GPUs and the Intel Xeon Many-Interconnected-Cores (MIC) Phi processors, in the infrastructure of DAS-5, virtualization was adopted as a high-priority design constraint in an effort to virtualize all computing, storage, networks, visualization, algorithms, and libraries components of distributed systems. These virtualized components enabled better service management from low bare metal hardware to high-level platforms. Higher level of service management was required to handle dynamic data streams in large data volume in the era of big data since the applications on DAS-5 were in the domains of multimedia (e.g., video and image retrieval), astronomy (e.g., radio signals and astrophysics), and climate (e.g., sea-level rise modeling).

As DAS-5 comprised heterogeneous hardware and accelerators, Docker version 18.06.0-ce was adopted to manage heterogeneous nodes, including the head one. In ASTRON's experiment, four hardware nodes in DAS-5 supercomputer were deployed by a docker daemon. These four nodes actually executed the LOFAR software stack and the DALiuGE execution framework.

Environment setup

To support ASTRON's experiments, LOFAR software suite version 2.21.4 and prefactor version 2.0.2 were extended with python wrappers to implement CALIB (calibration) pipeline and Prefactor pipeline.

The CALIB pipeline is a minimum example of a working image pipeline, each step of which operates on the same subband without additional fitting on the calibration solutions of different subband. Porting CALIB to DALiuGE will produce a logical graph of the CALIB pipeline in the form of DALiuGE graph component, e.g., a scatter block of DALiuGE graph component was adopted to define the whole pipeline within the scatter block to indicate that all steps in the workflow procedure are executed for every subband.

The prefactor pipeline was defined as the precursor to the factor pipeline. The prefactor pipeline was designed to produce initial calibrated images that further undergo facet calibration processing by the factor pipeline. The prefactor pipeline was composed of three major steps, i.e., calibration solution derivation, calibration solution application to the observed target, and calibrated target image generation. In ASTRON's experiment, only the step of calibration solution derivation of the Prefactor pipeline was ported to DALiuGE execution framework.

The data used in the CALIB pipeline experiment were the set of data from two subtarget bands and two calibrator subbands of LOFAR observation ID L182806. To reduce the file size, these time series data were averaged by a factor 64 to generate data files in the size of 2.5 GB.

Similar to the dataset used in CALIB pipeline, the data used for the experiment of Prefactor pipeline were based on data from the first 20 subbands of the LOFAR telescope with observation IDs L570741 (target) and L570745 (calibrator), respectively. To ensure a quick pipeline experiment, only the first 10 time slots of each subband data were taken into the experiment while the full version was available in archive.

Pipeline graph execution

Porting CALIB and Prefactor pipelines to DALiuGE started by describing the workflow of CALIB and Prefactor pipelines in an abstract logical graph of DALiuGE execution framework. A logic graph in general shows how data streams in the workflow move among connected drops, which represent a chunk of data or an executable object. The chunk of data can be either a file or a memory address while the executable can be a command, a python script, or C executable library. In a logical graph, dark blue boxes were used to indicate file drops, while light blue boxes were used to describe executable application drops. To differentiate from application executable, drops of python functions were indicated in cyan boxes.

Based on the logical graph and information on the available hardware, a physical graph can be generated to map processing units for multiple datasets. In case there is only one dataset as input to the workflow, the physical graph will be almost the same as the logical graph except for the dynamic status of execution, e.g., green units in the physical graph indicate completion of deployed work, dark yellow units indicate ongoing execution of deployed work, and light yellow units indicate deployed work to be executed.

Execution results

Porting the CALIB pipeline and the first part of the Prefactor pipeline to DALiuGE execution framework on four nodes of ASTRON's DAS-5 supercomputer was proven to be successful. The DALiuGE execution framework along with the CALIB and Prefactor software stack were capture in Docker images within the

Docker container of DAS-5 supercomputer to ensure the experiments were reusable and repeatable.

2.2.2 DALiuGE scalability test on Tianhe-2 supercomputer

While DALiuGE execution framework was evaluated on four nodes of DAS-5 supercomputer in ASTRON Netherlands in a small scale, DALiuGE execution framework was further tested to evaluate the scalability of DALiuGE [22] on Tianhe-2 supercomputer, which was the top one supercomputer in the world in 2013–15.

To ensure sufficient workload for scalability test, heavy workload at a scale similar to that of SKA-1 LOW array telescope was adopted. In addition to the heavy workload that is noticeably different from the workload in ASTRON's experiment, a complete pipeline, i.e., from the visibility data digestion to final image production, was described and executed in DALiuGE execution framework to evaluate the capability of DALiuGE.

Hardware

It was well known that Tianhe-2 supercomputer consists of 16,000 heterogeneous nodes, each of which contains 2 Intel Ivy Bridge-E Xeon E5 2692 CPUs and 3 Intel MIC Phi processors as accelerators. The total number of CPUs is 32,000 while total number of Phi processors is 48,000, all of which account for 3.12 million of computing cores.

Besides the 16,000 nodes containing Intel Xeon CPUs and MIC Phi processors, it is interesting to know that there are 16 GPU nodes on Tianhe-2 supercomputer. Each node is equipped with two Intel Xeon E5-2660v3 CPUs as well as two NVIDIA Tesla K80 accelerators. These nodes actually followed the architectural design of Tianhe-1A, which uses 7168 NVIDIA Tesla M2050 GPUs and 14,336 CPUs. These GPU nodes were available for big data applications that may have porting difficulties with Intel MIC Phi processors. On each GPU node, 2 T K80 GPUs support heavy double-precision computing workload with 4992 CUDA cores, 24 GB on board GDDR5 rams.

The network connecting these nodes are SFP+ 10 Gbps Ethernet and 56 Gbps FDR InfiniBand. LUSTER high-performance parallel distributed file system was adopted to support 1.1 PB storage capacity.

Environment setup

Red Hat Enterprise Linux Server release 6.5 was adopted as the operating system of Tianhe-2. The BIGDATA partition of the LUSTER file system was assigned to complete the DALiuGE scalability experiment. Batch job mode of classic scheduling tool SLURM was adopted to support the scalability test script.

Since SKA1 LOW array telescope is still being designed and prototyped, there are no observation data from SKA1. Instead of the observation data as used in ASTRON's experiment, simulated visibility data were adopted. These simulated data were produced with the Oxford's Square Kilometer Array Radio-telescope simulator (OSKAR) version 2.6.1. OSKAR is able to generate large-scale visibility data used in this experiment, whose scale is similar to that of SKA1 low array telescope. To emulate the scale of SKA1 LOW array telescope, the number of stations was set to 512, the number of simulated samples was set to 1000, while the sky model was generated by Centaurus A image from Murchison Widefield Array GLEAM with 23,811 point sources. Simulated observation phase center on the sky was set at (RA, Dec) = (199.61, −46.0) degrees with the field of view of [3, 4] degrees.

Pipeline development and execution

Similar to ASTRON's experiment on Porting CALIB and Prefactor pipelines to DALiuGE, DALiuGE scalability test also started by describing the workflow of an end-to-end pipeline in an abstract logical graph of DALiuGE execution framework.

As shown in Fig. 12.2, the logical graph shows major parts of the end-to-end pipeline, i.e., the OSKAR setting file generating, OSKAR interferometer simulation and data reduction with CASA described in light blue units. Dark blue units indicate File Drop to store files or data. The yellow unit of scattering operation indicates the workflow was expanded to multiple partitions under the control of *num_of_copies*.

DALiuGE execution results

Similar to ASTRON's experiment on porting CALIB and Prefactor pipelines to DALiuGE on DAS-5, DALiuGE execution framework on Tianhe-2 also underwent the standard process of DALiuGE execution steps by starting with unrolling the logical graph to a physical graph template, then translating the template into physical graph template partition along with generated physical graph specifications. These specifications and physical graph partitions will be submitted to drop manager who created sessions for these partitions. Further node/island managers further handle subgraph specifications if necessary. Once these managers are ready, graphs are pushed into these sessions for execution by triggering drops on some nodes to complete their tasks.

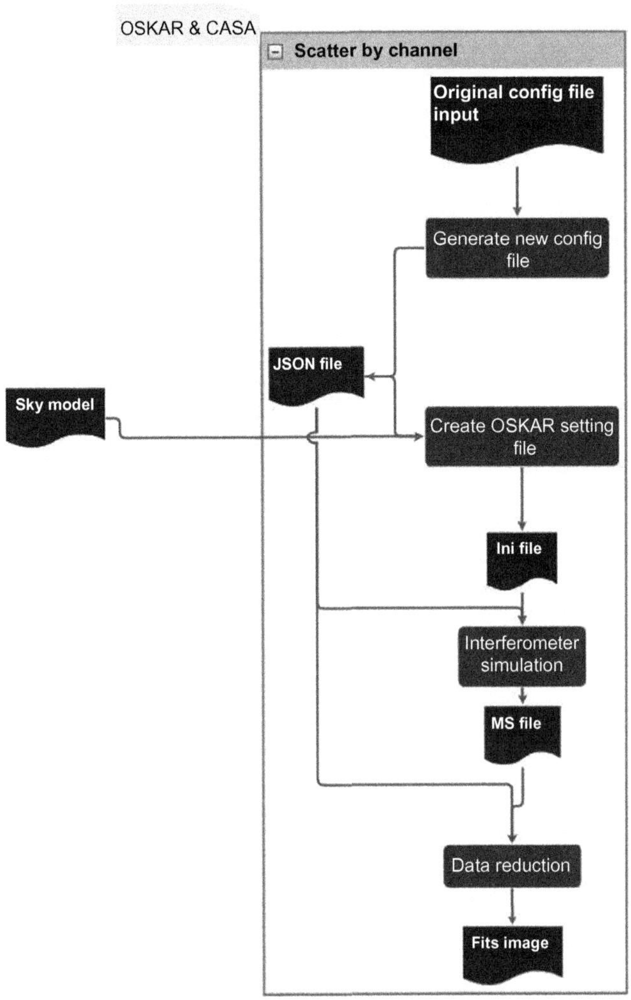

Fig. 12.2 logical graph of the pipeline [22].

The execution results are shown in Fig. 12.3, which were reduced by CASA library to generate four frequencies (100, 200, 300, and 400 MHz) dirty images whose channel averaged maps and spectral index maps were reported in [23].

To measure the scalability performance of DALiuGE, Fig. 12.4 shows the total completion time with DALiuGE and the execution time with MPI runs on Tianhe-2. The *X*-axis shows the number of the GPU nodes ranging from 1 to 16. It was clear that DALiuGE demonstrated stable scalability performance in terms of smaller variance in execution time in comparison with common

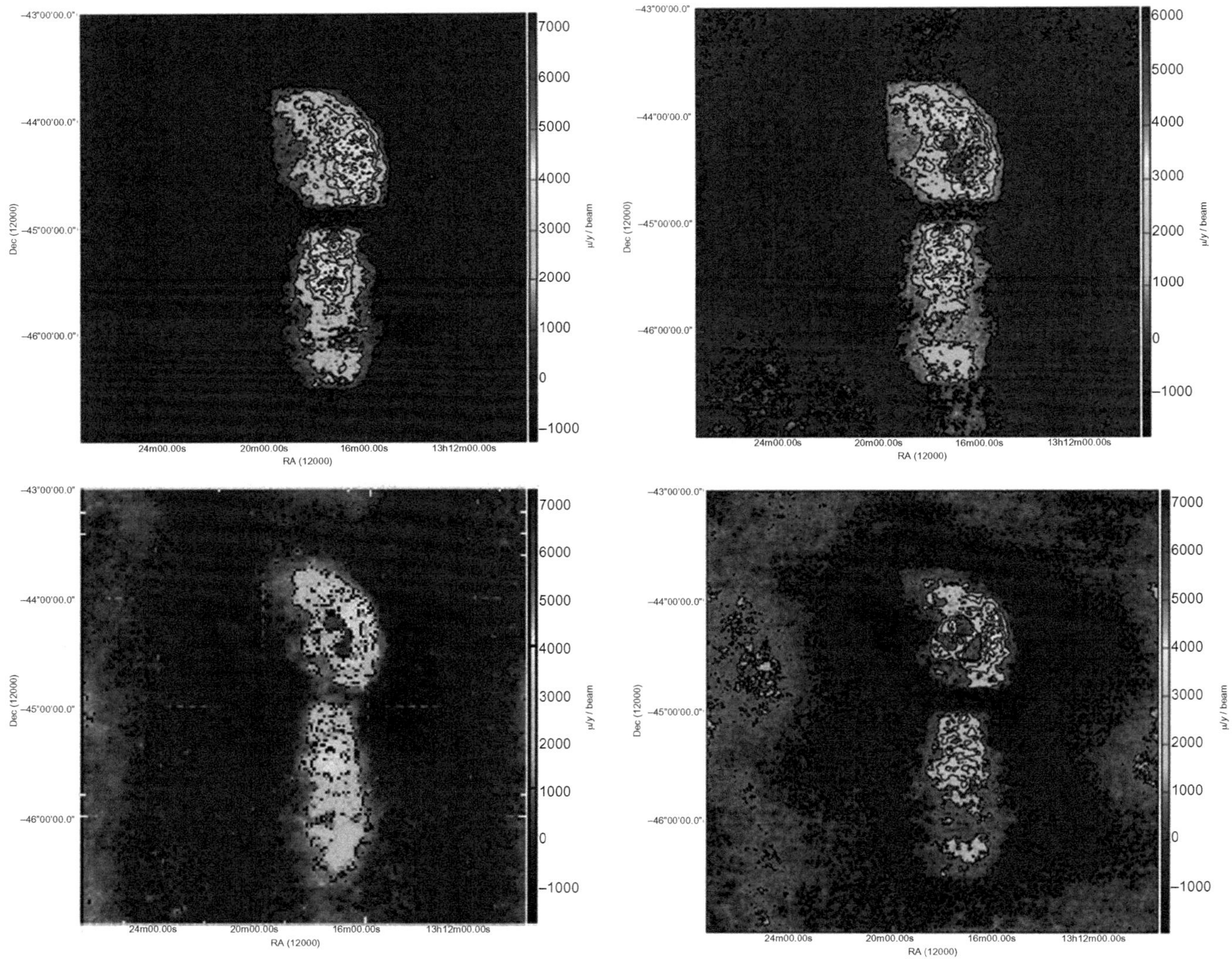

Fig. 12.3 The dirty images generated by CASA reduction cores library at four frequencies: (Top, Left) frequency is 100 MHz, (Top, Right) is 200 MHz, (Bottom, Left) frequency is 300 MHz, and (Bottom, Right) frequency is 400 MHz [22].

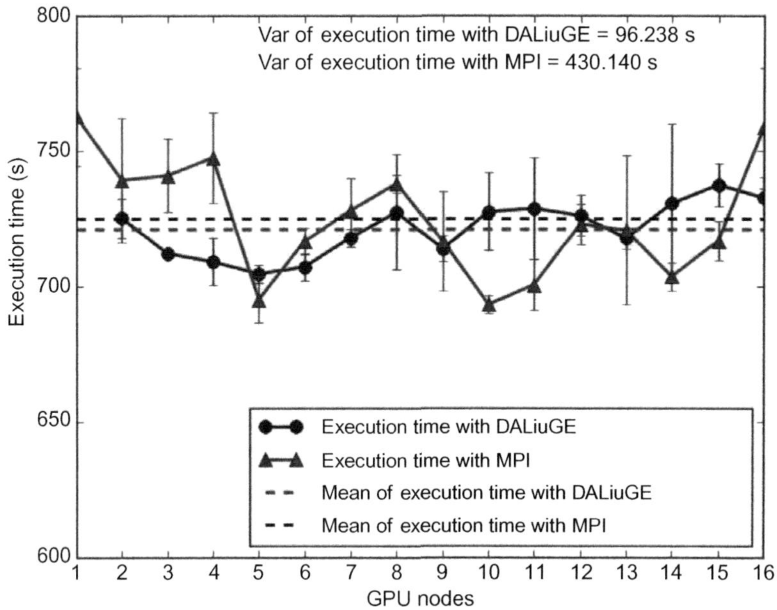

Fig. 12.4 Execution Time with DALiuGE and MPI. The blue solid line and green solid line are the pipeline execution time implemented with DALiuGE and MPI, respectively. The red dotted (gray dotted in print version) line and black dotted line are the average pipeline execution time implemented with DALiuGE and MPI, respectively. Variance of the execution time implemented with DALiuGE and MPI is shown in text [22].

MPI-based parallel implementation. DALiuGE execution framework-based implementation also has shorter average time than common MPI libraries-based implementation.

3 Improving SKA algorithm reference library on high-performance computing platform

ARL was a set of algorithms developed in Python for calibration and imaging pipelines. ARL consists of eight parts, i.e., CARL1.1 (data), CARL1.2 (image), CARL1.3 (visibility), CARL1.4 (Fourier transform), CARL1.5 (sky components), CARL1.6 (calibration), CARL1.7 (Utility), and CARL1.8 (Pipelines). As ARL was designed to illustrate the algorithm logic rather than performance, the ARL performance is well below requirement of astronomical data processing in a scale up to SKA telescope. In addition, major supercomputers, e.g., Tianhe-2, do not support high-performance execution of Python applications.

To handle insufficient performance issue of ARL on high-performance computing platforms, ARL was improved and the optimized by migrating the core and time-consuming algorithms in ARL, such as prediction, inversion, and cleaning, to C++ platform. A popular tool on high-performance computing platform, BLAS (Basic Linear Algebra Subprogram) library, was also used to improve performance. To minimize the impacts on astronomy data processing applications, these improvements are transparent to the upper-level applications since function calls are unchanged.

To further enhance the parallel execution of applications built on top of improved ARL, an automatic parallel optimization tool simulation tuning automatic runtime (STAR) was developed in shared memory environment to automatically tune applications in terms of fine granularity partition and dependency assignment as shown in Fig. 12.5, which can be exploited to achieve better dynamic scheduling of astronomical data processing tasks.

Experiments were conducted on our heterogeneous distributed platform BATON to evaluate the performance of our improved implementations of ARL-based astronomical imaging pipeline and algorithms. The results from an open access informal memo [23] of three type experiments: ARL computational kernels, simple imaging pipeline, and imaging algorithms are presented.

3.1 ARL computational kernel

By migrating ARL's underlying library *Numpy* library to C++ library, we managed to improve the performance of all the algorithms in ARL. To confirm our improvements, we evaluated the original performance and improved performance of gridding algorithm, degridding algorithm, FFT algorithm, and IFFT algorithm, which are major parts in ARL. The evaluation results are summarized in Tables 12.1 and 12.2.

It was shown that our improved implementation of degridding algorithm achieved 14 × speedup in performance while our improved implementation of gridding algorithm achieved 48 × speedup in performance. After further tuning with our STAR tool, the performance of improved degridding algorithm could achieve 24 × speedup, while the performance of gridding algorithm could achieve 54 × speedup.

We would like to highlight that our preliminary optimization results are well expected due to the low efficiency of python in handling heavy vector calculations in gridding and degridding algorithms. However, algorithm restructuring was required to achieve further optimization along with a significantly improved implementation since the nature of gridding algorithm implies

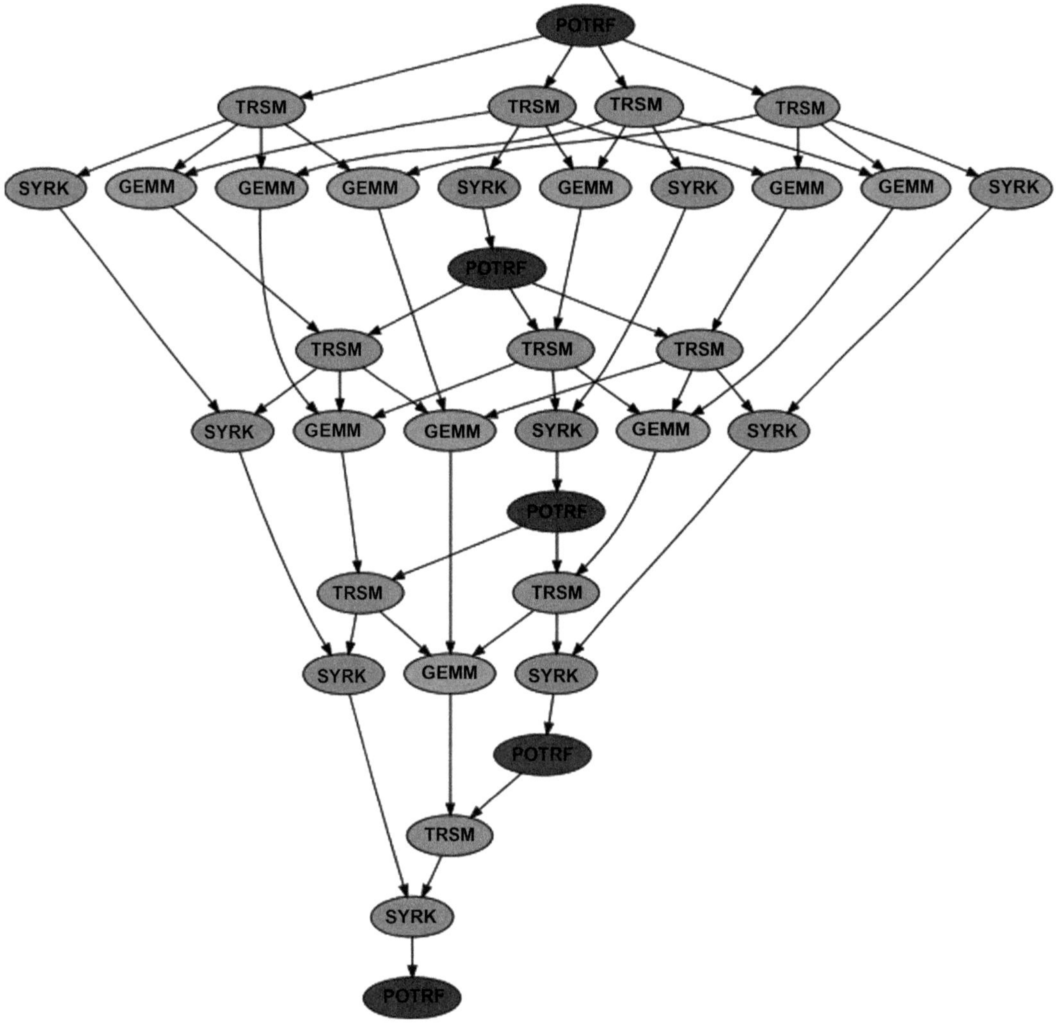

Fig. 12.5 Dependency assignment by STAR tool.

strong data interdependency, which is the major hurdle for massive parallelization on high-performance computing platform.

3.2 Simple imaging pipeline

The pipeline in our second experiment handled visibility data in the form of a simple imaging pipeline whose major stages included prediction, inversion, and cleaning. In this experiment, parameters settings were completed as 10 channels, 7 groups of

Table 12.1 The performance of gridding algorithm [23].

	Gridding(s)	Degridding(s)
Origin	8.08	4.38
Optimized	0.55	0.09
Speedup	×14.7	×48.7

Table 12.2 The performance of FFT algorithm [23].

	FFT(s)	IFFT(s)
Origin	0.63	0.37
Optimized	0.4	0.17
Speedup	×1.6	×2.2

time slots on top of basic configuration of 166 antennas as LOWDB2-CORE.

The ARL-based major stages of imaging pipelines were improved by migrating underlying Python library to C++ library. The original performance metrics and improved performance metrics of major stages in the pipeline are summarized in Table 12.3.

As shown in Table 12.3, overall performance in terms of total execution time was improved by 3.7 times of speedup. More specifically, the performance of prediction stage was improved by 5.3 times, the performance of inversion stage was improved by 11.2

Table 12.3 The performance of single image processing pipeline [23].

	Predict(s)	invert(s)	Deconv(s)	Total(s)
Origin	8.65	13.98	23.34	45.97
Optimized	1.64	1.25	9.37	12.26
Speedup	×5.3	×11.2	×2.5	×3.7

times, and the performance of deconvolution stage was improved by 2.5 times. This experiment confirmed the imaging pipeline built on top of improved ARL did produce significant performance improvement.

3.3 Functional imaging algorithms

After our experiment on simple imaging pipeline with basic imaging algorithms, we continued our experiments on more functional imaging algorithms, which were essential to realistic astronomical data processing pipeline regarding pattern predicting and pattern inversion. These algorithms include base, w-stack, timeslice, facets, and w-projection algorithms. The original performance in time and improved performance in time of these algorithms are listed in Tables 12.4 and 12.5.

It was reasonable to figure out that improved base algorithm and improved facets algorithm achieved significant performance enhancement since these algorithms contain frequent function calls to underlying gridding and FFT algorithms whose massive parallel versions exist. However, for w-stack algorithm and timeslice algorithm, the improved performance was dwarfed by frequent function calls to some underlying calculations whose current forms were difficult to be parallelized.

Table 12.4 The performance of algorithms regarding pattern predicting [23].

	2D(s)	W-stack(s)	Timeslice(s)	Facets(s)	W-project(s)
Origin	1.09	1.07	9.14	3.92	17.4
Optimized	0.28	0.3	8.64	0.77	9.22
Speedup	3.9	3.6	1.1	5.1	1.9

Table 12.5 The performance of algorithms regarding pattern inversion [23].

	2D(s)	Wstack(s)	Timeslice(s)	Facets(s)	Wproject(s)
Origin	1.52	0.88	12.79	8.56	8.78
Optimized	0.15	0.23	12.14	3.21	6.41
Speedup	10.1	3.8	1.1	2.7	1.4

4 Summary

For high-performance computing of astronomical big data, this chapter discusses the execution frameworks for astronomical big data processing and introduces the results of scalabilty test of DALiuGE execution framework on DAS-5 supercomputer as well as Tianhe-2 supercomputer. ASTRON's experiment on DAS-5 confirmed that porting CALIB pipeline and the first part of the Prefactor pipeline to DALiuGE execution framework was successful on 4 nodes of ASTRON's DAS-5 supercomputer. Further experiments on Tianhe-2 supercomputer indicated that while DALiuGE is able to execute GPU-accelerated pipelines in an end-to-end fashion, its performance is also comparable to other frameworks or libraries such as MPI in teams of performance stability and shorter average execution time. DaLiuGe-based execution seems to produce less latency than MPI-library-based implementation as the scale of parallel execution grows up.

In addition to experiments on DALiuGE execution framework, the optimized implementation of ARL was also described in order to enhance computing performance of ARL-library-based SKA imaging processing. The optimized implementation of ARL library improved the performance of the image processing pipeline through the optimization of the computational kernels. Additional performance gain can be brought by the fine tuning tool STAR.

References

[1] J.W. Romein, An efficient work-distribution strategy for gridding radio-telescope data on GPUs, in: Proceedings of the 26th ACM International Conference on Supercomputing, ACM, 2012, pp. 321–330.

[2] A. Wicenec, From antennas to multi-dimensional data cubes: the SKA data path, in: 2015 IEEE International Conference on Acoustics, Speech and Signal Processing (ICASSP), IEEE, 2015, pp. 5645–5649.

[3] R.T. Schilizzi, P.E.F. Dewdney, T.J.W. Lazio, The square kilometre array, in: Ground-Based and Airborne Telescopes II. International Society for Optics and Photonics, vol. 7012, 2008, p. 70121I.

[4] B. Nikolic, SDP consortium. square kilometre array science data processing, in: Exascale Radio Astronomy, 2014, p. 2.

[5] R. Chandra, L. Dagum, D. Kohr, et al., Parallel Programming in OpenMP, Morgan Kaufmann, 2001.

[6] W. Gropp, W.D. Gropp, A.D.F.E. Lusk, et al., Using MPI: Portable Parallel Programming With the Message-Passing Interface, MIT Press, 1999.

[7] J. Dean, S. Ghemawat, MapReduce: simplified data processing on large clusters, Commun. ACM 51 (1) (2008) 107–113.

[8] M. Zaharia, M. Chowdhury, M.J. Franklin, et al., Spark: cluster computing with working sets, HotCloud 10 (10 – 10) (2010) 95.

[9] K. Shvachko, H. Kuang, S. Radia, et al., The hadoop distributed file system, in: MSST, vol. 10, 2010, pp. 1–10.
[10] J.G. Shanahan, L. Dai, Large scale distributed data science using apache spark, in: Proceedings of the 21th ACM SIGKDD International Conference on Knowledge Discovery and Data Mining, ACM, 2015, pp. 2323–2324.
[11] M. Zaharia, R.S. Xin, P. Wendell, et al., Apache spark: a unified engine for big data processing, Commun. ACM 59 (11) (2016) 56–65.
[12] A.G. Shoro, T.R. Soomro, Big data analysis: apache spark perspective, Global J. Comp. Sci. Technol. 15 (1-C) (2015), https://computerresearch.org/index.php/computer/article/view/1137.
[13] The Apache Software Foundation. Welcome to Apache Hadoop. Online at: https://hadoop.apache.org/old/. (Accessed 22 April 2020).
[14] M. Malak, Data Locality: HPC vs. Hadoop vs. Spark, datascienceassn. org, 2014.
[15] J.E. Rodrigues, J.E. Rodriguez Bezos, A Graph Model for Parallel Computations, Technical Report, Massachusetts Institute of Technology, Cambridge, MA, USA, 1969.
[16] J.B. Dennis, First version of a data flow procedure language, in: Programming Symposium, Springer, Berlin, Heidelberg, 1974, pp. 362–376.
[17] J.B. Dennis, D.P. Misunas, A preliminary architecture for a basic data-flow processor, in: ACM SIGARCH Computer Architecture News, 3 ACM, 1975, pp. 126–132 (4).
[18] R.S. Nikhil, Executing a program on the MIT tagged-token dataflow architecture, IEEE Trans. Comput. 39 (3) (1990) 300–318.
[19] C. Wu, R. Tobar, K. Vinsen, A. Wicenec, D. Pallot, B. Lao, R. Wang, T. An, M. Boulto, I. Cooper, R. Dodson, M. Dolensky, Y. Mei, F. Wang, DALiuGE: a graph execution framework for harnessing the astronomical data deluge, Astron. Comput. 20 (2017) 1–15.
[20] Y. Grange, C. Wu, LOFAR pipeline DALiuGE, SDP Memo, http://ska-sdp.org/publications/released-sdp-memos-1, 2018.
[21] H. Bal, D. Epema, C. de Laat, R. van Nieuwpoort, J. Romein, F. Seinstra, C. Snoek, H. Wijshoff, C. de Laat, R. van Nieuwpoort, J. Romein, F. Seinstra, C. Snoek, H. Wijshoff, A medium-scale distributed system for computer science research: infrastructure for the long term, Computer 49 (5) (2016) 54–63.
[22] B. Lao, Full-Scale DALiuGE Data Simulation and Reduction on Tianhe-2, SDP Memo, http://ska-sdp.org/publications/released-sdp-memos-2, 2018.
[23] H. You, R. Yang, T. Wang, et al., Fast Implementation of SKA Algorithm Reference Library, SDP Memo, http://ska-sdp.org/publications/released-sdp-memos-2, 2018.

13

Spark and dask performance analysis based on ARL image library

Kaiyu Fu[a], Qiuhong Li[b], Siyu Fan[a], Ting Li[a], Tian Huang[c,d], Yuan Luo[a]

*[a]Department of Computer Science and Engineering, Shanghai Jiao Tong University, Shanghai, China. [b]School of Computer Science, Fudan University, Shanghai, China. [c]Astrophysics Group, Cavendish Lab, Cambridge University, Cambridge, United Kingdom. [d]Institute of High Performance Computing, Agency for Science, Technology and Research (A*STAR), Singapore, Singapore*

1 Introduction

With the development of data collection and storage technologies, humans have accumulated a large number of astronomical observation data. In this research field, an increasing portion of the data is about image. Thus, it is crucial to find the most suitable framework for large data processing in astronomy.

For example, the Square Kilometer Array (SKA) is a large-scale international scientific research cooperation project led by the United Kingdom in a consortium with more than ten countries. The original intention of the SKA project was that, in the 21st century, cosmology still had many fundamental problems to solve, such as the nature of elementary particles and basic forces, the formation and evolution of the universe, and the origin of dark matter. Therefore, in order to facilitate astronomers to further observe the structure of the early universe, it needs to build telescopes with higher sensitivity and survey speed, where SKA is one of them and SDP is of its science data processing part.

In the SKA large astronomical telescope network, the imaging data received by the telescope need to be processed by big data methods. However, the existing algorithms are serial and cannot be directly transferred to the big data frameworks where Spark and Dask are two important examples for current astronomers. Therefore, it is necessary to improve parallelization and optimize the computation performance. At the same time, the network

Big Data in Astronomy. https://doi.org/10.1016/B978-0-12-819084-5.00001-8

topology of the cluster needs to be studied in order to use the limited bandwidth resources to achieve maximal network performance. Therefore, the research content of this chapter is to analyze the performance of computation and network transmission on astronomical data processing and then propose better task scheduling models based on the computation topology, routing topology and their relations, respectively. One of the important benchmarks is suggested [1] to be SKA SDP Mid1 pipeline invoking the standard astronomy algorithm library (ARL).

In astronomical field, an expert team of engineers usually processed the collected images on a dedicated server, and then refined the results into a text for analysis by other astronomers. In addition, many data streaming jobs today are computationally intensive and data intensive. For example, astronomical data streaming applications generate large amounts of data and require huge computing power to process the data. Traditionally, high-performance computing (HPC) facilities have been optimized for handling such computationally intensive scientific applications. However, they are not suitable for handling data-intensive workloads. So our research on this aspect can first provide an easy-use tools for nonprofessionals, and secondly, it can provide an effective solution for data-intensive and computationally intensive operations.

In the structure of Spark, MapReduce [2] is the core of the paradigm, and resilient distributed datasets [3] (RDD) is the core of data structure. MapReduce makes fast and efficient parallel computing possible, while RDD can be stored in memory without the need for additional backups so that each query can quickly get the results. The authors in [4–9] described the usages of four functional modules in the Spark framework. The first module in Spark is core, which provides ordinary parallel batch processing. In addition, the second module "Spark Streaming" provides streaming parallel processing. Spark SQL, the third module, enables Spark's efficient parallelism to be applied into the massive data query domain. The last one, called MLlib, provides some APIs for distributed machine learning.

For the investigation of the benchmark, i.e., SKA SDP Mid1 pipeline, the data processing astronomical algorithm of SKA [1] is suggested to be multiscale multifrequency, which is abbreviated as MSMF [10]. In the transplant of this algorithm to Spark and Dask, this chapter investigates the difference of computation performance and network performance. The rest of the chapter is organized as follows. In Section 2, we introduce preliminaries and notations of astronomy and computer science in this paper. In Section 3, the performance of Spark and Dask on Mid1 pipeline is evaluated according to the variance of total cores, data size, and the number of threads per process. We find that the task scheduling models can be further improved by genetic algorithm in Section 4, which leads to a local optimal solution. Furthermore, Sections 4 and 5 introduce some basic ideas of the coordination between computation topology model, data transmission model of processors and physical machines, and also the routing model. Because of the limited space, more discussions will be in our future work.

2 Preliminaries and notations

This section introduces the main idea of Spark, Dask, and genetic algorithm. Spark introduces RDD model to make full use of memory resources and improve computational efficiency. Dask is a flexible parallel computing library based on Python and has been widely used in big data processing. The genetic algorithm is a method for solving both constrained and unconstrained optimization problems that is based on natural selection, the process that drives biological evolution.

2.1 Spark

The early big data parallel processing framework Hadoop was limited by a single point of failure and a relatively simple computational model. To overcome the shortcomings of MapReduce, Spark introduced RDD model to make full use of memory resources and improve computational efficiency. Unlike other large data processing frameworks, Spark can use ETL to SQL processes, machine learning and streaming data based on Shark, MLlib, GraphX, and Spark Streaming. So Spark has been widely used in the industry. Tencent, Yahoo, Taobao and other companies have integrated Spark into their own business successfully.

Apache Spark is a fast and general-purpose computing engine designed for large-scale data processing. Spark is a Hadoop MapReduce-like general-purpose parallel framework open sourced by UC Berkeley AMP lab (UC Berkeley's AMP Lab). Spark has the advantages of Hadoop MapReduce; but unlike MapReduce, the intermediate output of the job is stored in memory, so it is no longer necessary to read and write HDFS, so Spark can be more suitable for data mining and machine learning algorithms that need iterative MapReduce algorithms.

Spark is an open-source cluster computing environment similar to Hadoop, but there are some differences between the two. These useful differences make Spark perform better in certain workloads. In other words, Spark is enabled. In addition to the in-memory distributed dataset, it can optimize iterative workloads in addition to providing interactive queries.

Spark is implemented in the Scala language and uses Scala as its application framework. Unlike Hadoop, Spark and Scala are tightly integrated, where Scala can manipulate distributed datasets as easily as local collection objects.

Although Spark was created to support iterative jobs on distributed datasets, it is actually a complement to Hadoop that can run in parallel in the Hadoop file system. This behavior is supported through a third-party clustering framework called Mesos. Developed by the University of California, Berkeley's AMP Lab (Algorithms, Machines, and People Lab), Spark can be used to build large, low-latency data analysis applications.

RDD is one of the core concepts of Spark, with automatic fault tolerance, scalability, and location-aware scheduling. The creation of RDD can be achieved in two ways: stable physical storage of deterministic datasets and other deterministic operations performed by existing RDDs. Therefore, RDD contains information about how to calculate RDD from other RDDs. When some of the RDD partition data are lost, Spark can be reconstructed based on the log information.

Spark has three main features:

First, the high-level API strips away the focus on the cluster itself, and Spark application developers can focus on the calculations that the application has to do.

Second, Spark is fast and supports interactive computing and complex algorithms.

Finally, Spark is a general-purpose engine that can be used to perform a variety of operations, including SQL queries, text processing, machine learning, etc. Before Spark appeared, we generally needed to learn a variety of engines to handle these separately demand.

2.2 Dask

Dask is a flexible parallel computing library. Calculations are described as directed acyclic graphs (DAGs), using simple specifications built from dictionaries, tuples, and functions. These DAGs can be run using a variety of schedulers, from a single computer scheduler that uses threads to a distributed scheduler that extends across the cluster.

Above these core components, several collections have been established. These collections provide a parallel version of the mirrored familiar single-threaded collection—the dask.array for the NumPy array and the dask.dataframe for the Pandas DataFrame.

By simulating the existing interface, converting to a parallel version is very simple for user. For example, let's calculate the matrix multiplication of two random arrays *a* and *b* in NumPy and Dask:

```
# numpy version
import numpy as np
a=np.random.normal(size=(1000, 10,000))
b=np.random.normal(size=(10,000, 1000))
result=a.dot(b)
# dask version
import dask.array as da
a=da.random.normal(size=(1000, 10,000), chunks=1000)
b=da.random.normal(size=(10,000, 1000), chunks=1000)
result=a.dot(b).compute()
```

In order to convert NumPy code to Dask code, it needs to add the chunks keyword to tell Dask how to block the array, and it also needs to call the calculation method to tell Dask to execute the DAG. Otherwise, the NumPy and Dask code looks roughly the same. However, the Dask code can run in parallel and is not affected by the kernel, allowing user to perform calculations on datasets that are larger than memory.

Dask is a flexible library for parallel computing in Python and composed of two parts:

1. Dynamic task scheduling optimized for computation. This is similar to Airflow, Luigi, Celery, or Make, but optimized for interactive computational workloads.
2. "Big Data" collections like parallel arrays, dataframes, and lists that extend common interfaces like NumPy, Pandas, or Python iterators to larger-than-memory or distributed environments. These parallel collections run on top of dynamic task schedulers.

2.3 Genetic algorithm

Genetic algorithm was developed by Holland [11] to study the adaptive process of natural systems and to develop artificial systems that mimic the adaptive mechanism of natural systems.

The evolution usually starts from a population of randomly generated individuals, and is an iterative process, with the population in each iteration called a generation. In each generation, the fitness of every individual in the population is evaluated; the fitness is usually the value of the objective function in the optimization problem being solved. The new generation of candidate solutions is then used in the next iteration of the algorithm.

Genetic algorithm is a highly parallel, random, and adaptive optimization algorithm based on "survival of the fittest." The "chromosome" group represented by the problem solution is copied, crossed, and mutated. It has evolved from generation to generation, and finally converged to the most adaptive Group, so as to obtain the optimal or satisfactory solution to the problem. Its advantages are simple principle and operation, strong generality, unlimited constraints, and the ability of implicit parallelism and global solution search, which is widely used in combinatorial optimization problems. The earliest application of genetic algorithm to jb-shop scheduling problem is Davis genetic algorithm, which applies less knowledge of neighborhood when solving job-shop scheduling problem, which is more suitable for practical application. How to use genetic algorithm to solve job-shop scheduling problem efficiently has been regarded as a challenging problem and has become a research hotspot.

The crossover operator is the most important operator in the genetic algorithm, which determines the global convergence of the genetic algorithm. The most important criterion for crossover operator design is that children inherit the superior characteristics of their parents and the feasibility of their children. On the basis of in-depth analysis of the job-shop scheduling problem, Shao Xinyu et al. proposed a POX method based on process coding, and compared it with other process code-based crosses, proving that the cross method solves the job-shop scheduling problem Effectiveness. At the same time, in order to solve the premature convergence of the traditional genetic algorithm in solving the jb-shop scheduling problem, Shao Xinyu et al. also designed an improved offspring alternating pattern genetic algorithm, which significantly accelerated the convergence speed of the genetic algorithm. This principle is also applicable to other combinatorial optimization problems. The mutation of the genetic algorithm proposed by Shao Xinyu and others is different from the mutation operation (to maintain the diversity of the population) in the traditional genetic algorithm. It is to improve the performance of the offspring through local search.

2.4 Pipeline use case

We use SKA SDP Mid1 pipeline's first three stages—predict, calibration, and backwards—as our use case. This version of pipeline is abridged partially for the sake of simplicity, but it doesn't hurt its generalization.

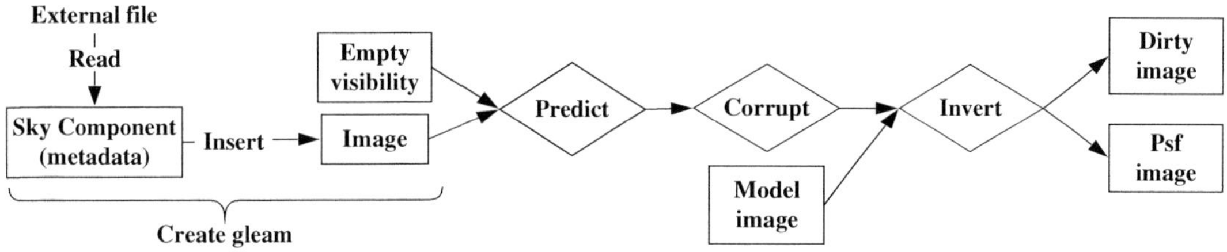

Fig. 13.1 SKA SDP Mid1 pipeline.

Data are extracted from *GLEAM_EGC. fits* file, which contains sky component information. With the metadata, we can produce as many data as possible by inserting it into the image of different channels. We use up to 512 channels in our experiment and there is an image on each channel. The size of each image is 256 * 256.

As Fig. 13.1 shows, this pipeline is consisted of 4 steps:

(1) Create-gleam: Step 1 extracts metadata from file and inserts this information into empty image, passing image results to the next step.

(2) Predict: Step 2 convolves image and inserts result into corresponding visibility data. In this step, we need to pair the image with corresponding visibility data, so there will be many shuffle operations, which may become a bottleneck in our program.

(3) Correct: Step 3 will calibrate visibility data coming from previous step with the randomly generated simulated gaintable. This step can be perfectly processed in parallel.

(4) Invert: Step 4 can be seen as the inverse of the step 2, which does deconvolution, converting visibility data back to the image. For the same reason, there are also many shuffle operations in this step.

The Stand-alone implementation and the Dask implementation is done by Tim Cornwell, which are both included in ARL.

3 Experiment

We evaluate the performance of the pipeline use case and compare efficiency of these two frameworks in different scenarios by tuning over the computing resources and data size.

3.1 End-to-end performance

We first evaluate the performance of running the Astronomy pipeline end to end on Spark and Dask.

Environment: All experiments are performed on a cluster composed of two machines, with 12 logical cores for one and 24 logical cores for another. Both machines have 120-GB memory.

Goal: Our main goal is to research the performance of Spark and Dask in different resources allocation and data size. During the experiment, we also find that even the total number of cores remains the same, the performance of both frameworks alters along with different process parallelism and thread parallelism.

Experiments:

- In the first experiment we use the largest data size and vary the total number of cores from 8 to 36, fixing the number of threads per process at four and memory per process at 2GB. The largest output data size is 537.6 MB, which fits the 2-GB memory well in our previous experiment. And in this chapter we will allocate adequate memory for most of program and mainly focus on number of cores and data size. Table 13.1 shows the data size of different n-frequency. N-frequency is the number of frequencies of the image, which determines the data size.
- In the second experiment, we fix the number of cores at 36 and number of threads per process at 4, varying n-frequency from 16 to 512.
- In the third experiment, we vary the number of threads per process, fixing the total cores at 36.
- In the last experiment, we focus on Spark framework and add the collect and cache operation in program to explore how to make the program faster. Collect operation will collect data distributed across all machines to the driver and cache operation will store the intermediate data on the machine for later use.

Experiment 1 Fig. 13.2 shows the runtimes for Dask and Spark as we increase the total cores.

From Fig. 13.2 we can see that with the number of cores increasing, the speedup is not proportional to it. Only when the number of cores increases from 8 to 16, the runtime decreases obviously. When further increasing number of cores, there is only neglectable improvement of performance. The reason is that while total number of logical cores is 36, there are actually only

Table 13.1 Data size in different n-frequency (MB).

n-frequency	16	32	64	128	256	512
Output	16.8	33.6	67.2	134.4	268.8	537.6

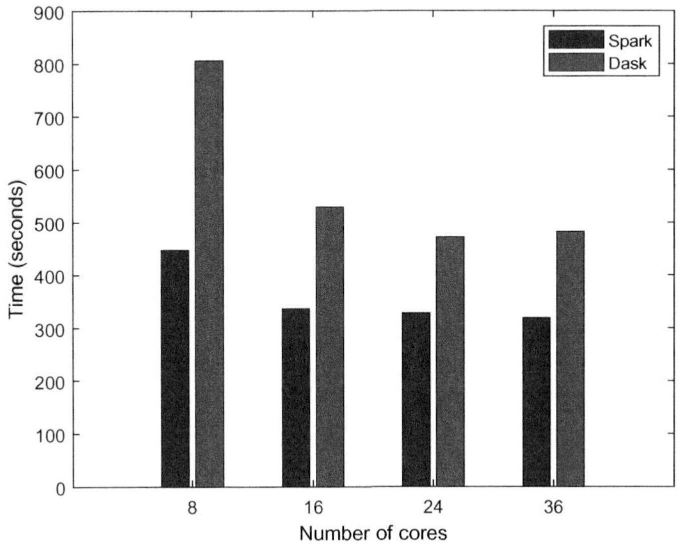

Fig. 13.2 Results for end-to-end experiments (Experiment 1).

18 physical cores in the cluster. Hyperthreading may be beneficial to some noncomputing-intensive applications, but when computation is so intensive that requires full attention of the processor, it may be not so effective and even negative. So when total cores are more than the de facto physical cores, program will not gain a significant improvement from Hyperthreading.

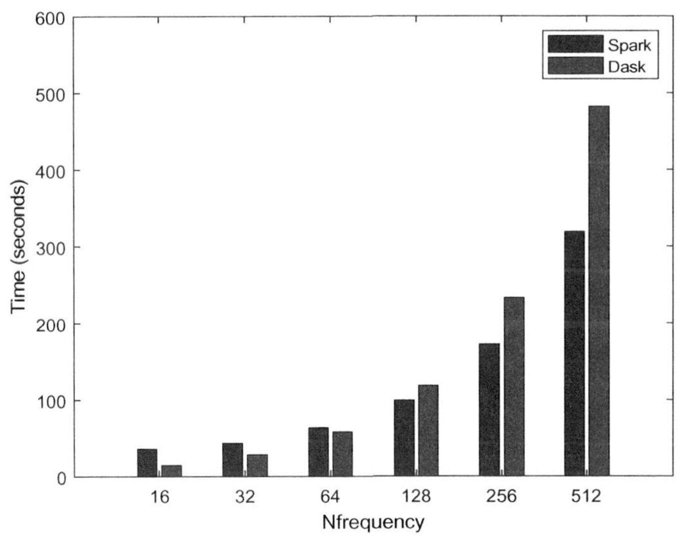

Fig. 13.3 Results for end-to-end experiments (Experiment 2).

Experiment 2 Fig. 13.3 shows the results as we increase the data size.

From Fig. 13.3, we can see that when n-frequency is smaller than 32, Dask is faster than Spark, and has similar runtimes when n-frequency is 64. However, with the data size increasing, Spark catches up with Dask gradually, and is 1.3 x faster than it when n-frequency is 512, according to which, we can believe that Spark will have better performance than Dask when processes big data in terms of this pipeline. Spark is slower in the start-up stage for that it is natively written in Scala rather than Python; therefore, all data that come to and from Python executor have to be passed through a socket and a JVM. This Overhead of JVM is not neglectable when data size is not large. As the data size becomes bigger and bigger, this side effect can be offset by its high efficiency and better resource control over Dask.

Experiment 3 Fig. 13.4 shows that while the total number of cores is constant, runtimes are changing with different number of threads per process. Furthermore, Spark and Dask have different tendencies in the face of increasing number threads per process. For Dask, performance declines with the increase of number of threads per process. The reason for this is that due to Python's global interpreter lock mechanism, which is a mechanism used in computer-language interpreters to synchronize the execution of threads so that only one native thread can execute at a time, thread parallelism can't function well on pure Python objects like dicts or lists, which take up most of our program. Therefore, this program is more compatible with multiprocessing mode. For Spark, performance is not influenced by number of threads per

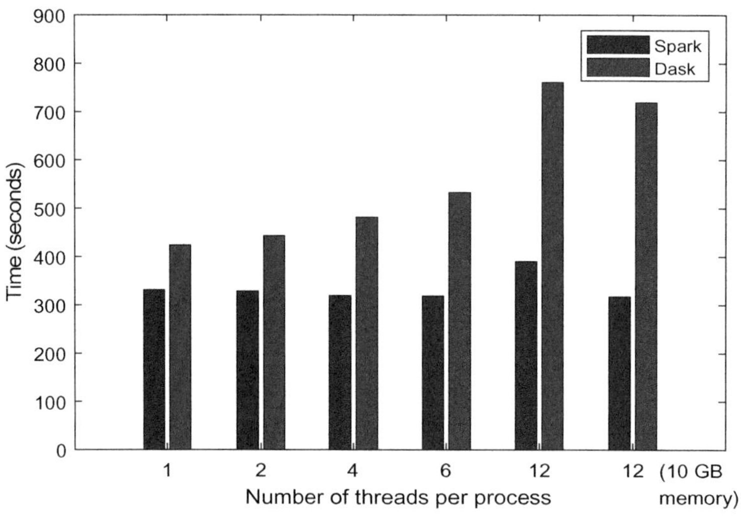

Fig. 13.4 Results for end-to-end experiments (Experiment 3).

process greatly and is only a little slower in the case of single thread per process because there are communication and data exchange between each executor and driver, the more processes, the more overheads. And we can see that this overhead is just a fraction of total runtime. However, when there are 12 threads for each process, performance suddenly becomes worse. The reason for this is that in the case of 12 threads per process, there are totally only three processes for our program, and each process only has been allocated 2-GB memory and the actual usable memory, which can be used to store RDD, the basic collection of elements that can be operated on in Spark, is just 970 MB, which is not much larger than the total size of intermediate data generated by our program. And as we know, JVM will do GC (Garbage collection) when memory is not so abundant, and executor even will spill data to disk when memory is full, both of which will worsen performance. So in this scenario, the bottleneck is memory. In order to verify our conclusion, we do an extra experiment, allocating 10-GB memory for each process with rest parameters remaining unchanged. Result well confirms our conjecture. When allocated adequate memory, Spark will perform better with as much as possible threads in one process by reducing communication overheads between different executors.

Experiment 4 Fig. 13.5 shows that collecting and caching data properly in the Spark program can improve performance.

From Fig. 13.5, we can see that if we don't collect every intermediate data into driver and cache-specific RDD which will be used once more, we can reduce runtimes effectively. The appropriately caching and collecting program can be 1.5 × faster than that without it.

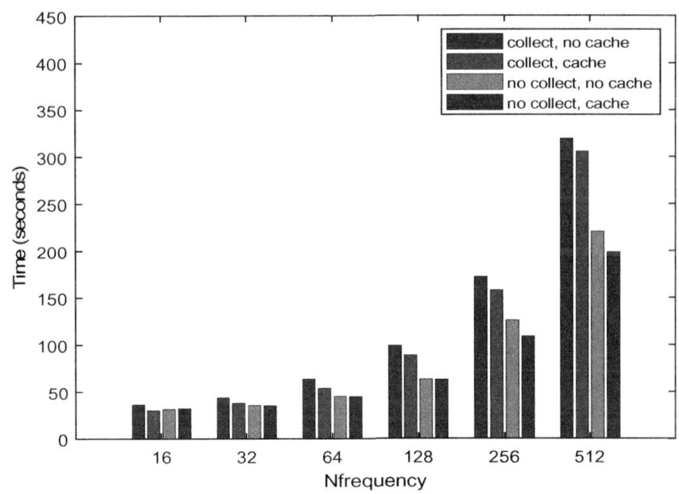

Fig. 13.5 Results for end-to-end experiments (Experiment 4).

3.2 Individual step performance

Then we will focus on individual steps and explore performance of Spark and Dask, respectively.

Creating gleam: In this step, our program will first read sky component information from FITS file, which is the most commonly used digital file format in astronomy. All these files are distributed by driver in Spark and scheduler in Dask, and its size is so small that we can ignore the overhead of its transmission. Then, this information of sky component will be inserted into empty image. Runtime of this step is shown in Fig. 13.6, from which we can see that while Spark performs well in all case, Dask performs badly in case of 8 cores. Despite this, Spark's performance keeps almost unchanged when we allocate more computing resources for it. The reason why Spark can't benefit from the increase of parallelism in this stage is that this stage is just a preprocessing stage and it spends most of the time on launching the task instead of doing computation, which is not the strengths of Spark (Fig. 13.7).

Predicting: In this Step, Image from previous step will be convolved, and the generated data will be put into the visibility data structure. Spark's performance is much better than that of Dask in this stage. From our perspectives, the reason is that this stage needs shuffle of data in order to make corresponding data locate in the same place. Spark and Dask have different mechanisms for this work. While Spark will keep larger data staying in place and smaller data shuffling between different processes as far as

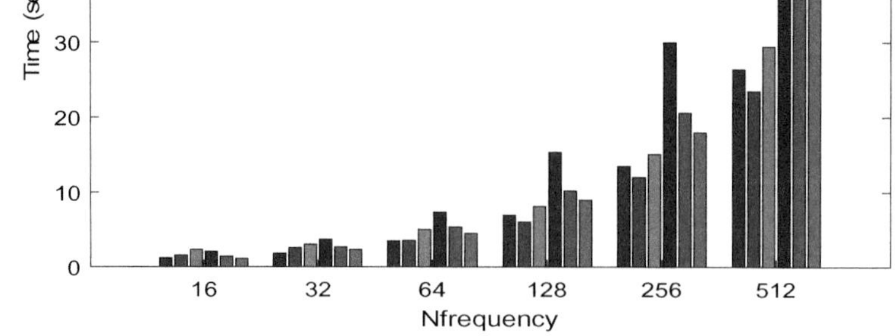

Fig. 13.6 Individual runtime of Spark and Dask in Creating gleam stage.

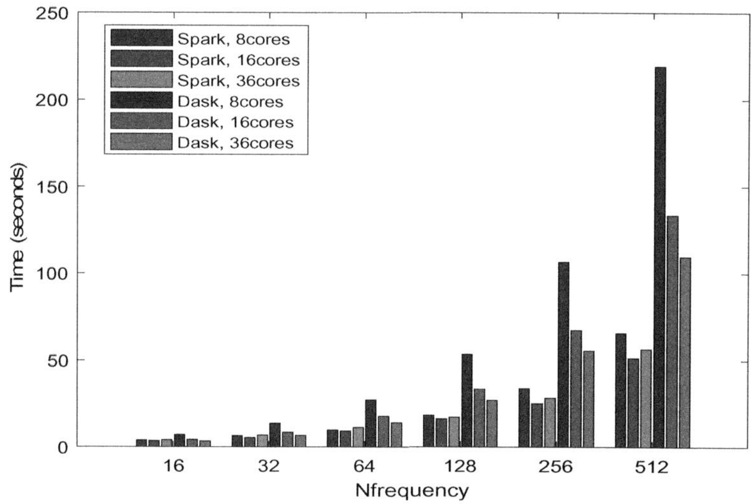

Fig. 13.7 Individual runtime of Spark and Dask in Predicting stage.

possible, Dask distributes all data from scheduler into workers, which is time consuming. In this scenario, the superiority of Spark's excellent resource scheduling is reflected (Fig. 13.8).

Correcting: In this step, visibility will be calibrated by many simulated gain tables. Spark also performs much better than Dask. Correcting offers a larger degree of data parallelism than any of other stages. For Correcting, the whole stage can be implemented with only a map operator in Spark, which is to its advantage. Here, map is the name of a higher-order function that applies a given function to each element of a collection (Fig. 13.9).

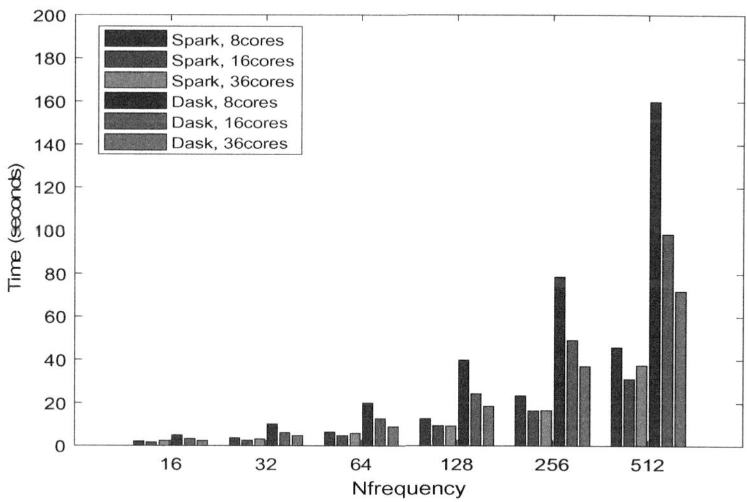

Fig. 13.8 Individual runtime of Spark and Dask in Correcting stage.

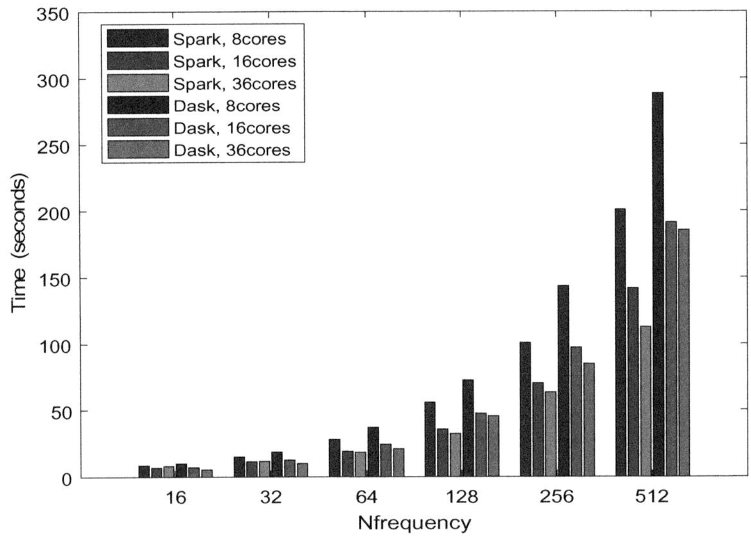

Fig. 13.9 Individual runtime of Spark and Dask in Inverting stage.

Inverting: Finally, Inverting stage deconvolves visibility back into image, which is computationally intensive, and will spend most of the time on computing. While Spark and Dask have different resource control and data transmission mechanism, both of them execute the same Python code from the ARL but wrapped in UDFs. Therefore, actual computing time for both frameworks is similar and this is the stage where the performance gap between Spark and Dask is the smallest.

In the previous article, we compared the performance of Spark and Dask on the Mid1 pipeline. After analyzing their task scheduling algorithms, we found that the task scheduling models of these two frameworks still have a lot of room for improvement. Therefore, we developed a new task scheduling model based on the genetic algorithm and network flow transmission. By simulation, we found that the theoretical performance of the model is much better than Spark. In the next part, we will illustrate the model in detail.

4 Task scheduling based on data processing capacity

We want to get a way of task scheduling, which can be represented by a 0–1 matrix A. Then, a system of linear equations can be established according to the relationship between the data processing capacity of the tasks X, the data processing capacity of the processors Y, and the matrix A, which is $Y=AX$. When X, Y

is known, some scheduling schemes A can be obtained according to the solution of the linear equations. At the same time, each processor has an upper bound Y_{max} of data processing capacity, and any of the solutions A should satisfy the inequality $Y_{max} \geq AX$.

However, there may be multiple matrices A that satisfy these equations. Although the number of matrices A is finite, it is still cumbersome to perform the exhaustive operation on matrix A. We cannot find the optimal solution of A through an exhaustive method; here, the optimal solution is a task scheduling plan that causes the processors to complete tasks with the shortest running time. In this part we will firstly propose a heuristic method to look for local optimal solution of matrix A and then analyze the system of equations $Y=AX$ according to the obtained local optimal solution to explore whether matrix A can be optimized. We will introduce how to apply GA on task scheduling in Section 4.1 and then describe the computation topology model and illustrate how to get a better solution using the computation topology model based on the GA in Section 4.2.

4.1 GA on task scheduling

In [12], authors proposed a method to apply genetic algorithm on multiprocessor scheduling problem. In this chapter, the representation of the search node is based on the order of the tasks being executed in each individual processor. The genetic operator proposed is based on the precedence relations between the tasks in the task graph.

Inspired by this article, we extended the algorithm and applied them to our astronomical data processing scenarios. The SKA SDP Mid1 pipeline can be represented by a DAG graph as shown in Fig. 13.10.

where the number of frequencies is 128 (listed as 2–129), but it can be scaled to arbitrarily large number without loss of generality. The circled number is the identifier of each subtask, ranging from 0 to 897. This graph refers to and all stages of the DAG graph are explained in detail in Section 2.4.

We perform the scheduling algorithm on the case where 898 subtasks to be scheduled on 12 processors with the number of frequencies is 128. Then we can get a solution to this task scheduling problem using the algorithm proposed. We set the probability of mutation to 0.05 and the probability of crossover to 1.0. And the whole program will terminate after running 10h. After we get a scheduling plan, we will represent it with a 0–1 matrix A^* and use it as a solution to the equation $Y=AX$ and do further analysis. The first row of the scheduling plan is as follows.

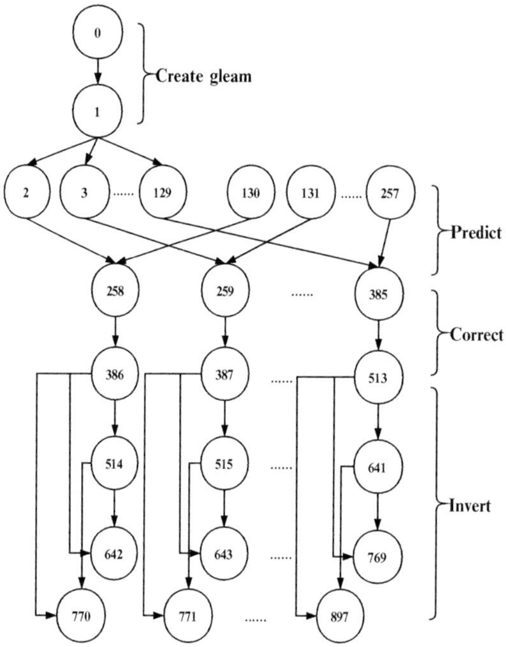

Fig. 13.10 Topology of the graph of the computation.

$$[41, 129, 74, 89, 76, \ldots, 583, 584, 585, 748, 865, 866, 867],$$

which means that the 41-st, ... until the 867-th subtasks will be assigned one after another to the first processor. Therefore, in the first row of the 0–1 matrix A, the value of the 41-st, ... and the 867-th columns will be 1s, and the rest will be 0s. The rest of the scheduling plan is omitted on account of the space limitations.

4.2 Computation topology model on task scheduling

A local optimal solution A^* can be got from the GA. In this section, we will describe the model in detail and illustrate how to get a better solution using the computation topology model based on the GA solution A^*.

Model description Suppose there is a serial and hybrid task processing system $\mathcal{N}_1$. A total of $\mathbf{n}$ pending tasks are in the system, use the vector $\mathbf{X} = (\mathbf{x}_1, \ldots, \mathbf{x}_n)^T$ to represent the data processing capacity of the tasks, and the data processing capacity of the k-th task is $\mathbf{x}_k$. The system has a total of $\mathbf{m}$ processors, each task can be assigned to one of the processors for processing, and the vector $\mathbf{Y} = (\mathbf{y_1}, \ldots, \mathbf{y_m})^{\mathbf{T}}$ is used to represent the processors' data processing capacity, where $\mathbf{y_i}$ represents the i-th processor's capacity of data processed. Let

$$a_{ij} = \begin{cases} 1, & \text{if the j} - \text{th task is assigned to the i} - \text{th processor,} \\ 0, & \text{else.} \end{cases}$$

denote $A = (a_{ij})_{m \times n}$ as the task scheduling matrix, then the relationship between the matrix A, the task's data processing capacity X, and the processor's data processing capacity Y can be expressed as $Y = AX$. It should be noted that data processing capacity here refers to the processing capacity over a period of time rather than the data processing capacity at a certain moment.

And every processor has its own upper bound of data processing capacity, denoted by $Y_{max} = (y_{1max}, \ldots, y_{mmax})^T$, where y_{imax} represents the upper bound of the i-th processor's data processing capacity. The data processing capacity of each processor should not exceed its upper bound, so the following inequality is generated:

$$Y_{max} \geq AX$$

Computation topology model combined with GA Usually, the data processing capacity of each task is known. When we get a scheduling plan for task processing, matrix $\boldsymbol{A}$ is determined accordingly, which means that the data processing capacity of each processor, that is, the matrix $\boldsymbol{Y}$ can be calculated by equations.

As we have seen, a task scheduling solution $\boldsymbol{A}^*$ can be obtained by GA, which can be easily transformed from the representation in Section 4.1. Let $\boldsymbol{Y}^* = \boldsymbol{A}^*\boldsymbol{X}$, which means $\boldsymbol{Y}^*$ is the data processing capacity of each task processor. We want to get all the task scheduling methods $\boldsymbol{A}$ by solving the equation $\boldsymbol{Y}^* = \boldsymbol{AX}$ and make $\boldsymbol{A}$ satisfy the inequality $\boldsymbol{Y_{max}} \geq \boldsymbol{AX}$ at the meantime, where $\boldsymbol{X}, \boldsymbol{Y}^*, \boldsymbol{Y_{max}}$ is known. Then analyze the matrix $\boldsymbol{A}$ to find out if there is a better task scheduling method than $\boldsymbol{A}^*$.

Suppose we have calculated an optimal task scheduling $\boldsymbol{A}$ now, as mentioned in Section 4.1, then the processor data transmission model can be generated according to the scheduling of the task (that is, $\boldsymbol{A}$) and the order of tasks' execution (as shown in Fig. 13.10). The number of each node in Fig. 13.10 will be replaced with the processor number assigned to this task, and the tasks processed by the same processor at the same stage will be merged into one point, finally the repeated directed edges will be deleted. According to these steps, we can get the topology graph of the processor data transmission. Since the graph is too complicated,

Table 13.2 Data transmission model of processors.

	1	2	3	4	5	6	7	8	9	10	11	12
1	0	0	0	1	0	0	0	1	0	0	0	0
2	0	0	0	0	0	1	0	0	0	0	0	1
3	1	0	0	1	0	1	1	1	0	1	0	1
4	0	1	1	0	0	0	1	1	0	0	1	0
5	0	0	0	0	0	0	0	0	0	0	0	0
6	0	0	0	0	0	0	1	0	0	0	0	0
7	0	0	1	0	0	1	0	1	1	0	0	0
8	0	1	1	1	0	1	0	0	1	1	1	0
9	0	0	0	0	0	0	0	0	0	1	0	0
10	0	0	0	0	0	1	0	0	0	0	0	1
11	0	0	0	0	0	0	0	0	0	0	0	1
12	0	1	1	1	0	1	1	1	0	0	0	0

Table 13.2 is used to represent the graph. If the number of the i-th row and the j-th column in the table is 1, it means that there is a directed edge from the i-th point to the j-th point in the graph, and each point represents a processor.

5 Network connection model and routing topology model

Two physical machines need time to do data transmission; this time can be calculated by routing topology model, which utilizes the linear equations. First, processor data transmission can be generated from the computation topology model. Second, machine data transmission, i.e., network connection model, can be got in accordance to the processor data transmission. Last, routing topology model can be brought according to the machine data transmission model.

In this section we will first introduce the routing topology model, then describe the specific process of generating a routing topology model.

Model description Let $\mathcal{N}_2$ be a network system with finite nodes. One source node and one destination node are called an OD pair. Let network $\mathcal{N}_2$ have $\boldsymbol{n}$ OD pairs, the flow of the k-th OD pair is x_k, vector $X=(x_1,x_2,\ldots,x_n)^T$ is the traffic matrix of $\mathcal{N}_2$ (actually a vector). Set network $\mathcal{N}_2$ has $\boldsymbol{m}$ links, and the link load of the i-th link is y_i, and the vector $\boldsymbol{Y}=(y_1, y_2, \ldots, y_m)^T$ is a link vector of $\mathcal{N}_2$. Let

$$b_{ij}=\begin{cases}1, & \text{if the traffic of the j}-\text{th OD pair goes through the i}-\text{th link,}\\ 0, & \text{else.}\end{cases}$$

$B=(b_{ij})_{m\times n}$ be the routing (information) matrix of network $\mathcal{N}_2$, then the relationship between routing matrix B, traffic matrix X and link vector Y can be expressed as $Y=BX$. Usually, the routing matrix B and the link vector $Y$of the network $\mathcal{N}_2$ are relatively easy to obtain. We hope to estimate the traffic matrix X, that is, the traffic transmitted between any two OD pairs.

Process of generate a routing topology model We suppose every machine has three processors, in order to reduce the load of network transmission between machines, we put the processors with more network transmissions on one machine. So here is the assignment scheme: 1-st, 5-th, 12-th processors on 1-st machine, 2-nd, 3-th, 4-th processors on 2-nd machine, 6-th, 7-th, 8-th processors on 3-rd machine, 9-th, 10-th, 11-th processors on 4-th machine. Then we use Table 13.2 to generate Table 13.3, a data transmission model of physical machine, i.e., an instance of network connection model. If the number of the i-th row and j-th column in the table is 1, it means that there is a directed edge from the i-th machine to the j-th machine. Every number 1 in Table 13.3 is an OD pair, so we have 9 OD pairs in our case.

Now consider there are 10 links in the network, and the routing path is known, which is shown in Table 13.4, the columns represent the OD pairs, which means the source node O transmits some data to destination node D (e.g., 1–2 means the 1-st machine transmit data to the 2-nd machine), the rows represent the physical links, which connect the machines or the routers. Then we can get the traffic matrix X according to

Table 13.3 Data transmission model of physical machine.

	1-st machine	2-nd machine	3-rd machine	4-th machine
1-st machine	0	1	1	0
2-nd machine	1	0	1	1
3-rd machine	0	1	0	1
4-th machine	1	0	1	0

Table 13.4 An instance of the matrix B in routing topology model.

	1–2	1–3	2–1	2–3	2–4	3–2	3–4	4–1	4–3
1-st link	0	1	0	0	1	0	0	1	0
2-nd link	1	0	1	1	0	1	1	1	1
3-rd link	0	0	0	0	1	0	1	1	1
4-th link	1	1	1	0	0	0	0	1	0
5-th link	1	0	1	1	0	1	0	1	0
6-th link	0	1	1	0	1	0	0	1	0
7-th link	1	0	1	0	0	0	1	1	1
8-th link	0	1	0	1	1	1	0	1	0
9-th link	1	0	1	0	1	0	0	1	0
10-th link	0	1	0	0	1	0	0	1	0

Note: Columns represent the OD pairs, rows represent the physical links.

B and *Y*. Now we can go back to the computation model. As the traffic transmission between two machines is known, time consumed by data transmission can be calculated, which can be added to the time spent according to scheduling method *A* in computation topology model.

6 Conclusion

This chapter presents a comprehensive study of large data processing of astronomical data on SKA SDP Mid1 pipeline systems. We first compare the performance of Spark and Dask using real-world use case from astronomical domain and then propose a local optimal task scheduling model based on the genetic algorithm. Dask can be more compatible with Python than Spark, and the migration and development of Dask program is relatively simple. However its performance is not as good as Spark. In Spark, although the transplant of the code is somewhat complicated, the user can use different functions such as reduce, combine, etc. In fact, according to the characteristics of the program, the data scheduling in Spark can be more accurate and efficient. Apart from this, we use genetic algorithm to get a local optimal solution of task scheduling, and then propose the corresponding computation topology model. The rest of the chapter focuses on the coordination between computation topology model, data transmission model of processors and physical machines, and also the routing model.

References

[1] A. Scaife, PDR.02.05.03 Imaging Pipeline. Science Data Processor Consortium, 2015.
[2] J. Dean, S. Ghemawat, MapReduce: simplified data processing on large clusters, Commun. ACM 51 (1) (2008) 107–113.
[3] M. Zaharia, M. Chowdhury, T. Das, A. Dave, J. Ma, M. McCauley, M.J. Franklin, S. Shenker, I. Stoica, Resilient distributed datasets: a fault-tolerant abstraction for in-memory cluster computing, in: Proceedings of the 9th USENIX Conference on Networked Systems Design and Implementation (NSDI'12), USENIX Association, Berkeley, CA, USA, 2012, p. 2.
[4] X. Meng, J. Bradley, B. Yavuz, E. Sparks, S. Venkataraman, D. Liu, J. Freeman, D. Tsai, M. Amde, S. Owen, D. Xin, R. Xin, M.J. Franklin, R. Zadeh, M. Zaharia, A. Talwalkar, MLlib: machine learning in apache spark, J. Mach. Learn. Res. 17 (1) (2016) 1235–1241.
[5] M. Zaharia, T. Das, H. Li, T. Hunter, S. Shenker, I. Stoica, Discretized streams: fault-tolerant streaming computation at scale, in: Proceedings of the Twenty-Fourth ACM Symposium on Operating Systems Principles (SOSP'13), ACM, New York, NY, USA, 2013, pp. 423–438.
[6] M. Armbrust, R.S. Xin, C. Lian, Y. Huai, D. Liu, J.K. Bradley, X. Meng, T. Kaftan, M.J. Franklin, A. Ghodsi, M. Zaharia, Spark SQL: relational data processing in spark, in: Proceedings of the 2015 ACM SIGMOD International Conference on Management of Data (SIGMOD'15), ACM, New York, NY, USA, 2015, pp. 1383–1394.
[7] M. Zaharia, M. Chowdhury, M.J. Franklin, S. Shenker, I. Stoica, Spark: cluster computing with working sets, in: Proceedings of the 2nd USENIX Conference on Hot Topics in Cloud Computing (HotCloud'10), USENIX Association, Berkeley, CA, USA, 2010, p. 10.
[8] M. Zaharia, T. Das, H. Li, S. Shenker, I. Stoica, Discretized streams: an efficient and fault-tolerant model for stream processing on large clusters, in: Proceedings of the 4th USENIX Conference on Hot Topics in Cloud Ccomputing (HotCloud'12), USENIX Association, Berkeley, CA, USA, 2012, p. 10.
[9] J. Fu, J. Sun, K. Wang, SPARK – a big data processing platform for machine learning, in: 2016 International Conference on Industrial Informatics-Computing Technology, Intelligent Technology, Industrial Information Integration (ICIICII), Wuhan, 2016, pp. 48–51.
[10] U. Rau, T.J. Cornwell, A multi-scale multi-frequency deconvolution algorithm for synthesis imaging in radio interferometry, Astron. Astrophys. 532 (2011) A71.
[11] J. Holland, Adaptation in Natural and Artificial Systems, first ed., University of Michigan Press, MI, 1975.
[12] E.S.H. Hou, N. Ansari, H. Ren, A genetic algorithm for multiprocessor scheduling, IEEE Trans. Parallel Distrib. Syst. 5 (2) (1994) 113–120.

14

Applications of artificial intelligence in astronomical big data

Yatong Chen[a], Rui Kong[b], Linghe Kong[b]
[a]Dalian University of Technology, Dalian, China. [b]Shanghai Jiao Tong University, Shanghai, China

1 Introduction

Astronomy has been developing stably in recent years, and the generated data have also been expanding both in diversity and complexity. So it is crucial to process these massive data. We stress two points. First, how do we efficiently and accurately archive these data collected every second? Second, how do we make full use of historical data for new discoveries? These challenges encourage some data-driven methods to cross with astronomy. In this chapter, we can see that machine learning (ML) methods can work well. ML utilizes these huge data in an appropriate way and use the date for preprocessing, feature, data mining, and the interpretation and evaluation of results. We will show some ML applications in some typical astronomy issues.

With the development of computing ability, ML gained great improvements in feature extraction, data compression, and spatial mapping. There are many categories of ML methods, including neural networks, fuzzy approximations, genetic algorithms, etc. Compared with traditional methods, ML has the following advantages: (1) it is a data-driven method, which is suited for a large amount of data in astronomy research; (2) it can automatically extract features from the training data according to the task target with little manual intervention; and (3) it has a good ability for learning transfer. We only need to use relevant data pairs when using a trained model to solve different problems, and then the model can be fine-tuned.

Big Data in Astronomy. https://doi.org/10.1016/B978-0-12-819084-5.00006-7

2 Machine learning for astronomical data calibration and repair

2.1 Clustering analysis algorithm for missing values: KSC

When researchers collect original data, some data are often missing. So, it is necessary to deal with these flaws. Some methods try imputation (estimating these missing values); some methods attempt marginalization (deleting all objects with missing values). They all rely on the assumption that missing values occur by random chance. However, this is a proper assumption in other areas. The truth is that values missed in an air astronomy catalog may be of great importance. Wagstaff and Laidler [1] presents an analysis of objects taken from the sloan digital sky survey (SDSS) to demonstrate how imputing the values can be misleading and why the KSC approach can produce more appropriate results. They demonstrate a clustering analysis algorithm, KSC, that does not impute values and does not discard the partially observed objects. KSC uses soft constraints defined by the fully observed objects to assist in the grouping of objects with missing values. This algorithm works well in this typical astronomy problem and avoids conclusion confusion. We know that some major discoveries occurred by random chance, so researchers need an overview of all collected data, including missing data.

2.2 PCA-based machine learning for classification of SDSS transient survey images

Transient image preprocessing is sometimes critical in astronomy problems. Occasionally, astronomy experiments need fast methods to preprocess data just collected so that researchers can move into the next work. For example, when observing planet collisions, instantaneous analysis is pretty crucial for the next step, or we are unable to get further discoveries. Du Buisson et al. [2] showed that multiple ML algorithms can match human performance in classifying transient imaging data from the SDSS supernova survey into real objects and artifacts. This is a first step in any transient science pipeline and is currently still done by humans, but future surveys such as the large synoptic survey telescope (LSST) will necessitate fully machine-enabled solutions. For example, using features extracted from eigen image analysis (principal component analysis, PCA) of single-epoch *g*, *r*, and *i* difference images, and can reach a completeness (recall) of 96%, while only incorrectly classifying at most 18% of artifacts as real objects, corresponding to a precision (purity) of 84%. In general, the random forest method performed best, followed by the k-nearest neighbor and the SkyNet artificial neural net algorithms, compared to other methods such as naive Bayes and kernel support vector machine (SVM). The results show that PCA-based ML can match human success levels and can naturally be extended by including multiple epochs of data, transient color, and host galaxy information, which should allow for significant further improvements, especially at low signal to noise.

2.3 CCD defect inspection with artificial neural network

When researchers do some experiments, there are usually some equipment failures. Nevertheless, the experiment's data sometimes cannot reflect where the problems are. We have to know how to find the reason. Artificial neural networks (ANNs) could often solve this difficulty. ANN establishes a mapping from inputs to outputs, and conversely it is able to construct a mapping from outputs to inputs, so we can make full use of most data. Rogers and Riess [3] has developed an ANN system that locates and classifies defects in CCDs. This system, based on a feed-forward neural network, was trained with a conjugate gradient training algorithm using observational data from an astronomical telescope. The network was tested with data from four large CCDs (2048×2048 pixels each), and it found defects with a higher efficiency and in a much shorter time than human inspectors. This method of detecting and classifying objects in images is quite general. In an appendix, the authors provide a recipe for neural computing to make this technique more accessible to the astronomical community.

3 Artificial intelligence algorithms in astronomy data classification and preprocessing

3.1 Morphological classification of galaxies

3.1.1 *Supervised learning method*

One advantage of supervised learning is its ability to construct and train a mapping model using lots of data with labels, and then predict data without labels. Thus, handwork is not so much required. In recent years, massive extragalactic galaxies have been seen. Many of them were provided by the Hubble Deep Field Telescope and the Large-Scale Sky Survey. How to identify and classify these galaxies in pictures is a challenge for researchers now, and supervised learning exactly makes it in its way.

Naim et al. [4] trained ANNs to classify galaxies based only on the morphology of the galaxy images that appeared on blue survey plates. The images were reduced, and morphological features such as bulge size and the number of arms were extracted, all in a fully automated way. The galaxy sample was first classified by six independent experts. Several definitions for each type of each galaxy were used, based on these training samples. Then the network was trained and tested on these features. They found that the rms error of the network classifications, as compared with the mean types of the expert classifications, was 1.8 Revised Hubble types. This is comparable to the overall rms dispersion between the experts. This result is robust and almost completely independent of the network architecture used in the experiment.

Strateva et al. [5] used visual morphology and spectral classification of the subsamples of 287 and 500 galaxies, respectively, to show that the two peaks corresponded roughly to early-type (E, S0, and Sa) and late-type (Sb, Sc, and Irr) galaxies, as expected from their different stellar populations. They also found that the colors of galaxies are often correlated with their radial profiles, which was measured by the concentration index and by the likelihood of exponential and de Vaucouleurs profile fits. While it is common that late-type galaxies are bluer than early-type galaxies, this is the first detection of a local minimum in their color distribution.

3.1.2 Unsupervised learning

Unsupervised learning methods do not need so many labels of data, so this reduces some requirements of the collected data. Unsupervised learning relies on its unique training system, and classifies original data with little or no already known label data. So unsupervised learning could find some potential quality that we humans could not such as when researchers observe and get some galaxies that have never been seen before, and they can't archive these galaxies because there is no knowledge of them. Therefore, unsupervised learning shows its power. Huertas-Company et al. [6] presented an automated morphological classification in four types (E, S0, Sab, Scd) similar to 700,000 galaxies from the SDSS DR7 spectroscopic sample based on SVMs.

3.2 Star/galaxy classification and detection

3.2.1 Supervised learning

In classification and detection tasks, supervised learning can utilize data with labels to construct models. And the more data we provide, the more accurate this is. Andreon et al. [7] presented NEXT (Neural Extractor), a new neural network (NN)-based package capable of detecting objects and performing both deblending and starting galaxy classification in an automatic way. Traditionally in astronomical images, objects are first distinguished from the noisy background by searching for sets of connected pixels having brightness above a given threshold. They are then classified as stars or galaxies through diagnostic diagrams having variables chosen according to the astronomer's taste and experience. In the extraction step, assuming that the images are well sampled, NEXT requires only the simplest a priori definition of "what an object is" (i.e., it keeps all structures composed of more than one pixel) and performs the detection via an unsupervised NN, approaching detection as a clustering problem that has been thoroughly studied in the artificial intelligence literature. The first part of the NEXT procedure consists of an optimal compression of the redundant information contained in the pixels via a mapping from pixel intensities to a subspace individualized through PCA. At magnitudes fainter than the completeness limit, stars are usually almost indistinguishable from galaxies. Therefore, the parameters characterizing the two classes do not lie in disconnected subspaces, thus preventing the use of unsupervised methods. The authors therefore adopted a supervised NN (i.e., an NN that first finds the rules to classify objects from examples and then applies them to the whole dataset). In practice, each object is classified depending on its membership of the regions mapping the input feature space in the training set. In order to obtain an objective and reliable classification, instead of using an arbitrarily defined set of features, the authors used an NN, shown in Fig. 14.1 (a three-layer fully connected neural network), to select the most significant features among the large number of measured ones. We then used these selected features to perform the classification task. In order to optimize the performance of the system, we implemented and tested several different NN models. The comparison of the NEXT performance with that of the best detection and classification package known to the authors (SEXTRACTOR) shows that NEXT is at least as effective as the best traditional packages.

Cross-correlating the XMM-Newton 2XMMi-DR3 catalog with the SDSS data release, Zhang et al. [8] obtained one of the largest

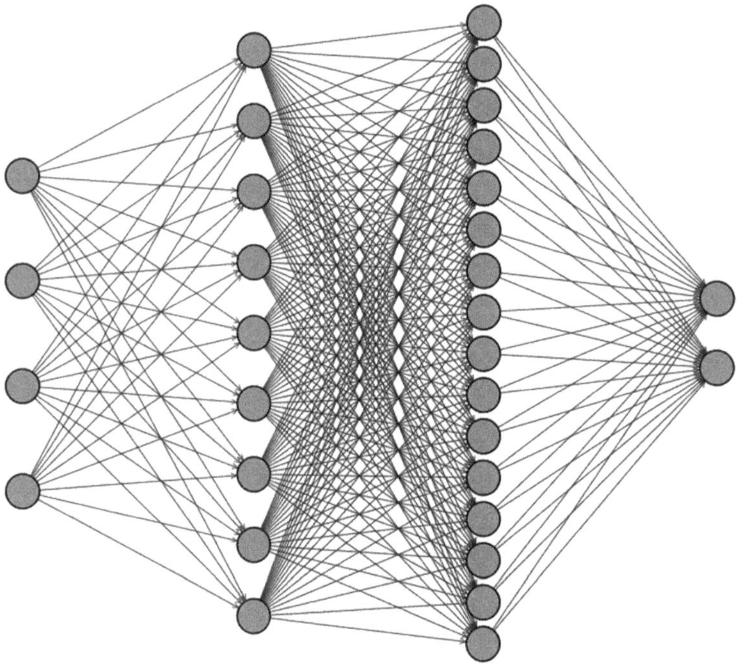

Fig. 14.1 The structure of a hierarchical NN.

X-ray/optical catalogs and explored the distribution of various classes of X-ray emitters in the multidimensional photometric parameter space. Quasars and galaxies occupy different zones while stars scatter in them. However, X-ray active stars have certain distributing rules according to spectral types. The earlier the type of star, the stronger its X-ray emitting. X-ray active stars have a similar distribution to most stars in the g–r versus r–i diagram. Based on the identified samples with SDSS spectral classification, a random forest algorithm for automatic classification was performed. The result showed that the classification accuracy of quasars and galaxies adds up to more than 93% while that of X-ray emitting stars only amounts to 45.3%. In other words, it is easy to separate quasars and galaxies, but it is difficult to discriminate X-ray active stars from quasars and galaxies. If we want to improve the accuracy of automatic classification, it is necessary to increase the number of X-ray emitting stars because the majority of X-ray emitting sources are quasars and galaxies.

Bertin et al. [9] presented the automated techniques we have developed for new software that optimally detects, measures, and classifies sources from astronomical images: SExtractor (Source Extractor), shown in Fig. 14.2. They show that a very reliable star/galaxy separation can be achieved on most images using

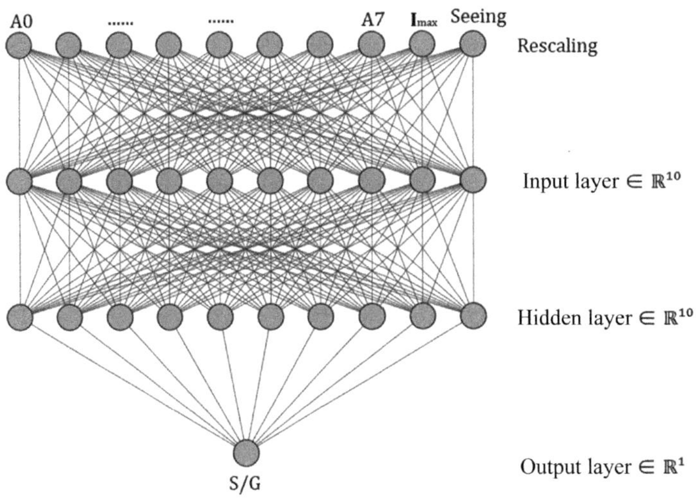

Fig. 14.2 Diagram of the neural star/galaxy classifier.

a neural network trained with simulated images. The salient features of SExtractor include its ability to work on very large images with minimal human intervention, and to deal with a wide variety of object shapes and magnitudes. It is therefore particularly suited to the analysis of large extragalactic surveys.

Gao et al. [10] introduced a decision tree method called random forest for multiwavelength data classification. The data was adopted from different databases, including SDSS data release five, USNO, FIRST, and ROSAT. They then studied the discrimination of quasars from stars and the classification of quasars, stars, and galaxies with the sample from optical and radio bands and with that from optical and X-ray bands. Moreover, feature selection and feature weighting based on random forest were investigated. The performances based on different input patterns were compared. The experimental results showed that the random forest method is an effective method for astronomical object classification and can be applied to other classification problems faced in astronomy. In addition, random forest will show its superiority due to its own merits such as classification, feature selection, feature weighting, and outlier detection.

Peng et al. [11] developed and demonstrated a classification system that is made up of several SVM classifiers, which can be applied to select quasar candidates from large sky survey projects such as SDSS, the UK Infrared Telescope Infrared Deep Sky Survey (UKIDSS), and the Galaxy Evolution Explorer (GALEX). Here, we present in detail a method for constructing this SVM classification system. When the SVM classification system works on the test set to predict quasar candidates, it acquires an efficiency of 93.21% and a completeness of 97.49%. In order to further prove the reliability and feasibility of this system, two chunks were randomly chosen to compare its performance with that of the XDQSO method used for the SDSS-III's Baryon Oscillation Spectroscopic Survey (BOSS). The experimental results showed that there is a distinct overlap between the quasar candidates selected by this system and those extracted by the XDQSO technique in the dereddened i-band magnitude range between 17.75 and 22.45, especially in the interval of dereddened i-band magnitude <20.0. In the two test areas, 57.38% and 87.15% of the quasar candidates predicted by the system are also targeted by the XDQSO method. Similarly, the prediction of subcategories of quasars according to redshift achieves a high level of overlap with these two approaches. Depending on the effectiveness of this system, the SVM classification system can be used to create an input catalog of quasars. In order to obtain a higher confidence of quasar candidates, the cross-results from the candidates selected by this SVM system and by the XDQSO method can be used.

Richards et al. [12] explored the multidimensional, multiwavelength selection of quasars from midinfrared (MIR) plus optical data, specifically from the Spitzer-Infrared Array Camera (IRAC) and the SDSS. Traditionally, quasar selection relies on cuts in two-dimensional color space, even though most modern surveys (optical and IR) are done in more than three bandpasses. In this paper, we apply modern statistical techniques to combined Spitzer MIR and SDSS optical data, allowing up to eight-dimensional (8D) color selection of quasars. Using a Bayesian selection method, we cataloged 5546 quasar candidates to an 8.0 μm depth of 56 μJy over an area similar to 24 deg(2). Roughly 70% of these candidates are not identified by applying the same Bayesian algorithm to four-color SDSS optical data alone. The 8D optical+MIR selection on this dataset recovers 97.7% of known type 1 quasars in this area and greatly improves the effectiveness of identifying $3.5<z<5$ quasars, which are challenging to identify (without considerable contamination) using MIR data alone. We demonstrate that even using only the two shortest wavelength IRAC bandpasses (3.6 and 4.5 μm), it is possible to use our Bayesian techniques to select quasars with 97% completeness and as little as 10% contamination (as compared to a similar 60% contamination using color cuts alone). We compute photometric redshifts for our sample; comparison with known objects suggests a photometric redshift accuracy of 93.6% (Delta $z \pm 0.3$), remaining roughly constant when the two reddest MIR bands are excluded. Even though our methods are designed to find type 1 (unobscured) quasars, as many as 1200 of the objects are type 2 (obscured) quasar candidates. Coupling deep optical imaging data with deep MIR data could enable the selection of quasars in significant numbers past the peak of the quasar luminosity function (QLF) to at least z similar to 4. Such a sample would constrain the shape of the QLF both above and below the break luminosity (L(Q)*) and enable quasar clustering studies over the largest range of redshift and luminosity to date, yielding significant gains in our understanding of the physics of quasars and their contribution to galaxy evolution.

3.2.2 Unsupervised learning

Mähönen and Frantti [13] introduced a simple method based on fuzzy set reasoning. The analysis presented here concentrates on separating point sources (stars) from extended ones. The tests show that the neural network approach is superior if compared to direct fuzzy classification. The paper shows that the inherent ability of neural networks to process complex nonlinear data justifies their use in astronomical classification. However, a combined fuzzy and neural network approach can be useful at least in special cases.

Miller and Coe [14] presented a classifier based on the Kohonen self-organizing map (SOM). The SOM was trained to classify by type using only a small number of manually classified objects. Galaxy classification accuracies of 98% were achieved for objects with a limiting magnitude of 20.

Mahonen and Hakala [15] reported progress in the development of an automatic star/galaxy classifier for processing images generated by large galaxy surveys such as APM. Their classification method is based on neural networks using the Kohonen SOM approach. Their method is novel because it does not need supervised learning, that is, the human factor, in training. The analysis presented here concentrates on separating point sources (stars) from extended ones. Using simple numerical experiments, we compared our method of image classification to the more traditional (PSF-fitting) approach of DAOFIND.

With the advent of surveys generating multiepoch photometry and the discovery of large numbers of variable stars, the classification of these stars has to be automatic. Eyer and Blake [16] developed such a classification procedure for about 1700 stars from the variable star catalog of the All-Sky Automated Survey 1–2 (ASAS 1–2) by selecting the periodic stars and by applying an unsupervised Bayesian classifier using parameters obtained through a Fourier decomposition of the light curve. For irregular light curves, we used the period and moments of the magnitude distribution for the classification. In the case of ASAS 1–2, 83% of variable objects are red giants. A general relation between the period and amplitude is found for a large fraction of those stars. The selection led to 302 periodic and 1429 semiperiodic stars, which are classified in six major groups: eclipsing binaries, sinusoidal curves, Cepheids, small amplitude red variables, SR, and Mira stars. The type classification error level is estimated to be about 7%.

Gonzalez et al. [17] presented a method for automatic detection that includes a novel data-augmentation procedure to make trained models more robust against the data taken from different instruments and contrast-stretching functions. Results are shown in Fig. 14.3, where they get for T1 a maximum recall ratio and IOU of 85.7% and 65.8%, respectively. In the case of T2, maximum recall ratio and IOU rise to 90.23% and 70.3%, respectively. T2 returns better results because higher contrast and brightness images built from FITS images give more information to training, in particular for disk galaxies where the disk can be seen more clearly with a higher contrast and brightness. However, excessive brightness may add too much noise to the background and excessive contrast tends to erase the central bulge profile.

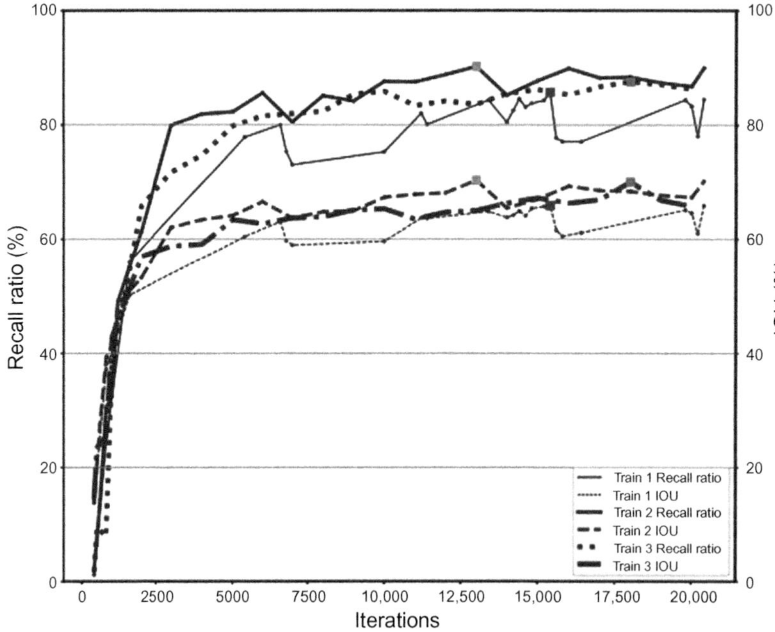

Fig. 14.3 Convergence of T1, T2, and T3 training sets. Dark gray, light gray, and gray squares indicate optimal recall ratio and IOU for the three trainings. From R.E. Gonzalez, R.P. Muñoz, C.A. Hernández, Galaxy detection and identification using deep learning and data augmentation, Astron. Comput. 25 (2018) 103–109, Fig. 2.

3.3 Spectral analysis

3.3.1 Artificial neural network

An ANN is a tool for fitting functions. When we know the existence of a certain model but it is not easy to directly solve it, we can try to artificially approximate the target model with an ANN. In fact, the universal approximation theory of neural networks has been proven correct, that is, we can use it to fit various mapping functions. The most commonly used neural network is a fully connected layer neural network: (1) it has only a forward propagation path; (2) it uses the back-propagation algorithm to update the model and iteratively; (3) it continuously reduces the error rate according to the supervised label, and finally achieves the effect of fitting various data. But many neural networks are prone to overfitting. These results can be improved through normalization and random drop operations. The advantages of ANN are huge because there are many laws in the universe that we can't figure out, and we can only observe the initial and results. The deep learning that appears in the subsequent sections is basically based on the neural network to further strengthen its fitting ability and reduce the phenomenon of overfitting.

Spectral classification is an important research topic in stellar and galaxy astronomy. The usual stellar spectral classification is performed by experts using artificial or semimanual methods to compare the stellar spectrum of an unknown spectrum with a standard stellar spectrum to obtain the stellar spectrum. This method is inefficient and subject to subjective factors, and it is difficult to form a unified standard. With the explosion of astronomical big data, there is an urgent need for an efficient and accurate automatic spectrum processing system. As an important artificial intelligence tool, the neural network has been widely used in this respect.

Vieira et al. [18] explored two automated classification methods: supervised classification using ANN and unsupervised classification using SOM. These methods are used to classify the IUE low-dispersion spectra of normal stars with spectral types ranging from O3 to G5.

Bailer-Jones et al. [19] investigated the application of neural networks to the automation of MK spectral classification. The dataset for this project consists of a set of more than 5000 optical (3800–5200 Å) spectra obtained from objective prism plates from the Michigan Spectral Survey. These spectra, along with their two-dimensional MK classifications listed in the Michigan Henry Draper Catalog, were used to develop supervised neural network classifiers. They show that neural networks can give accurate spectral type classifications across the full range of spectral types present in the data set (B2-M7). They show also that the networks yield correct luminosity classes for more than 95% of both dwarfs and giants with a high degree of confidence. Stellar spectra generally contain a large amount of redundant information. They investigate the application of PCA to the optimal compression of spectra. They show that PCA can compress the spectra by a factor of more than 30 while retaining essentially all the useful information in the dataset. Furthermore, it is shown that this compression optimally removes noise and can be used to identify unusual spectra.

Brett et al. [20] applied the SOM technique to the automated classification of singly periodic astronomical light curves. The algorithm they developed utilizes a form of neural network, namely the Kohonen SOM (1990). We are motivated in this choice by two desirable properties of Kohonen maps: the simplicity of implementation and their ability to learn in an unsupervised regime. They find that our maps readily distinguish between light curve types in both synthetic and real datasets, and that the resulting maps do not depend sensitively on the chosen learning parameters. Automated data analysis techniques are likely to become increasingly important as the size of astronomical datasets continues to increase, particularly with the advent of ultrawide-field survey telescopes such as WASP, RAPTOR, and ASAS.

3.3.2 Deep learning

Deep learning is the most popular model in recent years, and it has also become the most commonly used tool in industry. Mature frameworks such as TsensorFlow and PyTorch already support most of the deep learning algorithm models. Its popularity also benefits from the rapid increase in GPU computing power. The GPU produced by Nvidia has an advanced CUDA architecture, which can accelerate the calculation process of convolutional neural networks (CNNs). Lenet, Alexnet, GoogLenet, VGGnet Resnet, etc., are the most famous deep neural network models in recent years. They have shown great power in image classification, object detection, segmentation, and NLP as well as the hot AlphaGo, etc. It is more accurate in model prediction than traditional neural network models, and it is tens or even hundreds of times deeper in model size than neural network models.

In the paper, the authors [21] make the first attempt to employ the deep learning method for representation learning of the solar radio spectra. The original solar radio spectra are preprocessed, including normalization, enhancement, etc., to generate new images for the next processing. With the expertise of solar radio astronomy for identifying solar radio activity, the authors built a solar radio activity database that also contains solar radio spectra. The employed deep learning network is first pretrained based on the available massive amounts of unlabeled radio solar images. The model is shown in Fig. 14.4. Afterward, the weights of the network are further fined-tuned based on the labeled data. In the experiment, the pretraining process can help improve the classification accuracy.

Chen et al. [22] proposes the first attempt to utilize the multimodal learning method for the representation learning of the solar radio spectra. The solar radio signals sensed from different frequency channels, which present different characteristics, are regarded as different modalities. The authors employed the multimodal neural network shown in Fig. 14.5 to learn the representations of the solar radio spectrum, which can distinguish the differences and learn the interactions between different modalities. The original solar radio spectra are first preprocessed, including normalization, denoising, channel competition, etc., before being fed into the multimodal learning network. Experimental results have demonstrated that the proposed multimodal learning

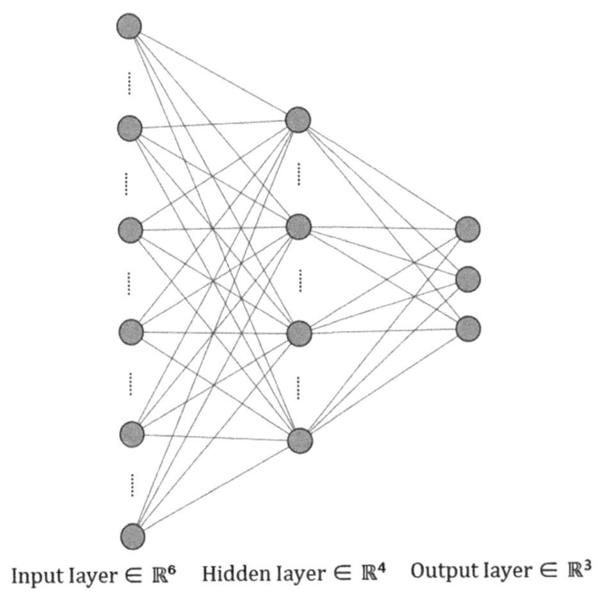

Fig. 14.4 Neural network for solar radio image classification.

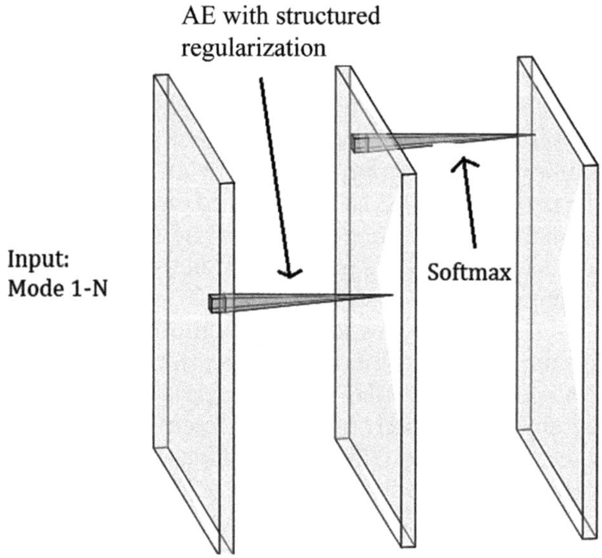

Fig. 14.5 The framework of multimodal learning.

network can learn the representation of the solar radio spectrum more effectively and improve the classification accuracy.

Chen et al. [23] made the first attempt to utilize a CNN for the classification of solar radio spectra. The solar radio spectrum is a two-dimensional gray-scale image with one dimension of frequency and the other of time. Taking advantage of CNN, we can efficiently learn the distinct characteristics of different types of spectra, and further classify them even more accurately. The proposed CNN-based network consists of four convolution layers, four pooling layers, and one fully connected layer. Its input is spectra of the size 120×120. The output gives the type of each spectrum among "burst," "nonburst," and "calibration." Experimental results demonstrate that the proposed CNN can achieve more classification accuracy of the solar radio spectrum beyond our previous efforts by employing a deep belief network (DBN) and an autoencoder (AE).

Yu et al. [24] made the first attempt to utilize long short-term memory (LSTM) to classify the solar radio spectrum. The vertical and horizontal dimensions of a spectrum correspond to frequency channel and time, respectively. Intrinsically, time dependence exists between columns of a spectrum, which indicates the slowly varying process of solar radio radiation. Thus, a spectrum can be treated as a time sequence instead of a general image - losing its time information. Inspired by the big success of LSTM for time sequence processing such as natural language recognition, it is reasonably believed that treating a spectrum as a time sequence would benefit its classification via LSTM. Thus, LSTM is employed to explore the sequential relations and interactions within a spectrum being treated as a time sequence. As such, LSTM learns the sequential properties and generates the representation of the solar radio spectrum for classification. The experimental results demonstrate that LSTM can well capture the characteristics of the solar radio spectrum, and thus achieve better classification accuracy.

3.4 Cosmic ray classification

3.4.1 Forward neural networks

Classification tasks are one of the specialties of neural network models. In cosmic ray classification tasks, feed-forward neural networks rely on their powerful prediction capabilities to predict unlabeled new data after they have some labeled datasets. The success of current and future air shower arrays in detecting point sources of cosmic rays above 10 TeV depends crucially on the possibility of finding efficient methods for separating gamma-induced air showers from the overwhelming background of hadron-induced showers. Westerhoff et al. [25] studied the application of computer-simulated feed-forward neural networks in the analysis of cosmic ray data taken with the Geiger towers of the HEGRA air shower array. The combination of these charged particle detectors with the neural net-based analysis was characterized by high background rejection and signal efficiency. In contrast to the often-quoted nontransparency of the net technique, the detailed analysis of the net performance gives insight into the physics involved and helps to assess the different information that allows gamma/hadron separation.

A large volume of low signal-to-noise multidimensional data is available from the CCD imaging spectrometers aboard the Chandra X-ray Observatory and the X-ray Multimirror Mission (XMM-Newton). To make progress analyzing this data, it is essential to develop methods to sort, classify, and characterize the vast library of X-ray spectra in a nonparametric fashion (complementary to current parametric model fits). Hojnacki et al. [26] developed a spectral classification algorithm that handles large volumes of data and operates independently of the requirement of spectral model fits. We use proven multivariate statistical techniques, including PCA and an ensemble classifier consisting of agglomerative hierarchical clustering and *K*-means clustering applied for the first time for spectral classification. The algorithm positions the sources in a multidimensional spectral sequence and then groups the ordered sources into clusters based on their spectra. These clusters appear more distinct for sources with harder observed spectra. The apparent diversity of source spectra is reduced to a three-dimensional locus in principal component space, with spectral outliers falling outside this locus. The algorithm was applied to a sample of 444 strong sources selected from the 1616 X-ray emitting sources detected in deep Chandra imaging spectroscopy of the Orion nebula cluster. Classes form sequences in N-H,A(V) and accretion activity indicators, demonstrating that the algorithm efficiently sorts the X-ray sources into a physically meaningful sequence. The algorithm also isolates important classes of very deeply embedded, active young stellar objects, and yields trends between X-ray spectral parameters and stellar parameters for the lowest mass, pre main sequence stars.

Baccigalupi et al. [27] implemented an independent component analysis (ICA) algorithm to separate signals of different origin in sky maps at several frequencies. Owing to its self-organizing capability, it works without prior assumptions on either the frequency dependence or the angular power spectrum of the various signal. Instead, it learns directly from the input data how to identify the statistically independent components on the assumption that all but, at most, one of the components have non-Gaussian distributions. We have applied the ICA algorithm to simulated patches of the sky at the four frequencies (30, 44, 70, and 100 GHz) used by the Low Frequency Instrument of the European Space Agency's Planck satellite. Simulations include the cosmic microwave background (CMB), synchrotron and thermal dust emissions, and extragalactic radio sources. The effects of the angular response functions of the detectors and of instrumental noise have been ignored in this first exploratory study. The ICA algorithm reconstructs the spatial distribution of each component with rms errors of about 1% for the CMB and 10% for the much weaker galactic components. Radio sources are almost completely recovered down to a flux limit corresponding to similar or equal to 0.7 sigma (CMB), where sigma (CMB) is the rms level of the CMB fluctuations. The signal recovered has equal quality on all scales larger than the pixel size. In addition, we show that for the strongest components (CMB and radio sources), the frequency scaling is recovered with percent precision. Thus, algorithms of the type presented here appear to be very promising tools for component separation. On the other hand, we have been dealing here with a highly idealized situation. Work to include instrumental noise, the effect of different resolving powers at different frequencies, and a more complete and realistic characterization of astrophysical foregrounds is in progress.

3.4.2 Clustering algorithm

The clustering algorithm belongs to unsupervised learning. As its name suggests, it does not need a model with artificially specified features. It only needs to give the number of classification results you want to train. The ray classification task is very similar to the galaxy classification task, and both use images to perform prediction classification.

Balastegui et al. [28] applied two different automatic classifier algorithms to the BATSE Current GRB Catalog data and we obtained three different classes of gamma ray bursts (GRBs). Our results confirm the existence of a third intermediate class of GRBs with mean duration similar to 25–50 s, as deduced from a cluster analysis and from a neural network algorithm. Our analyses imply longer durations than those found by Mukherjee et al. and Horvath, whose intermediate class had durations similar to 2–10 s. From the neural network analysis, no difference in hardness between the two longest classes is found, and from both methods we find that the intermediate-duration class constitutes the most homogeneous sample of GRBs in its space distribution while the longest-duration class constitutes the most inhomogeneous one with [V/V-max] similar to 0.1, being thus the deepest population of GRBs with z(max) similar to 10. The trend previously found in long bursts, of spatial inhomogeneity increasing with hardness, only holds for this new longest-duration class.

4 Artificial intelligence application in astronomy data analysis

4.1 Photometric redshift

4.1.1 Multilayer perceptron and artificial neural network

The redshift of the galaxy, the rate of retreat of the galaxy leaving the observer, is extremely important in the theory of observing cosmology. The redshift can be measured by both spectral data and photometric data. Although the measured redshift is accurate but time consuming, the photometric data is not accurate enough and has systematic errors, but it is easier to apply to large samples.

Therefore, for large samples, multiband photometric data from the same instrument or system can be used, and spectral redshift can be used as a subnet to evaluate the photometric redshift using a supervised neural network.

The multilayer perceptron (MLP) ANN is used to predict photometric redshifts in the HDF-S from an ultradeep multicolor catalog. The MLP can be trained on observed data, theoretical data, and mixed samples. Vanzella et al. [29] presented a technique for the estimation of photometric redshifts based on feed-forward neural networks. The prediction of the method is tested on the spectroscopic sample in the HDF-S (44 galaxies). Over the entire redshift range, $0.1 < z < 3.5$, the agreement between the photometric and spectroscopic redshifts in the HDF-S is good. In the evaluation phase, the computational speed is that 10(5) galaxies are estimated in a few seconds. This technique can be used in large databases.

Collister and Lahav [30] introduced ANNz to learn the relation between photometry and redshift from an appropriate training set of galaxies for which the redshift is already known. Where a large and representative training set is available, ANNz is a highly competitive tool when compared with traditional template-fitting methods. Nonideal conditions (spectroscopic sets that are small or brighter than the photometric set for which redshifts are required) are simulated, and the impact on the photometric redshift accuracy is assessed. ANNz is able to extrapolate with some success to around a magnitude fainter than is sampled by the training data.

4.1.2 Bayes algorithm

The Bayesian algorithm is part of the probability model. The results given by this algorithm are not determined classification targets but the probabilities of some classification targets, which can be processed artificially afterward.

Mortlock et al. [31] used the Bayesian method to report observations of a quasar (ULASJ112001.48+064124.3) at a redshift of 7.085, which is 0.77 billion years after the Big Bang. ULASJ1120+0641 has a luminosity of $6.3 \times 10(13)$L and hosts a black hole with a mass of $2 \times 10(9)$M (where L and M are the luminosity and mass of the Sun). The measured radius of the ionized near zone around ULASJ1120+0641 is 1.9 megaparsecs, a factor of three smaller than is typical for quasars at redshifts between 6.0 and 6.4. The near-zone transmission profile is consistent with a Ly alpha d damping wing (5), suggesting that the neutral fraction of the intergalactic medium in front of ULASJ1120+0641 exceeded 0.1.

4.1.3 Convolutional neural network

A CNN is more like a convolution operation in signal processing. It performs spatial mapping on existing data and processes these data in a new space. This makes it easier to observe new features.

Pasquet-Itam and Pasquet [32] applied a CNN to classify and detect quasars in the SDSS Stripe 82 and also to predict the photometric redshifts of quasars. The network takes the variability of objects into account by converting light curves into images. The CNN provides good results because its precision is higher than the same recall with a random forest classifier. For the redshift predictions, the CNN presents excellent results that are higher than those obtained with a feature extraction step and different classifiers (a *K*-nearest-neighbors, a SVM, a random forest, and a Gaussian process classifier). And the accuracy of the CNN within the Delta z vertical bar performed better than the KNN. So, the CNN successfully reduces the dispersion and the catastrophic redshifts of quasars. This new method is very promising for the future of big databases such as the LSST.

Fig. 14.6 shows blue and red histograms representing the distribution of absolute error for the estimation of photometric redshifts by using a KNN and a CNN classifier, respectively.

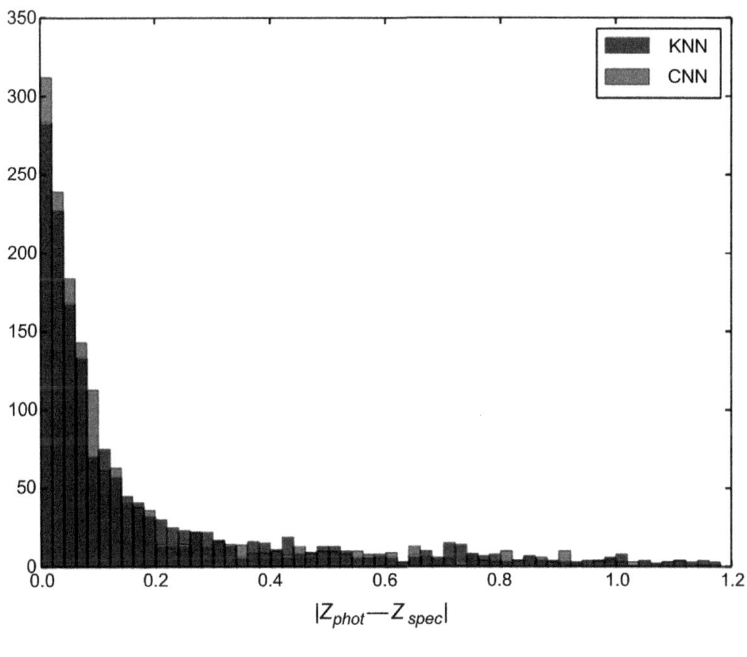

Fig. 14.6 Compare KNN with CNN. From Pasquet-Itam J, Pasquet J. Deep Learning approach for classifying, detecting and predicting photometric redshifts of quasars in the Sloan Digital Sky Survey stripe 82 [J]. Astronomy and Astrophysics, 2018, 611. Reproduced with permission © ESO.

The number of catastrophic redshifts is reduced with the CNN as the percentage of redshift estimation with an absolute error higher than 0.1 is about 45.78% for KNN and 38.03% for CNN.

4.2 Flare detection

4.2.1 Artificial neural network

Borda et al. [33] presented a new method for the automatic detection of flare events from images in the optical range. The method uses neural networks for pattern recognition and is conceived to be applied to full-disk H alpha images. The images are analyzed in real time, which allows for the design of automatic patrol processes able to detect and record flare events with the best time resolution available without human assistance. They use a neural network consisting of two layers, a hidden layer of nonlinear neurodes and an output layer of one linear neurode. The network was trained using a back-propagation algorithm and a set of full-disk solar images obtained by HASTA (H Alpha Solar Telescope for Argentina). This method is appropriate for the detection of solar flares in the complete optical classification, being portable to any H alpha instrument and providing unique criteria for flare detection independent of the observer.

4.2.2 Support vector machine

A SVM is one of the traditional ML methods. It also uses space mapping to map the original data to the new space, and performs easy and effective processing in the new space. It can perform various tasks such as classification and detection.

Nishizuka et al. [34] developed a flare prediction model using ML, which is optimized to predict the maximum class of flares occurring in the following 24 h. ML is used to devise algorithms that can learn from and make decisions on a huge amount of data. They used solar observation data during the period 2010–15, such as vector magnetograms, ultraviolet (UV) emissions, and soft X-ray emissions taken by the Solar Dynamics Observatory and the Geostationary Operational Environmental Satellite. We detected active regions (ARs) from the full-disk magnetogram, from which similar to 60 features were extracted with their time differentials, including magnetic neutral lines, current helicity, UV brightening, and flare history. After standardizing the feature database, they fully shuffled and randomly separated it into two for training and testing. To investigate which algorithm is best for flare prediction, they compared three machine-learning algorithms: the SVM, the k-nearest neighbors (k-NN), and the extremely randomized trees.

The prediction score, the true skill statistic, was higher than 0.9 with a fully shuffled dataset, which is higher than that for human forecasts. It was found that k-NN had the highest performance among the three algorithms. The ranking of the feature importance showed that previous flare activity is most effective, followed by the length of magnetic neutral lines, the unsigned magnetic flux, the area of UV brightening, and the time differentials of features over 24 h, all of which are strongly correlated with the flux emergence dynamics in an AR.

4.2.3 Deep learning

Solar flares originate from the release of the energy stored in the magnetic field of solar active regions. The triggering mechanism for these flares, however, remains unknown. For this reason, the conventional solar flare forecast is essentially based on the statistic relationship between solar flares and measures extracted from observational data. In the current work, the deep learning method is applied to set up the solar flare forecasting model, in which forecasting patterns can be learned from line-of-sight magnetograms of solar active regions.

The paper [35] uses the CNN network structure to perform solar flare prediction, which means we process the input magnetic map through the CNN and predict whether flares will occur. Therefore, the prediction of solar flares is transformed into a two-category problem. The testing results of the forecasting model indicate that the forecasting patterns can be automatically reached with the MDI data and they can also be applied to the HMI data; furthermore, these forecasting patterns are robust to the noise in the observational data. The performance of the deep learning forecasting model is not sensitive to the given forecasting periods (6, 12, 24, or 48 h). The performance of the proposed forecasting model is comparable to that of the state-of-the-art flare forecasting models, even if the duration of the total magnetograms continuously spans 19.5 years. Case analyses demonstrate that the deep learning-based solar flare forecasting model pays attention to areas with the magnetic polarity-inversion line or the strong magnetic field in magnetograms of active regions.

4.3 Galaxy parameter analysis

4.3.1 Machine learning algorithms

Extended statistical and cross-correlational analyses of the data time series of the parameters of the solar activity (the Wolf numbers, the radio-emission flux F10.7, and the global magnetic field of the Sun), the solar wind (velocity, density, and temperature), and the interplanetary magnetic field are performed. All the parameters involved have distributions that are close to the lognormal distributions. Integro-cross-correlational analysis with a floating averaging period showed that the maximum pair correlation is observed for all parameters for the averaging period of the order of one year. The attempt of complex simulation and forecast of annual smoothed values of the parameters with the use of ANNs led Veselovskii et al. [36] to conclude that a rapid degeneration of the forecast occurs at a time scale of the order of 1 year. This is due to the high degree of chaotization of the distributions of statistical parameters, which are characterized by rather high informational entropy, lying usually in the range of 0.8–0.9.

Fuentes [37] showed how ML methods can be effectively applied to the problem of automatically predicting stellar atmospheric parameters from spectral information, a very important problem in stellar astronomy. They applied feed-forward neural networks, Kohonen's SOM, and locally weighted regression to predict the stellar atmospheric parameters effective temperature, surface gravity, and metallicity from spectral indices. Our experimental results show that the three methods are capable of predicting the parameters with very good accuracy, as seen in Fig. 14.7. Locally weighted regression gives slightly better results than the other methods using the original dataset as input while SOM outperform the other methods when significant amounts of noise are added. They also implemented a heterogeneous ensemble of

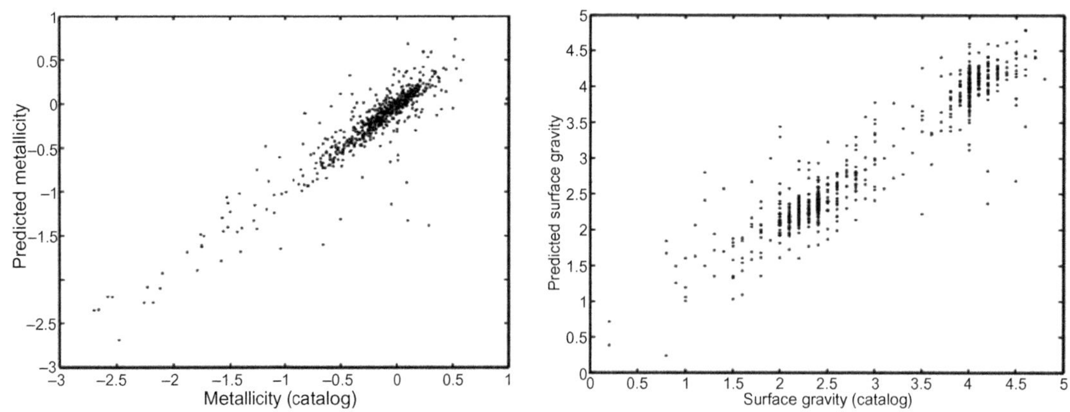

Fig. 14.7 Catalog versus predicted parameters using LWR with the original dataset. From O. Fuentes, Automatic determination of stellar atmospheric parameters using neural networks and instance-based machine learning, Exp. Astron. 12 (1) (2001) 21–31, Fig. 1.

predictors, combining the results given by the three algorithms. This ensemble yields better results than any of the three algorithms alone, using both the original and the noisy data.

Ntampaka et al. [38] presented a modern ML approach for cluster dynamical mass measurements that is a factor-of-two improvement over using a conventional scaling relation. Different methods are tested against a mock cluster catalog constructed using halos with mass $\geq 10^{14}$ M h^{-1} from Multidark's publicly available N-body MDPL halo catalog. In the conventional method, we use a standard M(sigma(nu)) power-law scaling relation to infer cluster mass, M, from line of sight (LOS) galaxy velocity dispersion, sigma(nu). The resulting fractional mass error distribution is broad, with width delta epsilon approximate to 0.87 (68% scatter), and has extended high-error tails. The standard scaling relation can be simply enhanced by including higher-order moments of the LOS velocity distribution. Applying the kurtosis as a correction term to log (sigma(nu)) reduces the width of the error distribution to delta epsilon approximate to 0.74 (16% improvement). ML can be used to take full advantage of all the information in the velocity distribution. We employ the support distribution machines (SDMs) algorithm that learns from distributions of data to predict single values. SDMs trained and tested on the distribution of LOS velocities yield delta epsilon approximate to 0.46 (47% improvement). Furthermore, the problematic tails of the mass error distribution are effectively eliminated. Decreasing cluster mass errors will improve the measurements of the growth of structure and lead to tighter constraints on cosmological parameters.

Miller et al. [39] reported the development of a ML framework capable of inferring fundamental stellar parameters (T-eff, log g, and [Fe/H]) using photometric-brightness variations and color alone. A training set is constructed from a systematic spectroscopic survey of variables with the Hectospec/Multimirror Telescope. In sum, the training set includes similar to 9000 spectra, for which stellar parameters are measured using the SEGUE Stellar Parameters Pipeline (SSPP). We employed the random forest algorithm to perform a nonparametric regression that predicts Teff, log g, and [Fe/H] from photometric time-domain observations. Our final optimized model produces a cross-validated rms error (RMSE) of 165 K, 0.39 dex, and 0.33 dex for T-eff, log g, and [Fe/H], respectively. Examining the subset of sources for which the SSPP measurements are most reliable, the RMSE reduces to 125 K, 0.37 dex, and 0.27 dex, respectively, comparable to what is achievable via low-resolution spectroscopy. For variable stars, this represents an approximately 12%–20% improvement in RMSE relative to models trained with single-epoch photometric colors. As an application of our method, we estimated stellar parameters for similar to 54,000 known variables. We argue that this method may convert photometric time-domain surveys into pseudo-spectrographic engines, enabling the construction of extremely detailed maps of the Milky Way, its structure, and history.

4.3.2 Deep learning algorithms

The ionosphere is an important space environment near the Earth. Its disturbance would result in severe propagation effects to radio information systems, thus causing bad influences on communication, navigation, radar, and so on. The total electron content (TEC) is an important parameter to represent the disturbance of the ionosphere, so the TEC forecast is very meaningful in scientific research. Sun et al. [40] proposed a LSTM-based model to predict the ionospheric vertical TEC of Beijing. The input of our model is a time sequence consisting of the vector of daily TECs and other closely related parameters. The output is TECs of the future 24 h. The result shows the root of mean square (RMS) error of the test data can reach 3.50 and the RMS error is less than this number during the period of low solar activity. Compared to a MLP network, LSTM is more promising and reliable to forecast ionospheric TEC.

Large-scale deep survey missions such as GAIA will collect enormous amounts of data on a significant fraction of the stellar content of our galaxy. These missions will require a careful optimization of their observational systems in order to maximize their scientific return, and will require reliable and automated techniques for parametrizing the very large number of stars detected. To address these two problems, Bailerjones [41] investigated the precision to which the three principal stellar parameters (T-eff, log g, [M/H]) can be determined as a function of spectral resolution and signal-to-noise ratio (SNR), using a large grid of synthetic spectra. The parametrization technique is a neural network that is shown to provide an accurate three-dimensional physical parametrization of stellar spectra across a wide range of parameters. It is found that even at low resolution (50–100 Å FWHM) and SNR (5–10 per resolution element), T-eff and [M/H] can be determined to 1% and 0.2 dex. Respectively. across a large range of temperatures (4000–30,000 K) and metallicities (−3.0 to +1.0 dex), and that log g is measurable to ±0.2 dex for stars earlier than solar. The accuracy of the results is probably limited by the finite parameter sampling of the data grid. The ability of medium band filter systems (with 10–15 filters) for determining stellar parameters is also investigated. Although easier to implement in an unpointed survey, it is found that they are only competitive at higher SNRs (greater than or equal to 50).

4.4 Periodicity analysis

4.4.1 Artificial neural network

Periodicity analysis of unevenly collected data is a relevant issue in several scientific fields. In astrophysics, we have to find the fundamental period of light or radial velocity curves that are unevenly sampled observations of stars. Classical spectral analysis methods are unsatisfactory to solve the problem. In this paper, Tagliaferri et al. [42] presented a neural network-based estimator system that performs the frequency extraction well in unevenly sampled signals. Pt uses an unsupervised Hebbian nonlinear neural algorithm to extract, from the interpolated signal, the principal components that, in turn, are used by the MUSIC frequency estimator algorithm to extract the frequencies. The neural network is tolerant to noise and works well with few points in the sequence. We benchmark the system on synthetic and real signals with the Periodogram and with the Cramer-Rao lower bound.

4.4.2 Clustering algorithm

Perlman and Java [43] discussed the development of a Java toolbox for astronomical time series data. Rather than using methods conventional in astronomy (e.g., power spectrum and cross-correlation analysis), we employed rule discovery techniques commonly used in analyzing stock market data. By clustering patterns found within the data, rule discovery allows one to build predictive models, allowing one to forecast when a given event might occur or whether the occurrence of one event will trigger a second. We have tested the toolbox and accompanying display tool on datasets (representing several classes of objects) from the RXTE All Sky Monitor. We use these datasets to illustrate the methods and functionality of the toolbox. We also discuss issues that can come up in data analysis as well as the possible future development of the package.

4.5 Supervised learning for detection of dispersed radio pulses

Devine et al. [44] presented a new, two-stage approach for identifying and classifying dispersed pulse groups (DPGs) in a single-pulse search output. The first stage identified DPGs and extracted features to characterize them using a new peak identification algorithm that tracks sloping tendencies around local maxima in plots of signal-to-noise ratio versus dispersion measure. The second stage used supervised ML to classify DPGs. We created four benchmark datasets: one unbalanced and three balanced versions using three different imbalance treatments. We empirically evaluated 48 classifiers by training and testing binary and multiclass versions of six ML algorithms on each of the four benchmark versions. While each classifier had advantages and disadvantages, all classifiers with imbalance treatments had higher recall values than those with unbalanced data, regardless of the ML algorithm used. Based on the benchmarking results, we selected a subset of classifiers to classify the full, unlabeled dataset of more than 1.5 million DPGs identified in 42,405 observations made by the Green Bank Telescope. Overall, the classifiers using a multiclass ensemble tree learner in combination with two oversampling imbalance treatments were the most efficient; they identified additional known pulsars not in the benchmark dataset and provided six potential discoveries, with significantly less false positives than the other classifiers.

4.6 Unsupervised learning for estimating extinction

Meingast et al. [45] presented a new technique, PNICER, that estimates intrinsic colors and extinction for individual stars using unsupervised ML algorithms. This new method aims to be free from any priors with respect to the column density and intrinsic color distribution. It is applicable to any combination of parameters and works in particular numbers of dimensions. It is not restricted to color space. Extinction toward single sources is determined by fitting Gaussian mixture models along the extinction vector to (extinction-free) control field observations. In this way, it becomes possible to describe the extinction for observed sources with probability densities, rather than a single value. PNICER effectively eliminates known biases found in similar methods and outperforms them in cases of deep observational data where the number of background galaxies is significant, or when a large number of parameters is used to break degeneracies in the intrinsic color distributions. This new method remains computationally competitive, making it possible to correctly de-redden millions of sources within a matter of seconds. With the ever-increasing number of large-scale high-sensitivity imaging surveys, PNICER offers a fast and reliable way to energy calculate extinction for arbitrary parameter combinations without prior information on source characteristics. The PNICER software package also offers access to the well-established NICER technique in a simple unified interface and is capable of building extinction maps, including the NICEST correction for cloud substructure.

5 Conclusion

Since the 20th century, benefits from the development of detectors and space technology and the deepening of research work and astronomical observations have further expanded from visible light and radio bands to various bands of electromagnetic waves including infrared, ultraviolet, X-ray, and gamma rays. Band astronomy provides a powerful means of observation for exploring the physical nature of various celestial and astronomical phenomena. Astronomy has reached a new stage: the astronomical big data period. In recent years, a large quantity of large-scale equipment has been put into use, such as the Hubble Space Telescope, the X-ray Space Observatory, etc. This reflects the vitality and broad development prospects of today's literature. The characteristics of astronomical data also show a diversified trend. Examples are the complexity, massiveness, and high dimensionality that make astronomical data gradually unable to be analyzed using traditional methods. Fortunately, the modeling and computing abilities of artificial intelligence solve this problem. Using the features of various knowledge methods in artificial intelligence (ANNs, deep learning algorithms, etc.) that their need for an enough training sets, these methods can solve the big data problem in astronomy perfectly. In conclusion, the rapid development of astronomical projects now and in the future requires more discipline penetration and cross-disciplinary applications.

References

[1] K.L. Wagstaff, V.G. Laidler, Making the most of missing values: object clustering with partial data in astronomy, in: 14th Astronomical Data Analysis and Systems, 2008.

[2] L. Du Buisson, N. Sivanandam, B.A. Bassett, et al., Machine learning classification of SDSS transient survey images, Mon. Not. R. Astron. Soc. 454 (2) (2015) 2026–2038.

[3] R.D. Rogers, A.G. Riess, Detection and classification of CCD defects with an artificial neural network, Publ. Astron. Soc. Pac. 106 (699) (1994) 532–541.

[4] A. Naim, O. Lahav, L. Sodre, et al., Automated morphological classification of APM galaxies by supervised artificial neural networks, Mon. Not. R. Astron. Soc. 275 (3) (1995) 567–590.

[5] I. Strateva, Z. Ivezic, G.R. Knapp, et al., Color separation of galaxy types in the Sloan digital sky survey imaging data, Astron. J. 122 (4) (2001) 1861–1874.

[6] M. Huertas-Company, J.A.L. Aguerri, M. Bernardi, et al., Revisiting the Hubble sequence in the SDSS DR7 spectroscopic sample: a publicly available Bayesian automated classification, Astron. Astrophys. (2010).

[7] S. Andreon, G. Gargiulo, G. Longo, R. Tagliaferri, N. Capuano, Wide Field Imaging. I. Applications of neural networks to object detection and star/galaxy classification, Mon. Not. R. Astron. Soc. 319 (3) (2000) 700–716.

[8] Y.X. Zhang, X.L. Zhou, Y.H. Zhao, et al., Statistical study of 2XMMi-DR3/SDSS-DR8 cross-correlation sample, Astron. J. 145 (2) (2013) 42.

[9] E. Bertin, S. Arnouts, SExtractor: software for source extraction, Astron. Astrophys. Suppl. Ser. 117 (2) (1999).

[10] D. Gao, Y. Zhang, Y. Zhao, Random forest algorithm for classification of multiwavelength data, Res. Astron. Astrophys. 9 (2) (2009) 220–226.

[11] N. Peng, Y. Zhang, Y. Zhao, et al., Selecting quasar candidates using a support vector machine classification system, Mon. Not. R. Astron. Soc. 425 (4) (2012) 2599–2609.

[12] G.T. Richards, R.P. Deo, M. Lacy, et al., Eight-dimensional mid-infrared/optical Bayesian quasar selection, Astron. J. 137 (4) (2009) 3884–3899.

[13] P. Mähönen, T. Frantti, Fuzzy classifier for star-galaxy separation, Astrophys. J. 541 (1) (2008) 261.

[14] A.S. Miller, M.J. Coe, Star/galaxy classification using Kohonen self-organizing maps, Mon. Not. R. Astron. Soc. 279 (1) (1996) 293–300.

[15] P.H. Mahonen, P.J. Hakala, Automated source classification using a Kohonen network, Astrophys. J. 452 (1) (1995).

[16] L. Eyer, C. Blake, Automated classification of variable stars for all-sky automated survey 1â2 data, Mon. Not. R. Astron. Soc. 358 (1) (2005) 30–38.

[17] R.E. Gonzalez, R.P. Muñoz, C.A. Hernández, Galaxy detection and identification using deep learning and data augmentation, Astron. Comput. 25 (2018) 103–109.

[18] E.F. Vieira, J.D. Ponz, Automated spectral classification using neural networks, Astron. Data Anal. Softw. Syst. 145 (145) (1998) 508.

[19] C.A.L. Bailer-Jones, M. Irwin, T.V. Hippel, Automated classification of stellar spectra. II. Two-dimensional classification with neural networks and principal components analysis, Mon. Not. R. Astron. Soc. 298 (2) (1998) 361–377.

[20] D.R. Brett, R.G. West, P.J. Wheatley, The automated classification of astronomical light curves using Kohonen self-organizing maps, Mon. Not. R. Astron. Soc. 353 (2) (2010) 369–376.

[21] Z. Chen, L. Ma, L. Xu, et al., Imaging and representation learning of solar radio spectrums for classification, Multimed. Tools Appl. 75 (5) (2016) 2859–2875.

[22] Z. Chen, L. Ma, L. Xu, et al., Multimodal learning for classification of solar radio spectrum, in: IEEE International Conference on Systems, IEEE, 2015.

[23] S. Chen, L. Xu, L. Ma, et al., Convolutional neural network for classification of solar radio spectrum, in: 2017 IEEE International Conference on Multimedia & Expo Workshops (ICMEW), IEEE, 2017.

[24] X. Yu, L. Xu, L. Ma, et al., Solar radio spectrum classification with LSTM, in: 2017 IEEE International Conference on Multimedia & Expo Workshops (ICMEW), IEEE, 2017.

[25] S. Westerhoff, B. Wiebelsooth, R.N. Sooth, et al., Separating gamma and hadron induced cosmic ray air showers with feed forward neural networks using the charged particle information, Astropart. Phys. 4 (2) (1997) 119–132.

[26] S.M. Hojnacki, J.H. Kastner, G. Micela, et al., An X-ray spectral classification algorithm with application to young stellar clusters, Astrophys. J. 659 (1) (2008) 585.

[27] C. Baccigalupi, L. Bedini, C. Burigana, et al., Neural networks and the separation of cosmic microwave background and astrophysical signals in sky maps, Mon. Not. R. Astron. Soc. 318 (3) (2000) 769–780.

[28] A. Balastegui, P. Ruiz-Lapuente, R. Canal, Reclassification of gamma-ray bursts, Mon. Not. R. Astron. Soc. 328 (1) (2001) 283–290.

[29] E. Vanzella, S. Cristiani, A. Fontana, et al., Photometric redshifts with the multilayer perceptron neural network: application to the HDF-S and SDSS, Astron. Astrophys. 423 (2) (2004) 761–776.
[30] A. Collister, O. Lahav, ANNz: estimating photometric redshifts using artificial neural networks, Publ. Astron. Soc. Pac. 116 (818) (2004) 345–351.
[31] D.J. Mortlock, S.J. Warren, B.P. Venemans, et al., A luminous quasar at a redshift of z = 7.085, Nature 474 (7353) (2011) 616–619.
[32] J. Pasquet-Itam, J. Pasquet, Deep learning approach for classifying, detecting and predicting photometric redshifts of quasars in the Sloan digital sky survey stripe 82, Astron. Astrophys. 611 (2018).
[33] R.A.F. Borda, P.D. Mininni, C.H. Mandrini, et al., Automatic solar flare detection using neural network techniques, Sol. Phys. 206 (2) (2002) 347–357.
[34] N. Nishizuka, K. Sugiura, Y. Kubo, et al., Solar flare prediction model with three machine-learning algorithms using ultraviolet brightening and vector Magnetograms, Astrophys. J. 835 (2) (2017) 156.
[35] H. Xin, W. Huaning, X. Long, et al., Deep learning based solar flare forecasting model. I. Results for line-of-sight magnetograms, Astrophys. J. 856 (1) (2018) 7.
[36] I.S. Veselovskii, A.V. Dmitriev, Y.V. Orlov, et al., Simulation of statistical distributions in the space of parameters of the solar wind and interplanetary magnetic field using artificial neural networks, Sol. Syst. Res. 34 (2) (2000) 116–123.
[37] O. Fuentes, Automatic determination of stellar atmospheric parameters using neural networks and instance-based machine learning, Exp. Astron. 12 (1) (2001) 21–31.
[38] M. Ntampaka, H. Trac, D.J. Sutherland, et al., A machine learning approach for dynamical mass measurements of galaxy clusters, Astrophys. J. 803 (2) (2015) 50.
[39] A.A. Miller, J.S. Bloom, J.W. Richards, et al., A machine learning method to infer fundamental stellar parameters from photometric light curves, Astrophys. J. 798 (2) (2015) 122.
[40] W. Sun, L. Xu, X. Huang, et al., Forecasting of ionospheric vertical total electron content (TEC) using LSTM networks, in: 2017 International Conference on Machine Learning and Cybernetics (ICMLC), 2017, pp. 340–344.
[41] C.A.L. Bailerjones, Stellar parameters from very low resolution spectra and medium band filters: Teff, logg and [M/H] using neural networks, Astron. Astrophys. 357 (1) (2000) 197–205.
[42] R. Tagliaferri, A. Ciaramella, L. Milano, et al., Spectral analysis of stellar light curves by means of neural networks, Astron. Astrophys. Suppl. Ser. 137 (2) (1999) 391–405.
[43] E. Perlman, A. Java, Predictive mining of time series data in astronomy, arXiv preprint astro-ph/0212413, 2002.
[44] T.R. Devine, K. Goseva-Popstojanova, M. Mclaughlin, Detection of dispersed radio pulses: a machine learning approach to candidate identification and classification, Mon. Not. R. Astron. Soc. 459 (2) (2016) 1519–1532.
[45] S. Meingast, M. Lombardi, J. Alves, Estimating extinction using unsupervised machine learning, Astron. Astrophys. 601 (2017).

PART

D

Future developments

Mapping the universe with 21 cm observations

Xuelei Chen
National Astronomical Observatories, Chinese Academy of Sciences, Beijing, China

Hydrogen is the most abundant element in the universe, and its 21 cm line is a very important observational probe for radio astronomy. The 21 cm cosmology is at the forefront of current research. I first briefly introduce the physics of the 21 line emission and absorption, and its observational signature during the cosmic evolution. I then describe some ongoing 21 cm tomography experimental efforts, and illustrate the typical experimental setup with the Tianlai pathfinder array; the global spectrum experiment is also briefly introduced. The basics of synthesis imaging are briefly introduced and the challenge posed by foreground presented, with an example of foreground subtraction using the PCA method. The foreground wedge in the 21 cm power spectrum is explained. Finally, I conclude with an outlook to future development.

1 The neutral hydrogen and 21 cm line

The ground state of the neutral hydrogen atom is split into two states by the interaction between the magnetic moments of the proton and electron. Transitions between these hyperfine split states produce or absorb radio waves with a frequency of 1420 MHz corresponding to a wavelength of 21 cm; this spectral line is usually called the 21 cm line. Historically, this was the first spectral line predicted and detected in radio astronomy, and it has been widely used in the observation of neutral hydrogen (often referred to as HI) in the galactic interstellar medium as well as nearby galaxies.

Big Data in Astronomy. https://doi.org/10.1016/B978-0-12-819084-5.00009-2

With the expansion of the universe, a radio wave emitted in an earlier epoch is redshifted to a longer wavelength, but customarily it is still referred to as the (redshifted) 21 cm radiation for its origin. As the space is filled with photons from the cosmic microwave background (CMB) or more generally some continuum radio background, when we speak of observing the 21 cm radiation, we are actually speaking of observing the excess (emission) or dearth (absorption) features produced by the neutral hydrogen on top of such background radiation.

The neutral hydrogen atoms first appeared right after the epoch of recombination, which marked the end of the Big Bang. As the baryonic matter in the early universe was almost entirely made up of hydrogen and helium atoms with very little heavy elements, the 21 cm line provides perhaps the only means to probe the ensuing dark age. As structures formed, the ionizing photons from stars and accreting black holes began to ionize the gas of the universe, and through this epoch of reionization (EoR), the 21 cm line can be used to map neutral hydrogen distribution, thus revealing the process of reionization. After the EoR, the neutral hydrogen atoms are found primarily within galaxies where the gas density is sufficiently high, such that the recombination rate (proportional to density squared) can compensate the ionization rate from the extragalactic ultraviolet background.

Although the redshifted 21 cm radiation is in principle observable up to the Dark Ages, at present it is mostly detected at low redshifts ($z < 0.2$) because its signal is much smaller than some foreground radiations, such as the synchrotron emission of cosmic ray electrons moving in the galactic magnetic field, the Bremsstrahlung radiation from electrons in the interstellar medium, and the radio galaxies and quasars. The galactic synchrotron radiation is particularly severe, as its strength is about 10^5 that of the EoR 21 cm signal. At least in principle, the foreground can be identified and removed as their spectra are smooth, whereas the 21 cm spectrum varies as it follows the underlying HI density, which on large scales traces the matter density and fluctuates accordingly. The potential of 21 cm observation is great as it can be used to probe a large portion of the comoving volume of the observable universe. Once a precision observation technique is mastered and the signal extracted, this will supply an unprecedented amount of precious information for cosmological studies.

1.1 The physics of spin temperature

The 21 cm line is produced by the transition between the singlet ($F=0$) and triplet ($F=1$) hyperfine levels of the hydrogen atom at the electronic ground state (1s) (Fig. 15.1). These two

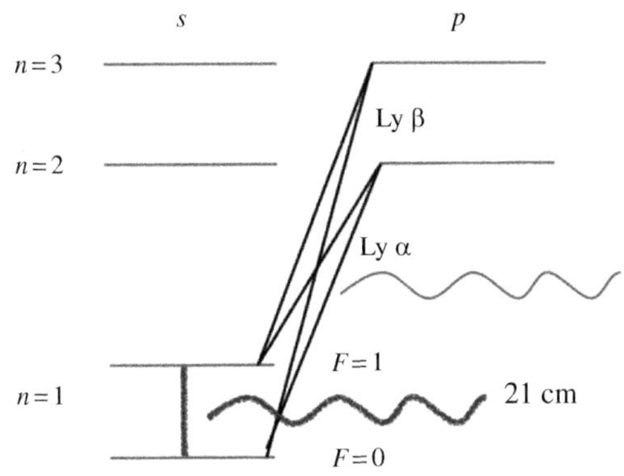

Fig. 15.1 The neutral hydrogen energy levels.

states are distinguished by the different spin of the electron with respect to the proton. A spin temperature can be defined according to the occupation numbers in the two states,

$$\frac{n_1}{n_0} = 3e^{-h\nu/kT_s}$$

where h is the Planck constant, ν the frequency, so hν gives the energy of the 21 cm photon, k is the Boltzmann constant, and T_s the spin temperature. If the spin temperature is much higher than $h\nu$, then the difference in energy level is negligible and the two states are nearly equally occupied except for the degeneracy factor of 3. But if the spin temperature is low, then the lower state is much more occupied than the higher one.

In a cosmological setting, the line can be observed in absorption or emission against the cosmic microwave background (CMB) with the brightness temperature given by [1].

$$\begin{aligned}\delta T &= \frac{T_s - T_r}{1+z}(1-e^{-\tau}) \\ &\simeq (0.025K)\left(\frac{\Omega_b h_0}{0.03}\right)\left(\frac{0.3}{\Omega_{m0}}\right)^{1/2}\left(\frac{1+z}{10}\right)^{1/2}\frac{\rho_{HI}}{\bar{\rho}_H}\frac{T_s - T_r}{T_s}.\end{aligned}$$

where T_r is the radio background temperature at the frequency 1420 MHz at that particular redshift. In the absence of other strong backgrounds, it would be the CMB temperature, and τ is the optical depth. The second line is the approximation at the epoch of reionization. ρ_{HI} is the density of neutral hydrogen and $\bar{\rho}_H$ is the mean hydrogen density of the universe (we have assumed the optically thin case and approximated $\Omega_{m0}(1+z)^3 \gg 1$.

The hydrogen spin temperature plays a key role in determining the sign and amplitude of the 21 cm emission or absorption signal against the radio background. To study the evolution of the spin temperature, we may regard the two hyperfine split energy levels of the ground state hydrogen atoms as a thermal system in contact with other thermal systems (Fig. 15.2). The systems in contact include the radio background photons (below we shall take this as the CMB, the radio emission from galaxies and stars may also contribute some part but it is just a small fraction in the conventional scenario), the kinetic motion of the atoms, and the Lyman series photons. According to the zeroth law of thermodynamics, two contacting thermal systems will approach a common equilibrium temperature. As the spin system has a very small heat capacity, it has little impact on either of the other two systems, and if it is in contact with only one of them, then the spin temperature would approach the temperature of that system. As it is in contact with several systems, the spin temperature is a weighted average of the temperature of these, with the weights given by the corresponding coupling strength.

The coupling between the radio background and the spin system depends only on the transition oscillator strength and the flux of the radio background, and in the case of a CMB-dominated background, it is entirely predictable. The kinetic motion is directly coupled to the spin by atomic collisions. In an atomic collision, both the velocity and the spin can be changed simultaneously. As the collision rate is proportional to the density squared, it is ineffective in thin gas.

The atom on the ground state 1s can jump to a higher excited state by absorbing a photon, ending up on the $2p$, $3p$, ... states for the Lyman alpha (Ly α), Lyman beta (Ly β), ... Lyman series photon. The most important among these are the Ly α photons. In this case, after a short time the atom will fall back to the ground state

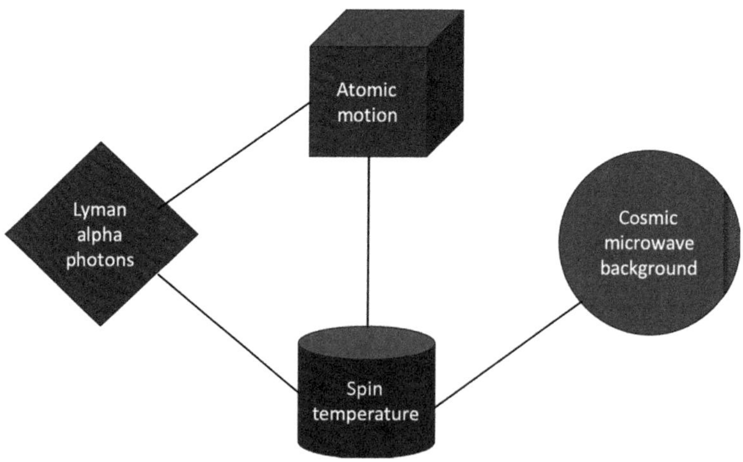

Fig. 15.2 The atomic motion, Lyman alpha photons, CMB, and spin as interacting thermal systems.

with the spontaneous emission of a Ly α photon. We may also regard this as the resonant scattering of the Ly α photon. The higher Lyman series are less important because they do not always fall back to the ground state, but can instead cascade down the energy levels. Due to the large scattering cross-section, each Ly α photon will be resonantly scattered repeatedly. This whole process can also be seen as the atom spin system exchanging energy with the Ly α photons, so that the spin temperature should approach the color temperature of the Ly α photons. The color temperature is defined by the slope of the radiation spectrum at the line center. Normalized to the predictable radio background temperature, the spin temperature is then given by

$$T_S = \frac{T_r + y_c T_k + y_\alpha T_\alpha}{1 + y_c + y_\alpha}$$

where T_r, T_k, T_α are the radio background temperature, kinetic temperature of the gas, and color temperature of the Ly α photons, respectively, and the collisional and Wouthuysen-Field coupling weights are given by

$$y_\alpha \equiv \frac{P_{10} T_*}{A_{10} T_k}; y_c \equiv \frac{C_{10} T_*}{A_{10} T_k}$$

where $T_* = 0.0682\,\mathrm{K}$ is the hyperfine energy splitting, $A_{10} = 2.87 \times 10^{-15} s^{-1}$ is the spontaneous emission coefficient of the 21 cm line, $C_{10} = \kappa_{10} n_H$ is the collisional deexcitation rate, κ_{10} ranges from $2 \times 10^{-14} \sim 2.5 \times 10^{-10} \mathrm{cm}^3 \mathrm{s}^{-1}$ and has been tabulated.

The indirect deexcitation rate P_{10} of the hyperfine structure levels is related to the total Ly α scattering rate $P_{10} = 4P_\alpha/27$, with

$$P_\alpha = \int d\nu c n_\nu \sigma(\nu)$$

where n_ν is the number density of photons per unit frequency, and $\sigma(\nu)$ is the cross-section for Ly α scattering. The interaction between the Ly α photons and the hydrogen gas, on the other hand, tends to make the color temperature to that of the gas kinetic temperature. Thus, to a good approximation,

$$T_\alpha = T_k.$$

1.2 The evolution of 21 cm signal through cosmic history

We now consider the evolution of the spin temperature and 21 cm brightness temperature throughout the history of the universe. The neutral hydrogen first appeared after recombination. Initially, its temperature would be nearly equal to that of the CMB. In adiabatic expansion, the nonrelativistic gas temperature drops as $T \propto a^{-2}$ while the CMB photons drops as $T \propto a^{-1}$ where a is the scale factor, so the gas temperature would soon fall below that of the CMB if there is no heat change between the photons and the gas. In fact, there is indeed some heat exchange—there are still residue-free electrons that can scatter with the CMB photons, during which energy is transferred from the photons to gas. However, as the universe expands, this heat transfer rate falls quickly, such that during the late Dark Ages ($z < 200$), the gas temperature fell well below that of the CMB. The gas density is sufficiently high, so the collisional coupling is important, and the gas spin temperature approaches that of the kinetic temperature. As a result, we will see a broad negative 21 cm brightness temperature or absorption feature in the average spectrum corresponding to this redshift. On the other hand, the primordial fluctuations began to grow during the dark age, forming nonlinear structures such as halos and filaments. In such systems, the gas is compressed and shock heated and has a significantly higher temperature, though as such the structures are still very rare and occupy a very small fraction. However, as the density drops further, the collisional coupling becomes less effective, the spin temperature tends to that of the CMB temperature, and the 21 cm brightness temperature approaches 0 (Fig. 15.3).

Then, as sufficiently large halos formed, the gas within can collapse to form stars. These first stars supply Ly α photons in two ways: the continuum radiation can be redshifted to the Ly α wavelength, and ionization and deexcitation in the gas

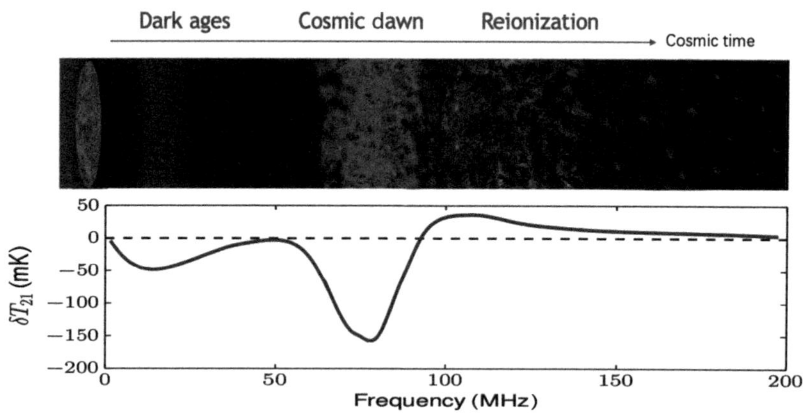

Fig. 15.3 The cosmic evolution history and the 21 cm brightness temperature.

surrounding the star can produce the Ly α photons. This first form a Ly α halo, in which the spin temperature couples to the kinetic temperature by the Wouthuysen-Field effect of the Ly α photons. Then, a Ly α background is built up in the whole space as more emitters are formed. If the majority of the gas is not heated by x-ray photons, its kinetic temperature will remain below the CMB temperature, so that a 21 cm absorption trough will be produced again [3,4].

Eventually, the gas would be heated by the ionizing photons, so some sort of 21 cm emission would be seen. However, as more and more stars, galaxies, and accreting blackholes (quasars) are formed, the gas would all be ionized, except that within the galaxies or dense clumps inside the dark matter halos where the recombination rate is higher, which can keep a self-screened neutral region in the presence of the ionizing background. The 21 cm signal from the intergalactic gas would again drop to zero, but of course there are still the 21 cm signals from the neutral hydrogen in the galaxies [2].

2 The 21 cm experiments

There are generally three types of 21 cm experiments: 21 cm tomography, 21 cm forest observation, and 21 cm global spectrum measurement.

In the 21 cm tomography experiment, one surveys the 21 cm variation in both the spatial direction and frequency, thus yielding a three-dimensional data cube. As different frequency corresponds to different redshift, this gives a three-dimensional map of the universe along the light cone, revealing spatial distribution as well as time evolution. This is the primary mode of observation. It can either be carried out in the traditional galaxy survey mode, where one tries to observe the distribution of individual objects (galaxies), or it can be carried out in the intensity mapping mode, where the focus is on obtaining the large-scale distribution.

The 21 cm forest is analogous to the well-known Lyman alpha forest in the quasar absorption line. One tries to observe the spectra of high redshift bright continuum spectra radio sources, such as quasar, gamma ray burst radio afterglow, or fast radio burst. The neutral hydrogen absorbers at different redshifts along the line of sight produce absorption lines at different part of the spectrum. A great advantage of the 21 cm forest is that it is a good probe of the spin temperature. Finding the usable bright radio source is the challenge for this kind of observation [5] (Fig. 15.4).

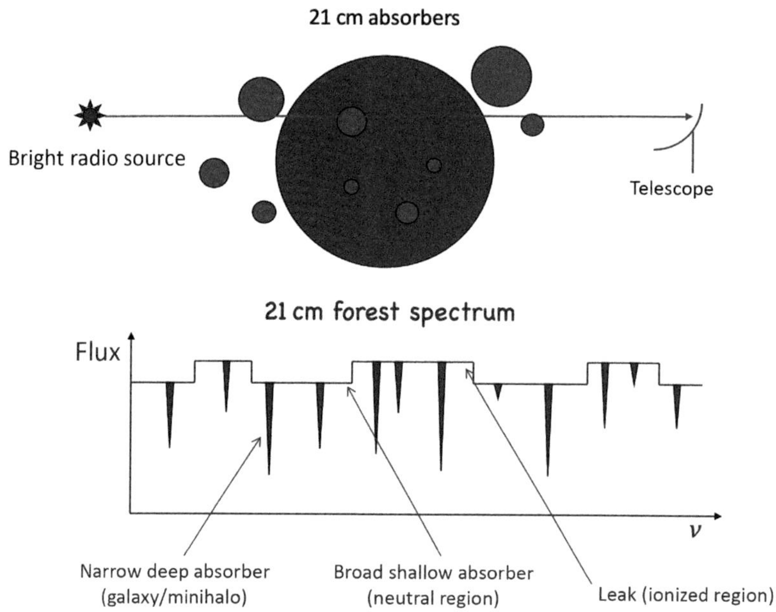

Fig. 15.4 Illustration of the 21 cm forest concept *(top)* and the absorption spectrum *(bottom)*.

The global averaged 21 cm spectrum gives the mean brightness temperature of the 21 cm emission as a function of redshift. In such an experiment, the flux of the whole sky or a large portion of it is integrated, so it can be done with a single dipole-type antenna that has a very wide beam. However, on top of the 21 cm signal, there is the much stronger foreground radiation. Nevertheless, the cosmology theory predicts some features in the global spectrum, such as the absorption trough at the cosmic dawn and Dark Age, and the sharp drop of the mean brightness temperature at the completion of the reionization. The global spectrum experiments can look for such features.

Below, we shall discuss mostly the 21 cm tomography experiments, as these are the primary efforts in 21 cm cosmological observation. This is more relevant for the big data problem, although we will also briefly review the progress in the 21 cm global spectrum experiment.

2.1 HI galaxy survey

The traditional way of cosmological observation with the 21 cm line is the HI galaxy survey, where one detects the 21 cm line of individual galaxies, just as in the typical optical galaxy redshift survey. Although the 21 cm line has long been detected in the

Milky Way and nearby galaxies, it is not easy to detect in the distant universe. For an interferometer array, the sensitivity is

$$\Delta S = \frac{2kT_{sys}}{\eta A\sqrt{2N_p \Delta\nu t}}$$

where T_{sys} is the system temperature, $\eta < 1$ is a factor describing quantization loss, A the effective area of one antenna, $N_p = n_a(n_a - 1)/2$ is the number of antenna pairs, $\Delta\nu$ the bandwidth, and t the integration time. For large n_a, we may approximate

$$\Delta S = \frac{2kT_{sys}}{\eta A_{eff}\sqrt{\Delta\nu t}} \approx 2.74\left(\frac{1000}{\eta A_{eff}/T_{sys}}\right)\left(\frac{\Delta\nu t}{MHzs}\right)^{-1/2} mJy$$

As a reference, the typical A_{eff}/T_{sys} values for currently operating or upcoming arrays are, for example, eMERLIN(60), ASKAP(65), GMRT(250), JVLA(265), MeerKAT(321), and SKA1-mid (1650) [6].

The 21 cm line flux from the galaxy is given by

$$S_\nu = 4.25 \times 10^{-2} \frac{(1+z)^2}{(d_L/Mpc)^2}\left(\frac{M_{HI}}{10^6 M_\odot}\right)\left(\frac{\Delta V}{100kms^{-1}}\right)^{-1} Jy$$

where d_L is the luminosity distance, M_{HI} is the mass in HI in the volume pixel (voxel), and ΔV is the velocity spread in the rest frame of the galaxy. For optimal detection, the voxel should be just the size of the galaxy. The limiting mass of the galaxy in terms of the flux limit is then

$$M_{HI} = 2.36 \times 10^{10}\left(\frac{D_L}{Gpc}\right)^2 \frac{S_i}{mJy} \frac{\Delta v}{100kms^{-1}}(1+z)^{-2} M_\odot$$

We see a massive galaxy in cosmological distance, sub-mJy sensitivity is required, and even for the SKA-1mid array, an integration time of several hours is needed.

Furthermore, in order to distinguish the different galaxies, one needs to achieve an angular resolution of a few arcseconds. This can be achieved with an array distributed over several tens of kilometers. Many interferometer arrays such as JVLA, ASKAP, MeerKAT, and SKA1-mid have baselines on such scales. However, if one wants to survey a sizable fraction of the sky (a few thousand square degrees or more) within a plausible time frame (a few years), and at each pointing integrate for a few hours as discussed above, then a large field of view is also required. For SKA1-mid, which is made up of antennas of 15 m aperture plus the 13 m MeerKAT antennas, the field of view is 0.8 degrees at 1420 MHz. This means that within each field of view, there would be at least

10^6 pixels, and then combined with 10^{3-4} frequency bins, each image cube would have at least 10^{10} voxels. For the whole survey, this would be 10^{14} voxels, posing a daunting task for imaging synthesis computation and data storage.

2.2 Intensity map observations

Another mode of observation is the intensity mapping observation, in which one maps the radiation intensity of the sky at a low resolution without attempting to resolve individual objects but to try to obtain the distribution on the larger scale (Fig. 15.5). This mode of observation was first considered for the epoch of reionization, where the main aim has been observing the reionization of the intergalactic medium, which is expected to happen on scales much larger than individual galaxies. Later, it is also considered for the HI survey at low- or mid-redshift surveys, which can be used to measure the baryon acoustic oscillation (BAO) features and to detect the dark energy. It is noted that such a survey can be conducted with a dedicated array of ~100 m size [7,8].

The flux is related to the brightness temperature by

$$S = \frac{2k\Omega_b}{\lambda^2} T_b$$

where Ω_b is the solid angle of the synthesized beam while λ is the wavelength, so the temperature noise per pixel in intensity mapping is given by

$$\Delta T_b = \frac{\lambda^2 T_{sys}}{\Omega_b \eta A \sqrt{2N_p \Delta \nu t}}$$

For the same number of antennas and integration time, a smaller error on the temperature can be achieved with larger pixels.

Intensity mapping is carried out with a compact, almost aperture-filled array. This is necessary for the observation in the presence of a strong foreground. The foreground, especially the galactic synchrotron radiation, is always much larger (by 4–5 orders of magnitude) than the 21 cm signal. In order to extract the 21 cm signal, one has to use the property that along the line of sight, the foreground radiation has a relatively smooth spectrum while the 21 cm flux fluctuates randomly to separate them. However, at a given wavelength, the interferometry of each baseline probes a particular spatial mode of the sky radiation intensity [9], and as the wavelength varies, the interferometer array output gives different spatial modes. It is necessary to have a compact array for which the coverage in the baseline (u,v,w) space is dense to subtract the foreground modes.

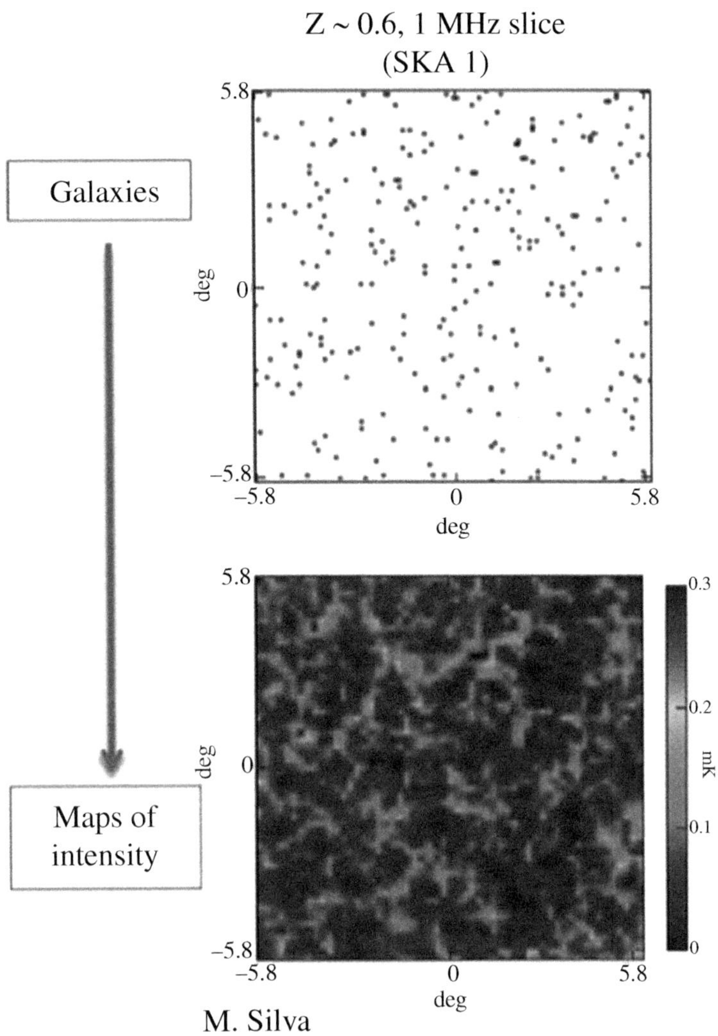

Fig. 15.5 An illustration of intensity mapping. Made by M. Silva.

A potential problem in the intensity mapping observation is the confusion of other spectral lines. In the galaxy survey, it is rare for two galaxies at two different redshifts to happen to be on the same line of sight and get confused with each other. In the intensity mapping observation, two spectral lines with wavelengths λ_1, λ_2 will be confused if they are located at redshifts z_1, z_2 if $(1+z_1)\ \lambda_1=(1+z_2)\lambda_2$. Fortunately, at the low frequency of the 21 cm line, there are not many other lines of comparable strength, though for higher frequency molecule lines this may be a problem [10].

2.3 The 21 cm tomography experiments

The intensity mapping technique has been tried on existing telescopes, such as GBT (United States) and Parkes (Australia), and the results show that there is a correlation between the signal and the density of optical galaxies, though so far the signal has proved to be too weak to be detected positively in the autopower spectrum [11–14]. It has also been suggested to apply this technique to the new large multipurpose telescopes, such as the FAST (China) [15] and the SKA1-mid (South Africa) [16]. The layout of the SKA1-mid does not have a complete covering in the (u,v) plane, but it has been proposed that the many dishes in the array may be used individually as single dishes for this task.

All the EOR or low-frequency radio experiments such as the 21 cmA (China), LOFAR (Netherland and European countries), MWA (Australia), PAPER (United States-South Africa), HERA (United States-South Africa), and the SKA1-low (Australia) can make intensity mapping mode observations, though some of them also have longer baselines for other imaging observations. These experiments make observations in the meter-wave band, as the waves are not very sensitive to the details of the antenna much smaller than the wavelength, so most of them use some form of dipole antennas. HERA, however, uses dishes made of cheap structural material. As single dipoles are still relatively small, multiple dipoles are put together to form a station. The signals received by the dipoles are combined after amplification. The station serves as interferometer units. In some cases, multiple beams can also be formed for each station (Figs. 15.6 and 15.7).

Here, as a concrete example, we discuss the Tianlai experiment [17,18] in some detail. This experiment aims to detect the baryon acoustic oscillations of the Big Bang, so it is named Tianlai or "heavenly sound," a phrase that appeared in the classical works of the ancient Taoist philosopher Zhuangzi (Chuang-Tzu). The goal of the Tianlai experiment is to observe the 21 cm signal at redshift 0–3. At present, it is a pathfinder experiment to verify the basic principles (e.g., array design, calibration method, foreground subtraction techniques, etc.) and key technologies (e.g., the analog receiver, signal transportation system, and digital correlators). The pathfinder consists of three $15\,\text{m} \times 40\,\text{m}$ cylinders lined in the north-south direction and adjacent to each other as well as 16 dishes of 6 m aperture. Cylinder reflectors offer a low-cost way to build a telescope array with a large effective area. For the Tianlai cylinder array, the feeds are put on the focus line that lies along the N-S line, so that a narrow strip crossing the zenith could be observed at any time. The reflector is fixed on ground, as Earth rotates, the array maps out the whole observable

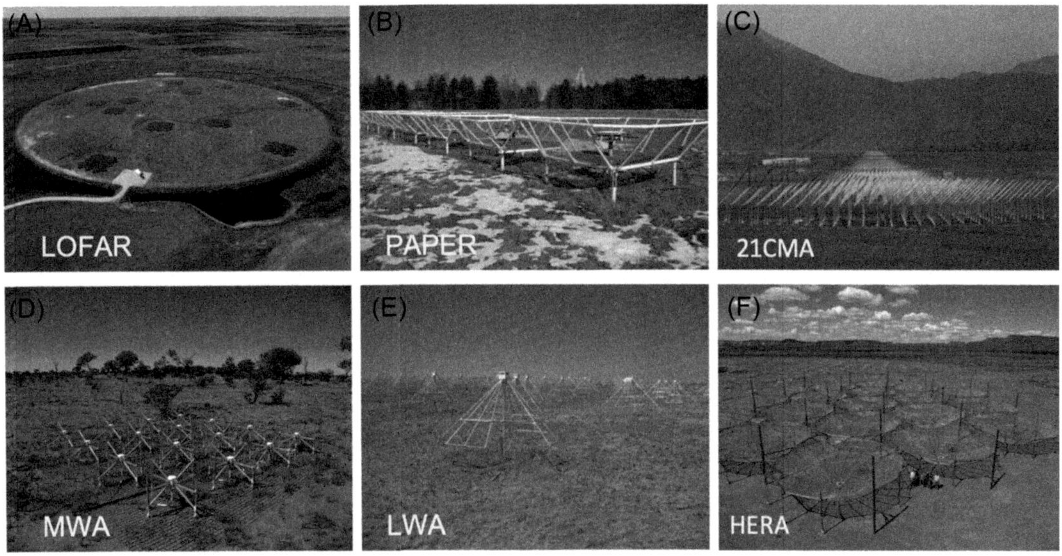

Fig. 15.6 Some EOR experiments. (A) From https://en.wikipedia.org/wiki/Low-Frequency_Array_(LOFAR)#/media/File:LOFAR_Superterp.jpg. (D) From https://commons.wikimedia.org/wiki/File:MWA_32T_Tile.jpg. (F) HERA Collaboration, from http://reionization.org/.

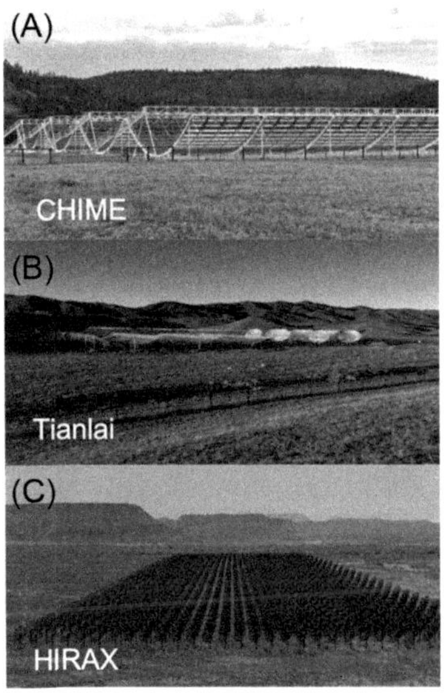

Fig. 15.7 Some mid-redshift 21 cm experiments. (A) From https://commons.wikimedia.org/wiki/File:Canadian_Hydrogen_Intensity_Mapping_Experiment_-_overall.jpg. (C) Used with permission from Cynthia Chiang.

part of the sky (see Fig. 15.8). As there is no movable part, the array is stable and can be constructed at low cost.

In the dish array, the instant field of view is a circular area defined by the primary beam of the dish. It can also make

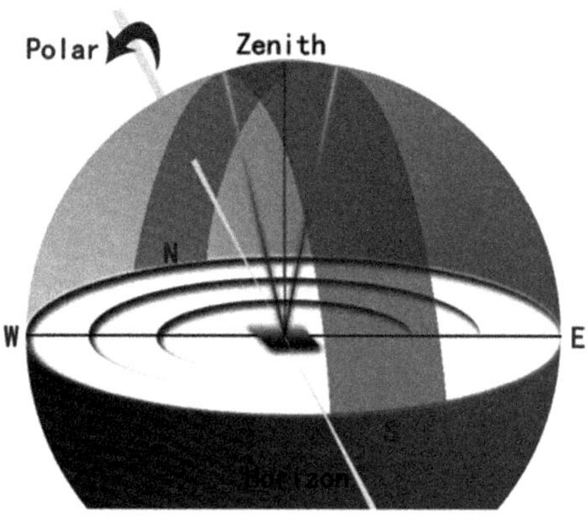

Fig. 15.8 The instant view of the sky by the cylinder array; the whole northern sky is surveyed as Earth rotates.

observations with the drift scan mode, that is, use the rotation of the Earth to scan over the area to be surveyed. Because of the finite size of the primary beam, one needs to adjust the declination angle of the dish in order to scan over the whole observable sky [19]. The dishes, however, also have certain advantages. For each receiver, the effective antenna area is large even for small dishes, so it is more sensitive. With a driver mechanism, the dishes can be pointed to different directions and track targets, making it easier for test and calibration.

The signals received by the feeds on the cylinder reflectors or dishes are first amplified with a low noise amplifier (LNA), then sent via a coaxial cable to an optical transmitter below the antenna. Here, the radio signal is used to modulate the optical signal, which is sent via a multifiber optical cable to the station house that is located in the nearby village about 7 km away. The optical signal is converted back to radio there, then downconverted to an intermediate frequency (IF), then further amplified and digitized. The digitized time series signal is fast Fourier transformed (FFTed), then cross-correlated in the FX correlator. The integrated cross-correlations (interferometry visibilities) are stored on hard drives for offline processing (Fig. 15.9).

2.4 The 21 cm global spectrum experiments

Besides tomographic observations, the global average 21 cm brightness temperature can also be measured. As the measurement is for the whole sky, angular resolution is not required, and a single antenna can be used for such a measurement. The EDGES (Arizona State and MIT) experiment is the first to make

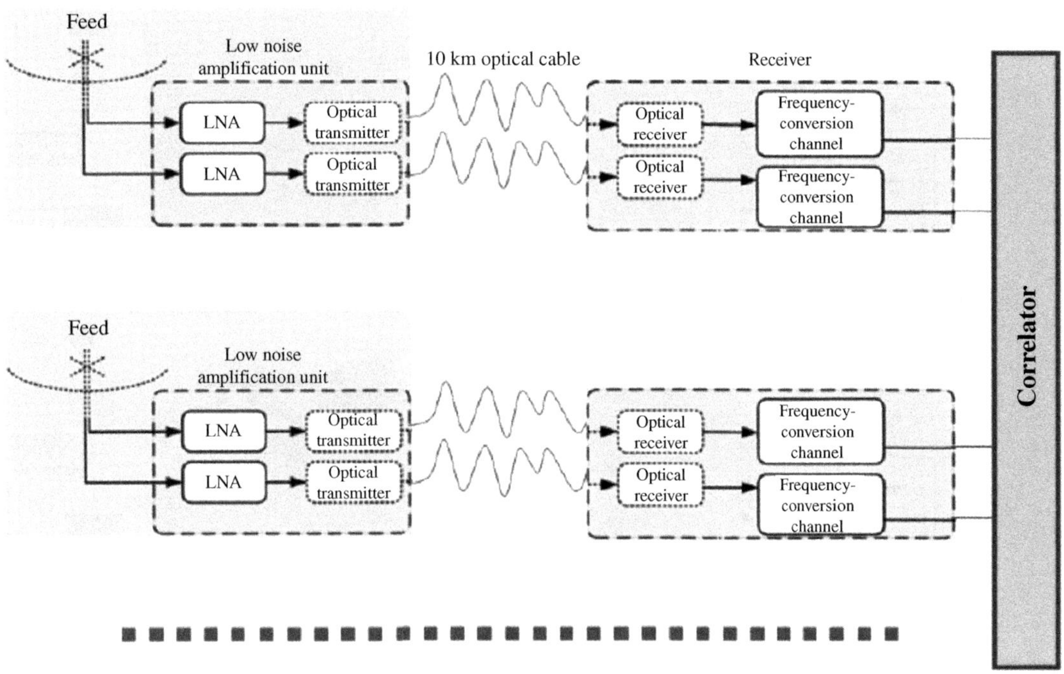

Fig. 15.9 Schematic view of the analog signal path of the Tianlai pathfinder array.

such experiments [20]. Subsequently, similar experiments have also been carried out by the SCI-HI (United States-Mexico) experiment, the PRIZM (South Africa and Canada) experiment, the SARAS (India) experiment, the Big Horn experiment (Australia), and LEDA (Harvard-Berkeley). There are also upcoming ones such as the REACH (Cambridge University) and Tianlai-Global (China) (Fig. 15.10). In addition, several lunar orbit mission concepts have been studied, including the DARE, DAPPER, and DSL [21].

In the global spectrum experiment, the whole sky average is to be measured. It does not require high angular resolution. In low frequency, the sky is quite bright, but the 21 cm signal is small so the experiment requires very high precision and dynamic range. Some usually ignored effects, such as the drifting of amplifier gain and the reflection in the circuit due to impedance mismatch, must all be considered.

In order to achieve such precision, the EDGES experiment receiver includes internal calibration mechanisms. As shown in Fig. 15.11, there is a switch at the base of the antenna that is connected to the circuitry for the measurement of antenna reflection, which can be used to measure the impedance of the antenna. The

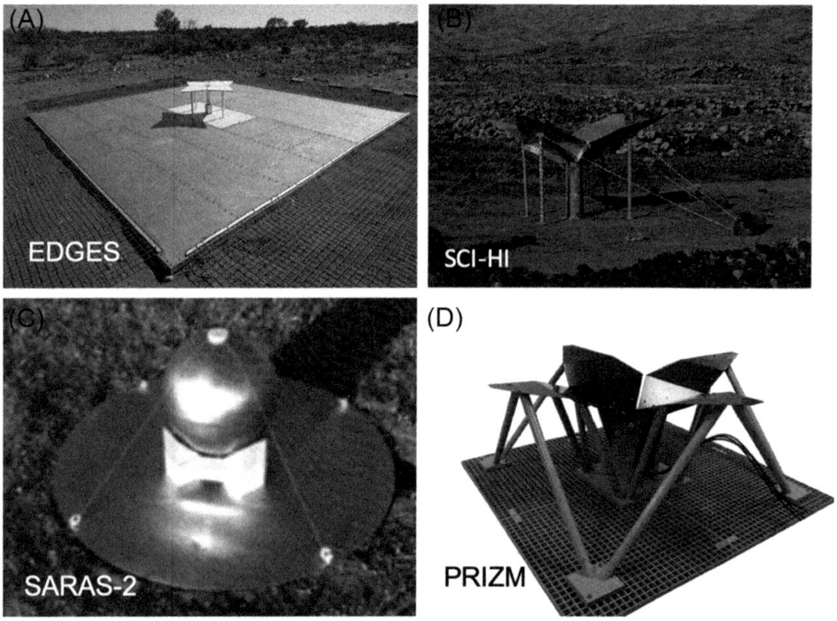

Fig. 15.10 Some global spectrum experiments. (A) From https://commons.wikimedia.org/wiki/File:EDGES_antenna.JPG. (B) From Tabitha C. Voytek et al. 2014, Probing the Dark Ages at z ~20: The SCI-HI 21 cm All-Sky Spectrum Experiment, ApJL 782 L9. (C) From Saurabh Singh et al. 2017, First results on the Epoch of Reionization from First Light with SARAS 2, ApJL 845 L12. (D) Used with permission from Cynthia Chiang.

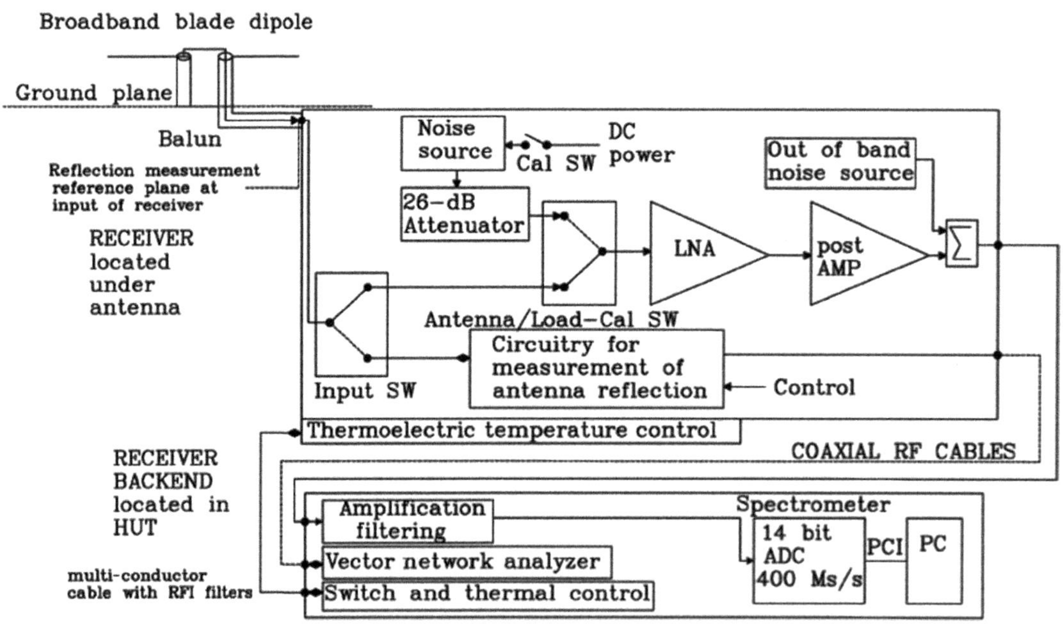

Fig. 15.11 Schematic of the EDGES receiver system [22].

input of the low noise amplifier (LNA) can also be switched between the antenna input and a noise source with an adjustable attenuator so that the gain of the LNA can be measured. The output is then digitized. The system will be automatically controlled to switch between measuring the sky signal and calibration with the internal noise source.

The other experiments have used principles similar to EDGES, though in detail the designs are each different. For example, different types of antennas have been employed. The antenna should be as independent of the frequency as possible because a frequency-dependent beam would mean that the antenna will see slightly different skies at the different frequencies, thus mixing the spatial pattern of the sky with the frequency feature to be measured. The SCI-HI and PRIZM experiments used a "flower petal" shaped antenna while the SARAS used a cone-sphere antenna. These experiments also designed the system to have a higher characteristic frequency so that in the band of observation, the system response is smoother. The price paid is that due to the imbalance of impedance, only a small fraction of the antenna power can be fed in to the amplifier, and hence the amplifier noise is more significant.

In 2018, the EDGES experiment announced that an absorption feature was detected around 78 MHz, corresponding to redshift $z \sim 18$, which is exactly the expected cosmic dawn time [22]. However, the detected absorption of 550 mK is much stronger than the prediction of most models—indeed, the trough is about twice as deep as the maximum allowed by the standard model. Various ideas have been proposed to explain such strong absorption, such as the gas could be cooler than usually assumed by an exotic interaction with dark matter that is generally colder than the gas due to earlier decoupling from the cosmic fluid. Alternatively, a strong radio background radiation other than the cosmic microwave background might be present in the early universe, so that the absorption signal is stronger. On the other hand, the required precision of the experiment is extremely high, so it might suffer from various contaminations or unexpected system errors. Further experiments are required to produce convincing results.

3 Data processing

3.1 Imaging and beam forming

The angular resolution of an observation is given by

$$\Theta \approx \frac{\lambda}{D}$$

where λ is the wavelength and D is the aperture of the instrument. Due to the long wavelength of the radio waves, in order to obtain a good angular resolution, large D is required. Even with the 500 m spherical telescope (FAST) that has a total size of 500 m and an aperture of 300 m, its angular resolution is merely 2.9 arcmin at the 21 cm wavelength.

In order to achieve higher angular resolution, an interferometer array is often used. The voltage ε_i induced on the antenna i can be written as

$$\varepsilon_i = \int d^2\hat{\mathbf{n}} A_i(\hat{\mathbf{n}}) E(\hat{\mathbf{n}}) e^{-i2\pi \vec{u_i}\cdot\hat{n}}$$

where $E(\mathbf{n})$ is the electric field induced by the radiation from direction $\mathbf{n}$ and $A_i(\mathbf{n})$ is the voltage response by antenna i. Under the assumption that the time-varying electric field is uncorrelated for any two different directions, as one would expect for astrophysical sources, the interferometer visibility, that is, the short time averaged correlation between the voltage signal of the two elements, is related to the sky temperature $T(\hat{n})$ by

$$V_{ij} = <\varepsilon_i{}^*\varepsilon_j> = \int d^2\hat{\mathbf{n}} A_{ij}(\hat{n}) T(\hat{\mathbf{n}}) e^{-i2\pi\vec{u}\cdot\hat{n}}$$

where $\mathbf{u}$ is the baseline vector between the elements i and j in wavelength units and A_{ij} is the beam response for the antenna pair (i,j). Roughly speaking, each pair of interferometry units (antennas) measures one Fourier component of the sky intensity [9]. The sky temperature can be recovered from the measured visibility data by inverting the above relation. If we define a reference point in the celestial sphere, and use it to set up a coordinate system (u,v,w), the vector pointing to an arbitrary point is defined in terms of the direction cosines (l, m, n) with respect to the (u,v,w) axes (see Fig. 15.12). Then, the above integration can be reduced to

$$V_{ij} = \int \frac{dl\,dm}{\sqrt{1-l^2-m^2}} A_{ij}(l, m)\, T(\mathbf{l}, \mathbf{m})\, e^{-i2\pi\left(ul+vm+w\left(\sqrt{1-l^2-m^2}-1\right)\right)}$$

where we have used $l^2 + m^2 + n^2 = 1$.

With small angle approximation, the w-term can be neglected. The above is then reduced to a two-dimensional Fourier transform, and can be easily inverted as efficiently as an inverse two-dimensional Fourier transform. Of course, in reality the array baselines can only sample part of the (u,v) space, so one has to make interpolations from the measurement to obtain the visibility at regular grids. This procedure is called "gridding." In many interferometer arrays, the antenna distribution is sparse and the spacing between the antennas is far larger than the size of the antenna,

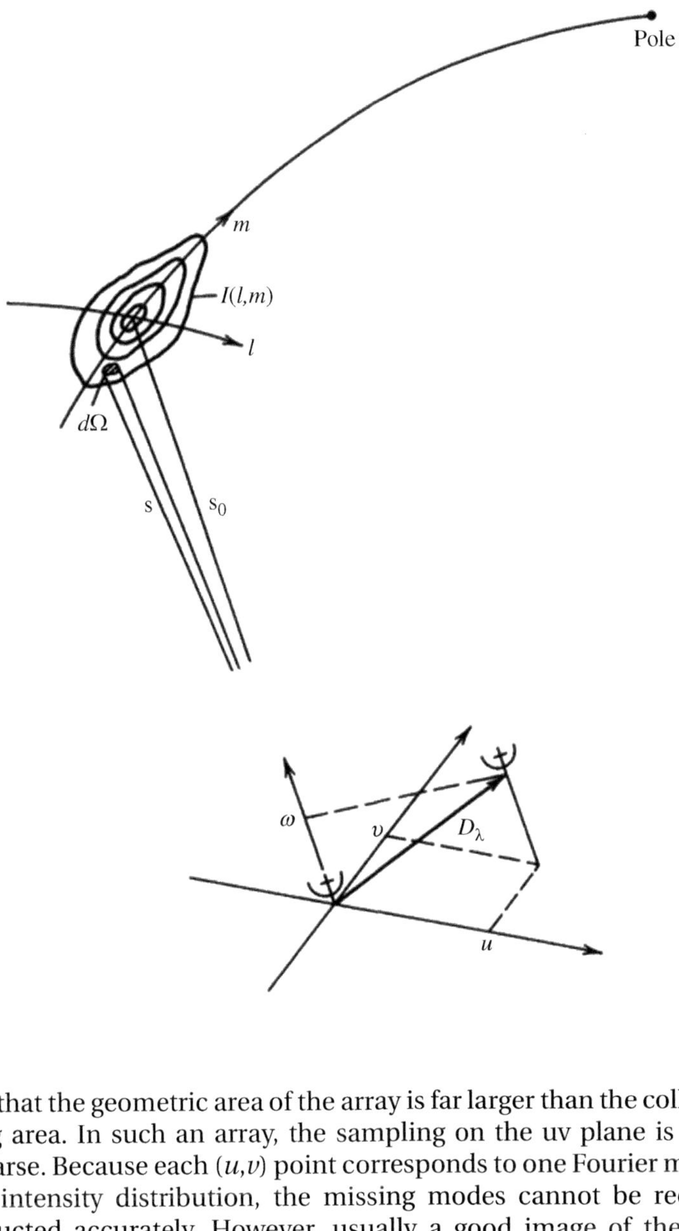

Fig. 15.12 A coordinate system often used in radio interferometry. From A.R. Thompson, J.M. Moran, G. Swenson Jr., Interferometry and Synthesis in Radio Astronomy, 3rd ed., Springer, Cham, Switzerland, 2017.

so that the geometric area of the array is far larger than the collecting area. In such an array, the sampling on the uv plane is also sparse. Because each (u,v) point corresponds to one Fourier mode of intensity distribution, the missing modes cannot be reconstructed accurately. However, usually a good image of the sky can still be obtained, thanks to the sparse nature of sky source distributions. Such an image often has strong side lobes. This can be improved by a deconvolution procedure, using either the CLEAN or maximum entropy method (MEM). The general procedures of such processing are reviewed in the usual radio astronomy textbooks [9].

An alternative to visibility-based image synthesis is beam forming. In this case, one can form a beam by

$$I(\mathbf{n}) = |\sum_j w_j(\mathbf{n})\varepsilon_j|^2$$

where the weight for array element j is given by

$$w_j(\mathbf{n}) = e^{2\pi i\,\mathbf{u}_j\cdot\mathbf{n}}$$

which compensates the phase difference between the element j and the array origin point for direction $\mathbf{n}$, so that for waves coming from direction $\mathbf{n}$, the voltages from the different array elements add constructively. Given sufficient computing power, multiple beams toward different directions can be formed simultaneously. Note that for an array with N elements, at most N independent beams can be formed simultaneously. More beams will be a linear combination of the first N independent beams. The beam forming is equivalent to the visibility measurement. From the definitions, the short time average of the beam output is

$$< I(\mathbf{n}) >= \sum_{ij} w_i^* w_j V_{ij}$$

It would be possible to form the time-averaged beam output by using the visibilities, as shown by the equation above. However, the beam-forming procedure allows using all the array elements to check for the radiation from a particular direction without time averaging, which is particularly suitable for transient source search or observation.

3.2 Foreground

In the intensity mapping observation mode, there is strong foreground radiation from other sources. The diffuse galactic synchrotron emission is the dominant foreground at low frequencies. It is approximately given by

$$T(\nu) = T_0\left(\frac{\nu}{\nu_0}\right)^{-\gamma}$$

with $\gamma \approx 2.7$, and for $\nu_0 = 150$ MHz, $T_0 \approx 150$ K at high galactic latitude while greater near the galactic plane. There are also radio point sources, galactic free-free emissions, etc. Fortunately, these foreground signals are all smooth in the spectrum, so that the 21 cm signal can be extracted by removing the smooth components. This is illustrated in Fig. 15.13.

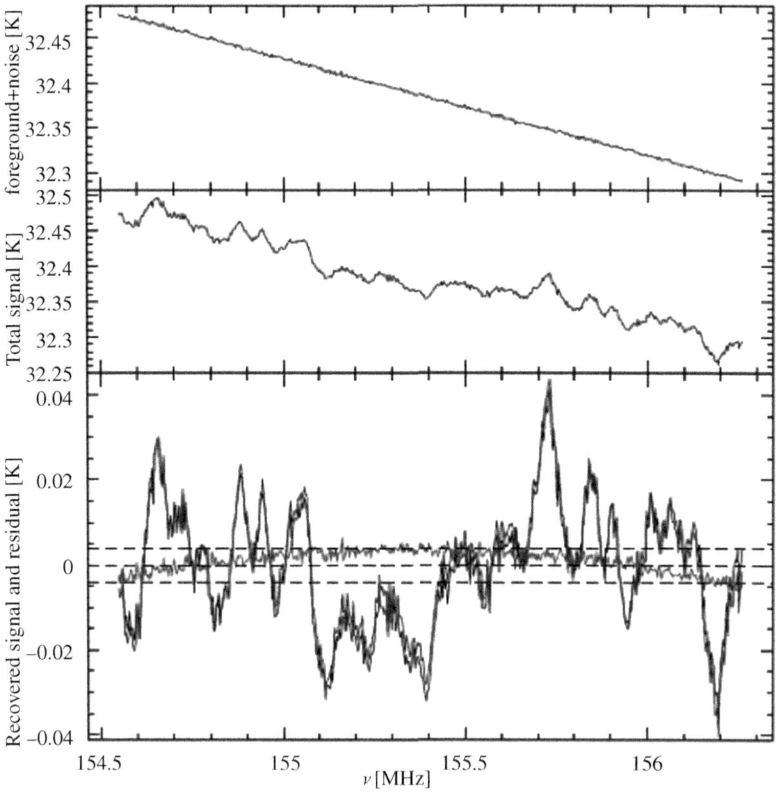

Fig. 15.13 The smooth foreground (top panel), foreground plus 21 cm signal *(middle panel)*, the original 21 cm signal *(black)*, the recovered signal after foreground removal (*blue*; dark gray in print version), and the residue difference (*red*; dark gray in print version) *(bottom panel)* [23].

The simplest method of foreground removal is polynomial fitting. We may assume the foreground brightness temperature is given in the form of

$$\log\frac{T}{T_0} = a_1 \log\frac{\nu}{\nu_0} + a_2\left(\log\frac{\nu}{\nu_0}\right)^2 + \ldots$$

then the smooth component can be removed.

However, in reality, the telescope response is neither known nor smooth, and the beam size generally depends on the frequency. This all makes the foreground removal task very difficult. Nevertheless, blind or semiblind methods have been proposed to remove the largely unknown foreground and extract the 21 cm signal.

As a concrete example, we shall consider the PCA/SVD method for foreground removal in the single dish survey with the GBT. The observation data are a mixture of the 21 cm signal and the foregrounds convolved with the unknown beam response function, plus the noise. The telescope response function is apparently not smooth, for the data spectrum is far from smooth and the nonsmoothness is much higher than the estimated noise level.

Without knowing the precise response of the telescope, can we subtract out this nonsmooth "apparent foreground?" Suppose the data are related to the sky temperature linearly, so in discrete form, we can write down the matrix equation

$$\mathbf{y}=\mathbf{A}\,(\mathbf{sf}+\mathbf{s21})+\mathbf{n},$$

where $\mathbf{s_f}$ and $\mathbf{s_{21}}$ are the foreground and 21 cm components, respectively, $\mathbf{A}$ is the telescope response matrix, and $\mathbf{n}$ the noise. Here, we have omitted indices, but the data have indices of angular position and frequencies. Taking the autocorrelation, we have

$$<\mathbf{y}\mathbf{y}^{\dagger}>=\mathbf{A}\left(<s_f s_f{}^{\dagger}>+<s_{21}s_{21}{}^{\dagger}>+<\mathbf{nn}^{\dagger}\right)\mathbf{A}^{\dagger}$$

or

$$\mathbf{C}_y=A(\mathbf{C_f}+\mathbf{C_{21}}+\mathbf{C_n})\,\mathbf{A}^{\dagger}$$

where we assume that the foreground, 21 cm signal, and noise are all uncorrelated. For different frequencies,

$$\mathbf{C_{21}}(\nu_1,\nu_2)\approx\mathbf{0}$$

while the foreground part is nonzero. This suggests that we subtract the components that have nonzero correlations.

This can be done by performing the principal component analysis (PCA). Solving the eigenvalue equation

$$\mathbf{C}\mathbf{x}=\lambda\mathbf{x}$$

one would obtain a series of eigenvalues and associate eigenvectors. Ordering them by the magnitude of the eigenvalue, we may subtract the largest ones (principal components), so then a large amount of foreground would be removed. In practice, the matrix involved is sometimes a nonsquare matrix. In such cases, the singular value decomposition (SVD) analysis can be performed instead.

The singular values for some GBT HI intensity mapping data are shown in Fig. 15.14. As one can see, the first few eigenvalues are quite high. The first few modes in frequency are shown in Fig. 15.15. Except for the first one, they are not smooth in frequency. With the subtraction of the first 10 or 20 modes, the variance of the data is significantly reduced.

The blind foreground subtraction procedure will also remove some 21 cm signal. This is unavoidable, but the amount of signal loss can be computed statistically. This is done by generating a mock 21 cm signal, which is generally much smaller than the data.

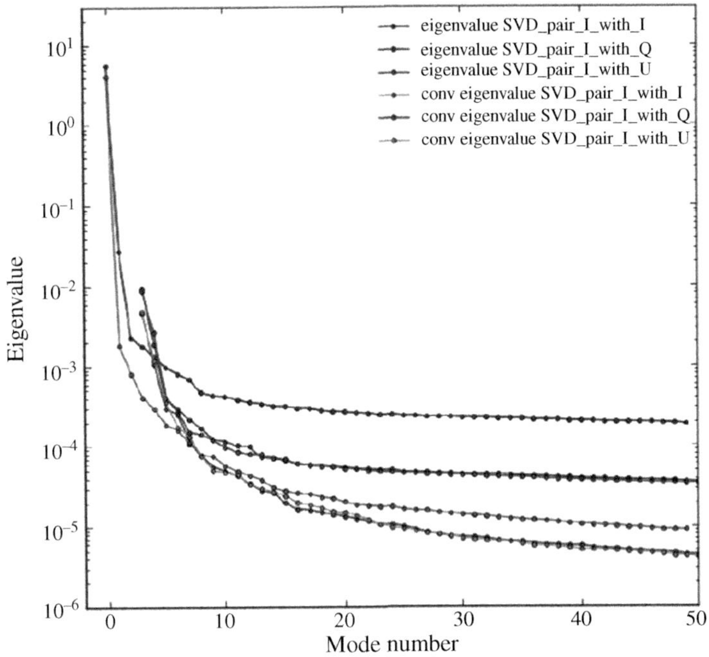

Fig. 15.14 The SVD eigenvalues in an analysis of GBT HI intensity mapping survey data.

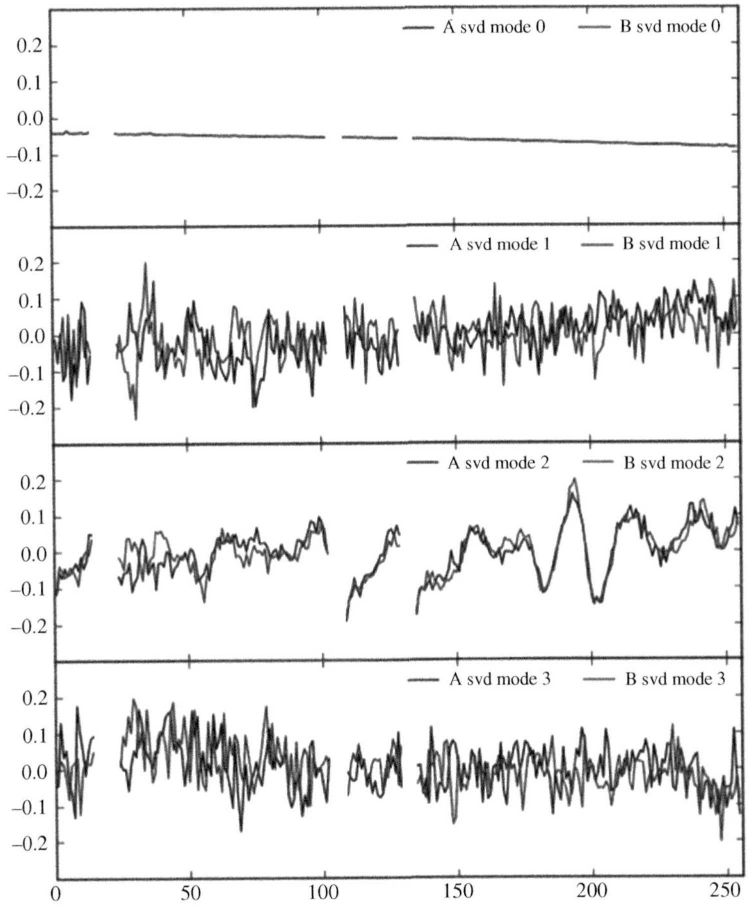

Fig. 15.15 The first few SVD modes of the GBT observation data.

We add such a mock signal into the dataset, perform the foreground subtraction procedure, and compute the amount of 21 cm signal loss. Repeat this simulation many times with different random numbers, and we will be able to estimate the signal loss transfer function [13].

Other semiblind foreground subtraction methods have also been proposed, including the independent component analysis (ICA), which uses the non-Gaussianity of the foregrounds to subtract them [24]. Generally speaking, the foreground subtraction can be put into a Bayesian framework to extract the 21 cm signal as much as possible by utilizing the known information.

3.3 The foreground wedge

For cosmology, the power spectrum is the observable to be measured. The cosmological 21 cm signal can be Fourier transformed,

$$\widetilde{T}\left(k_{\perp}, k_{\parallel}\right) = \int d^2 r_{\perp} dr_{\parallel}\, e^{-i\left(k_{\perp} r_{\perp} + k_{\parallel} r_{\parallel}\right)}\, T\left(r_{\perp}, r_{\parallel}\right).$$

Similarly, we can also Fourier transform the visibilities,

$$\widetilde{V}\left(\mathbf{b}, \eta\right) \approx \int d^2\mathbf{u}\ \widetilde{T}\left(\mathbf{u}, \boldsymbol{\eta}\right) \widetilde{\mathbf{A}}_{\boldsymbol{p}} \left(\frac{\mathbf{b}}{\lambda} - \mathbf{u}, \nu_0\right) \approx \widetilde{T}\left(\frac{\mathbf{b}}{\lambda}, \eta\right)$$

where η is the Fourier dual of frequency ν. In the last step of the above, we have made the approximation of ignoring the primary beam of the antenna, which ideally is smooth or even constant on the scale of interest. However, the same physical baseline probes different scales at different frequencies. The cosmological scales are related to the instrumental scales in the visibility by

$$k_{\perp} = \frac{2\pi\nu_0\, b}{c\, D_c}; k_{\parallel} = \frac{2\pi\nu_{21} H_0 E(z)\eta}{c(1+z)^2}$$

where D_c is the comoving distance, and

$$E(z) = \sqrt{\Omega_m (1+z)^3 + \Omega_\Lambda}.$$

The finest angular scales or maximum $k_{\perp}$ are determined by the angular resolution, which is determined by the longest baselines of the array. The largest angular scales (lowest $k_{\perp}$) are determined by the survey area. Along the radial direction, the finest resolution is limited by the spectral resolution of the instrument while the lowest $k_{\parallel}$ is limited by the frequency range of observation.

The foregrounds with a smooth or even flat spectrum will induce some foreground power that contaminates the 21 cm signal. Naively, one might think this sets a lower limit in $k_\parallel$, above which the 21 cm signal can be measured. However, the fact that the same baseline measures different angular scales at different frequencies induces a mode-mixing effect, which tilts the limit in power spectrum scales. To see this, note that the measured power spectrum is given by

$$\hat{P}(\mathbf{k}) \propto \left\langle \left| \widetilde{V}_b(\tau) \right|^2 \right\rangle$$
$$= \int \frac{d^3k}{(2\pi)^3} P(\mathbf{k}) \left| \int dy e^{ik_y y} A_p \left[\frac{cD_c}{2\pi b} (\alpha k_\parallel + 2\pi\tau), y \right] \right|^2.$$

The squared term in the integrand corresponds to the window function of the power spectrum estimator. For unclustered diffuse point sources with flat spectra, the foreground power has the form $P(k) \propto \delta(k_\parallel)$, that is, strong noise at low $k_\parallel$ as expected. Plugging this foreground power into the above expression, we would obtain

$$\hat{P}(\mathbf{k}) \propto \int dy A_p^2 \left[\frac{cD_c\tau}{b}, y \right] \equiv \overline{A_p^2} \left[\frac{k_\parallel}{k_\perp} \frac{c}{H_0} \frac{(1+z)}{E(z)} \right]$$

where $\overline{A}_p^2$ is the squared primary beam integrated over the direction perpendicular to the baseline. Even though the initial foreground is entirely at $k_\parallel = 0$, the power is leaked out, contaminating a region with $k_\parallel \propto k_\perp$. This is the so-called foreground wedge [25], as shown below in Fig. 15.16. The primary beam response falls off away from the pointing center, at an angle θ_0. In some cases, the primary beam is very wide, but it is still limited by the horizon. Thus, the function $\overline{A}_p^2$ drops to 0 at

$$k_\parallel = k_\perp \frac{H_0 D_c E(z) \theta_0}{c(1+z)},$$

This is the horizon wedge in the figure. Above the horizon wedge is the so-called EoR window where the 21 cm signal can be extracted without contamination.

4 Conclusion

The 21 cm line offers an observational probe that covers most of the observable cosmic volume, from the Dark Ages down to the modern universe. At present, a number of experiments are running or are being planned to measure the cosmological 21 cm signal at different redshifts, as shown in Fig. 15.17.

The low- and mid-redshift experiments are primarily designed to map the large-scale structure and measure the baryon acoustic

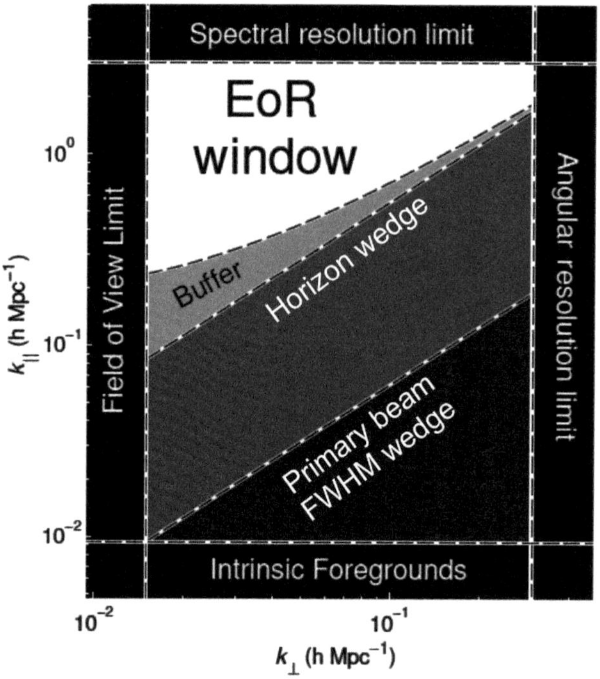

Fig. 15.16 The power spectrum in ($k_{\perp}$, $k_{||}$), showing the EoR wedge.

oscillation features to probe the dark energy equation of state. The existing 21 cm experiments, such as Tianlai, CHIME, HIRAX, and the intensity mapping survey of SKA-mid, are mostly dedicated to $z<3$. It has been proposed that a Stage 2 experiment could map the postreionization universe at $2<z<6$ [26]. The higher redshift experiments such as LOFAR, MWA, HERA, and the future SKA-low are looking into the epoch of reionization and cosmic dawns.

Ideas have also been advanced to go beyond the cosmic dawn into the cosmic Dark Ages. Due to the limitation of the Earth's ionosphere, such experiments may need to be conducted from the far side of the moon [27]. If fully explored, the Fourier modes that can be measured with the 21 cm line are about 10^6 that of the CMB modes [28]. Therefore, this contains the largest known wealth of information about the primordial fluctuations, which we can use to probe the cosmic origin. However, as the synchrotron foreground scales as $\nu^{-\gamma}$ where $\gamma \approx 2.5$, the foreground is stronger at lower frequencies, so it is also increasingly harder to make the measurement at the higher redshifts. This requires increasingly larger arrays to satisfy the basic sensitivity requirement.

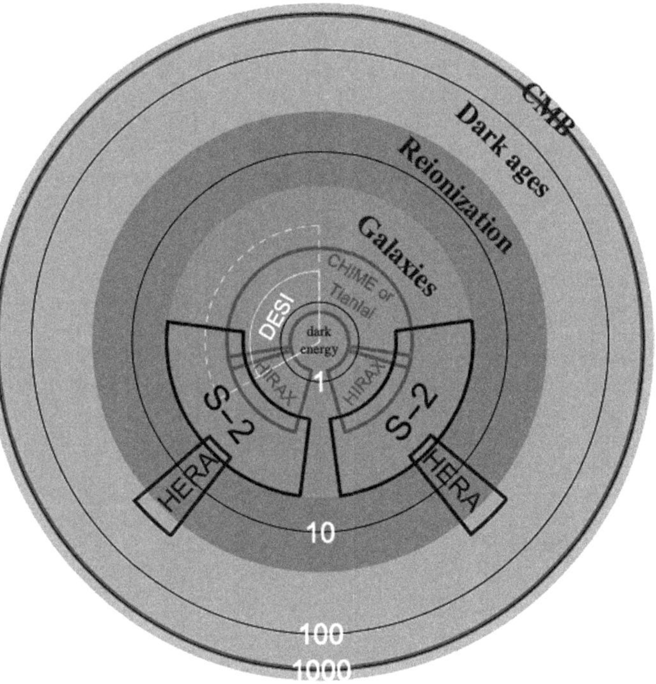

Fig. 15.17 A schematic two-dimensional representation of the observable universe where the area is proportional to the comoving volume, with the covered redshift range and survey field of the various experiments [26].

The real world 21 cm experiment also has to deal with many complications, especially the complicated telescope response and the contamination of the various foregrounds. Nevertheless, with the increase of sensitivity and growing computational capability, these problems will be overcome, and we expect 21 cm cosmology will come of age in the next decade.

References

[1] S.R. Furlanetto, S.P. Oh, F.H. Briggs, Cosmology at low frequencies: the 21cm transition and the high redshift universe, Phys. Rep. 433 (2006) 181.

[2] J.R. Pritchard, A. Loeb, 21 cm cosmology in the 21st century, Rep. Prog. Phys. 75 (2012) 086901.

[3] X. Chen, J. Miralda-Escude, The spin-kinetic temperature coupling and the heating rate due to Lyman alpha scattering before reionization: predictions for 21cm emission and absorption, Astrophys. J. 602 (2004) 1.

[4] X. Chen, J. Miralda-Escude, The 21cm signature of the first stars, Astrophys. J. 684 (2008) 18.

[5] B. Ciardi, S. Inoue, K.J. Mack, Y. Xu, G. Bernardi, 21 cm forest with the SKA, arxiv:1501.04425. (2015).

[6] P. Dewdney, SKA1 System Baseline Design V2, SKA-TEL-SKO-0000002, 26 February (2016).

[7] T. Chang, U.-L. Pen, J.B. Peterson, P. McDonald, Baryon acoustic oscillation intensity mapping as a test of dark energy, Phys. Rev. Lett. 100 (2008) 091303.

[8] H.J. Seo, et al., A ground-based 21cm baryon acoustic oscillation survey, Astrophys. J. 721 (2010) 164.
[9] A.R. Thompson, J.M. Moran, G. Swenson Jr., Interferometry and Synthesis in Radio Astronomy, third ed., Springer, Cham, Switzerland, 2017.
[10] Y. Gong, et al., The OH line contamination of 21 cm intensity fluctuation measurements for z=1∼4, Astrophys. J. Lett. 740 (2011) 20.
[11] T. Chang, et al., Hydrogen 21-cm intensity mapping at redshift 0.8, Nature 466 (2010) 463.
[12] K.W. Masui, et al., Measurement of 21 cm brightness fluctuations at z ∼ 0.8 in cross-correlation, Astrophys. J. Lett. 763 (2013) 20.
[13] E.R. Switzer, et al., Determination of z∼0.8 neutral hydrogen fluctuations using the 21 cm intensity mapping auto-correlation, Mon. Not. R. Astron. Soc. Lett. 434 (2013) 46.
[14] C.J. Anderson, et al., Low-amplitude clustering in low-redshift 21-cm intensity maps cross-correlated with 2dF galaxy densities, Mon. Not. R. Astron. Soc. 476 (2017) 3382.
[15] W. Hu, et al., Forecast for FAST: from galaxies survey to intensity mapping, (2019). arxiv:1909.10946.
[16] D.J. Bacon, et al., Cosmology with Phase 1 of the Square Kilometre Array, Red Book 2018: Technical specifications and performance forecasts, (2018) arxiv:1811.02743.
[17] X. Chen, The Tianlai project: a 21cm cosmology experiment, Int. J. Phys. Conf. Ser. 12 (2012) 256 Proceeding of the 2nd Galileo-Xu Guangqi Meeting, arxiv:1212.6278.
[18] Y. Xu, X. Wang, X. Chen, Forecasts on the dark energy and primordial non-Gaussianity observations with the Tianlai cylinder array, Astrophys. J. 798 (2015) 40.
[19] J. Zhang, et al., Sky reconstruction from transit visibilities: PAON-4 and Tianlai dish array, Mon. Not. R. Astron. Soc. 461 (2016) 1950.
[20] J.D. Bowman, A.E.E. Rogers, J.N. Hweitt, Toward empirical constraints on the global redshifted 21 cm brightness temperature during the epoch of reionization, Astrophys. J. 676 (2008) 1.
[21] X. Chen, et al., Discovering the sky at the longest wavelengths with small satellite constellations, (2019). arxiv:1907.10853.
[22] J.D. Bowman, et al., An absorption profile centred at 78 megahertz in the sky-averaged spectrum, Nature 555 (2018) 67.
[23] X. Wang, et al., Twenty-one centimeter tomography with foregrounds, Astrophys. J. 650 (2006) 529.
[24] E. Chapman, et al., Foreground removal using FASTICA: a showcase of LOFAR-EoR, Mon. Not. R. Astron. Soc. 423 (2012) 2518.
[25] M.F. Morales, et al., Four fundamental foreground power spectrum shapes for 21 cm cosmology observations, Astrophys. J. 752 (2012) 137.
[26] R. Ansari, et al., Cosmic visions dark energy: inflation and early dark energy with a stage II hydrogen intensity mapping experiment, (2018)arxiv:1810.09572.
[27] L. Koopmans, et al., Peering into the dark (ages) with low-frequency space interferometers, (2019). White paper submitted to ESA voyage 2050, arxiv:1908.04296.
[28] A. Loeb, M. Zaldarriaga, Measuring the small-scale power spectrum of cosmic density fluctuations through 21cm tomography prior to the epoch of structure formation, Phys. Rev. Lett. 92 (2004) 1301.

Further reading

A. Liu, R. Shaw, Data analysis for precision 21cm cosmology, (2019) arxiv:1907.08211.

Index

Note: Page numbers followed by *f* indicate figures, *t* indicate tables, and *np* indicate footnotes.

F

M

N

O

S

9780128190845